FLUID MECHANICS SOURCE BOOK

THE McGRAW-HILL SCIENCE REFERENCE SERIES

Acoustics Source Book
Communications Source Book
Computer Science Source Book
Meteorology Source Book
Nuclear and Particle Physics Source Book
Optics Source Book
Physical Chemistry Source Book
Solid-State Physics Source Book
Spectroscopy Source Book

FLUID MECHANICS SOURCE BOOK

Sybil P. Parker, *Editor in Chief*

McGRAW-HILL BOOK COMPANY

New York   St. Louis   San Francisco
Auckland   Bogotá   Caracas   Colorado Springs   Hamburg
Lisbon   London   Madrid   Mexico   Milan   Montreal
New Delhi   Oklahoma City   Panama   Paris   San Juan
São Paulo   Singapore   Sydney   Tokyo   Toronto

Cover: Laminar and turbulent flow in a heated smoke plume. The smoke originally rises in a smooth stream (blue region), but this buoyancy-driven flow is unstable and a transition (red region) to turbulent flow (purple region), characterized by irregular motion, takes place.

Cover illustration by E. T. Steadman.

This material has appeared previously in the McGRAW-HILL ENCYCLOPEDIA OF SCIENCE AND TECHNOLOGY, 6th Edition, copyright © 1987 by McGraw-Hill, Inc. All rights reserved.

FLUID MECHANICS SOURCE BOOK, copyright © 1988 by McGraw-Hill, Inc. All rights reserved. Printed in the United States of America. Except as permitted under the United States Copyright Act of 1976, no part of this publication may be reproduced or distributed in any form or by any means, or stored in a data base or retrieval system, without prior written permission of the publisher.

1 2 3 4 5 6 7 8 9 0   DOC/DOC   8 9 5 4 3 2 1 0 9 8

ISBN 0-07-045502-3

Library of Congress Cataloging in Publication Data:

Fluid mechanics source book / Sybil P. Parker, editor in chief.
    p. cm. — (McGraw-Hill science reference series)
  "This material has appeared previously in the McGraw-Hill encyclopedia of science and technology, 6th edition"— CIP t.p. verso.
  Bibliography: p.
  Includes index.
  ISBN 0-07-045502-3
  1. Fluid mechanics—Dictionaries.   I. Parker, Sybil P. II. McGraw-Hill Book Company.   III. Title: McGraw-Hill encyclopedia of science and technology (6th ed.) IV. Series.
TA357.F582 1988   87-36638
620.1'06'0321—dc19

TABLE OF CONTENTS

Introduction	1
Fluid Statics	5
Fluid Flow	13
Fluid Dynamics	53
Nonviscous Flow	105
Viscous Flow	113
Waves and Disturbances in Fluids	133
Similitude	151
Physics of Fluids	177
Measurement and Display of Properties	193
Contributors	263
Index	269

FLUID MECHANICS SOURCE BOOK

INTRODUCTION

FLUID MECHANICS is the science concerned with fluids, either at rest or in motion. It deals with pressures, velocities, and accelerations in the fluid, including fluid deformation and compression or expansion. Fluid mechanics may be divided into two branches, fluid statics and fluid dynamics; the first deals with pressure intensities and forces exerted by a fluid at rest, and the second with forces exerted on fluids and their resulting motions.

The laws of fluid mechanics control a great portion of natural phenomena. The flight of an insect or bird, the motion of a fish through water, the relative movement of air masses as in frontal weather systems, and the eruption of a volcano are examples of flow that follow the laws of fluid mechanics. The science of fluid mechanics is involved in many phases of aeronautical, chemical, civil, and mechanical engineering. It requires the combination of theoretical analysis and orderly experimentation. In the theoretical approach, assumptions are made as needed to keep the resulting mathematical expressions manageable, such as assuming an incompressible fluid, or one without viscosity. The experimental work must be planned in such a way that similitude relations are used to a maximum advantage.

Definitions. Flow may be classified in many ways, such as turbulent, laminar; real, ideal; isothermal, isentropic; steady, unsteady; and uniform and nonuniform.

Turbulent-flow situations are most prevalent in engineering practice. The fluid particles move in irregular paths, causing an exchange of momentum from one portion of fluid to another. The fluid regions involved in the transfer of momentum due to turbulence can range in size from large-scale turbulence with many cubic miles of fluid participating in a single tornadic eddy to a small-scale turbulence with only a few thousand molecules interacting. Turbulent flow causes the conversion of mechanical energy into thermal energy at a rate varying roughly as the square of the velocity.

In laminar flow, particles move along smooth paths in layers, or laminae, with one layer gliding over an adjacent one. Laminar flow is governed by Newton's law of viscosity. It may be considered as flow in which all turbulence has been damped out by the action of viscosity.

A real fluid always has viscosity and, whether a liquid or a gas, has compressibility. An ideal fluid is considered to be frictionless (nonviscous) and incompressible. There is no means by which an ideal fluid can convert mechanical energy into thermal energy. When ideal fluid particles have no rotation, the fluid is irrotational and a velocity potential exists.

When gas flows without change in temperature, the flow is isothermal. When flow occurs such that no heat is added or subtracted at the boundaries, the flow is adiabatic. Reversible adiabatic (frictionless adiabatic) flow is called isentropic flow.

Steady flow occurs when conditions at any point in the fluid do not change with time. In unsteady flow one or more quantities, such as density or velocity, change with time at a point. Uniform flow occurs when the velocity vector throughout the fluid is everywhere the same at any instant, and nonuniform flow occurs when the velocity vector has varying values throughout the fluid at any instant. The strict definitions of steady flow and uniform flow are relaxed in practical flow situations, due to turbulent fluctuations in the first case and to variations over a cross section in the second case. For example, liquid flow through a long pipe at constant rate is steady-uniform flow; liquid flow through a long pipe at a decreasing rate is unsteady-uniform flow; flow through an expanding tube at a constant rate is steady-nonuniform flow; and flow through an expanding tube at an increasing rate is unsteady-nonuniform flow.

A streamline is a continuous line through the fluid that has the direction of the local velocity vector at every point. A stream tube is composed of all streamlines through a small closed curve.

Basic equations of fluid flow. In any fluid-flow situation, three conditions exist: (1) Newton's laws of motion hold for every particle at every instant; (2) the continuity relationship holds, that is, net mass inflow into any small volume per unit time equals its time rate of increase of mass; (3) at a boundary, the velocity component normal to the boundary equals the velocity component of the boundary normal to itself. For real fluids, in addition, the tangential component of fluid velocity at the boundary is zero relative to the boundary.

Integration of Newton's second law of motion may lead either to the Bernoulli equation or to the momentum equation. By considering the steady flow of a frictionless, incompressible fluid, the Bernoulli equation states that the mechanical energy remains constant along a streamline. By considering steady, irrotational flow of a frictionless incompressible fluid, a form of the Bernoulli equation is obtained that shows that the energy is constant everywhere throughout the fluid. For steady flow the momentum equation states that the resultant force acting on any free body of fluid is just equal to its time rate of change of momentum or, for a fixed control volume, that the resultant force equals the difference between the momentum per unit time leaving and the momentum per unit time entering.

The continuity equation for steady flow states that the mass per unit time flowing along a stream tube is everywhere constant.

Use of the basic equations permits many fluid-flow situations to be analyzed, provided that energy losses are small and can be neglected. In certain special situations, application of the continuity, Bernoulli, and momentum equations permits the energy losses to be approximately computed.

For the vast majority of real fluid-flow situations, experimental information is required to determine the amount of mechanical energy converted to thermal energy. The effect of losses in steady flow of a fluid along a streamline may be expressed by including one or more loss terms in the energy equation. The equation states that the energy per unit weight at one point is equal to the energy remaining at a downstream point plus all the losses between the two points. For example, in turbulent pipe flow the losses due to wall friction tend to vary almost directly as the length of pipe, inversely as the diameter, and directly as the square of the velocity. Fluid properties and condition of the pipewall surface enter in a more complicated manner. Experimentally the losses may be determined and expressed in a dimensionless form by a chart so that the results apply to other fluids and to other sizes of geometrically similar pipe. Losses in pipe flow due to changes in cross section or direction, such as elbows and valves, tend to vary about as the square of the velocity, or as a constant times the kinetic energy per unit weight.

Similitude. The performance of hydraulic structures or fluid machines may be studied by utilizing similitude relations. Two systems are said to be dynamically similar if (1) they are geometrically similar, (2) they have the same boundary conditions, and (3) either their streamline configurations are geometrically similar or their dynamic pressure ratios at corresponding points are the same.

The particular method of interpreting the data from model studies depends upon the types of forces that predominate. For example, with hydraulic structures such as dams, spillways, and canal transitions, the important forces are those due to gravity and inertia of the liquid, with viscous forces of lesser importance. The ratio of inertial forces to gravity forces is a dimensionless parameter, known as the Froude number. By adjusting flows and depths so that the Froude number is the same in model and prototype, measurements in the model can be made and converted to corresponding prototype values. If the ratio of linear dimensions of the prototype to the model is λ, then the velocity in the prototype is $\sqrt{\lambda}$ times the corresponding velocity in the model, and (for the same liquids in model and prototype) the pressure intensity in the prototype is λ times the corresponding pressure intensity in the model.

Model studies of fluid machinery. In making a test on a model of a turbomachine, special relationships must be observed. For geometrically similar streamlines, the ratio of velocity of flow at some point in the machine to the peripheral velocity of the runner or rotor must be the same in model and prototype. Also, since inertial forces are of great importance in the machine, there must be a definite relation between head on the machine and velocity at some point within the machine, this relation being the same for model and prototype. In addition, for compressible flow the ratio of velocity of fluid to local acoustic velocity must be the same at corresponding points in model and prototype. The above conditions, together with geometric similitude, permit tests on a model to be used to predict performance of the prototype. The above relations, referred to as homologous relationships, do not permit viscous forces to be scaled properly; hence there is a slight difference in efficiency of the various sizes. The larger the machine, the more efficient it is, but with the change being usually not more than 2 or 3%.

In flow through pipes and other closed conduits, the controlling forces are inertial and viscous for velocities that are small compared with acoustic velocity. The ratio of inertial to viscous forces is expressed by a dimensionless parameter known as Reynolds number. When two geometrically similar closed-flow systems have the same Reynolds number at corresponding points, the dimensionless flow and loss coefficients will be the same. When closed-channel flow occurs at velocities near acoustic velocity, the Mach number (ratio of velocity to acoustic velocity) becomes a controlling parameter.

VICTOR L. STREETER

1

FLUID STATICS

Fluid statics	6
Pascal's law	7
Buoyancy	7
Archimedes' principle	8
Hydromechanics	8
Hydrostatics	9
Gas mechanics	10
Aeromechanics	11
Aerostatics	11

FLUID STATICS
Victor L. Streeter

The determination of pressure intensities and forces exerted by liquids and gases at rest. Hydrostatics, although implying the statics of water alone, applies to liquids in general. By definition, a fluid at rest cannot sustain a shear stress; therefore, a fluid force exerted on an element of boundary area must act normal to the area. For example, consider the small free body of fluid shown in the **illustration**. Equilibrium requires that $dp = -\gamma dz$, in which p is the absolute pressure, γ is the specific weight (weight of a unit volume of fluid), and z is the elevation, measured vertically upward. For liquids, γ is substantially constant, and the equation shows that pressure decreases linearly as the elevation increases. *See* Hydrostatics.

With gases, the variation of γ with pressure or elevation must be known in order to integrate the equation. For an isothermal gas ($p/\gamma = p_0/\gamma_0$ = constant), the pressure variation with elevation is given by Eq. (1), in which p_0 and γ_0 are the values of p and γ at elevation z_0.

$$p = p_0 \exp[-(z - z_0)/(p_0/\gamma_0)] \quad (1)$$

On the average, the absolute temperature T of the atmosphere tends to decrease linearly with elevation within the troposphere, $T = T_0 - \beta z$. By using the general gas law $p/\gamma = RT$, in which R is the gas constant, the specific weight is given by Eq. (2). Use of this expression for γ in terms of p and z yields the variation of pressure with elevation in a standard atmosphere, as in Eq. (3), in which p_0 and T_0 are the values at $z = 0$.

$$\gamma = p/R(T_0 - \beta z) \quad (2) \qquad p = p_0(1 - \beta z/T_0)^{1/R\beta} \quad (3)$$

The determination of fluid pressure is of importance in many flow-measuring devices. The manometer is one method which makes use of transparent tubes with one or more liquids contained in them. By measuring the difference in elevation of the fluid menisci, the desired pressure is determined from the laws of fluid statics.

Fluid forces on plane surfaces may be determined by integration of $p\,dA$ over the surface, with p the pressure intensity and dA an element of the surface area. For liquids, the magnitude of the force is the product of the area and the pressure at the centroid of the area. The line of action of the resultant force is normal to the surface and acts at a point termed the center of pressure. For horizontal submerged surfaces, the centroid of the area and the center of pressure coincide, but for all other orientations of the surface, the center of pressure is below the centroid (at less elevation). *See* Pressure measurement.

To determine the resultant fluid force exerted on a curved surface, it is convenient to consider horizontal component of force exerted on a curved surface is equal to the force exerted

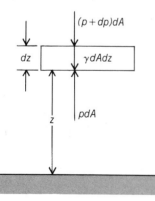

Free-body diagram for vertical forces acting on fluid element.

on a projection of the surface onto a plane normal to the direction of the component. The line of action for the horizontal force will be through the center of pressure of the projected area. The vertical component of force exerted on a curved surface by the fluid pressure equals the weight of liquid vertically above the curved surface, and acts through the centroid of this volume of liquid. The resultant force is thus the sum of three vectors, two horizontal components at right angles and the vertical component. SEE ARCHIMEDES' PRINCIPLE; BUOYANCY.

PASCAL'S LAW
KARL ARNSTEIN AND ROBERT S. ROSS

A law of physics which states that a confined fluid transmits externally applied pressure uniformly in all directions. Blaise Pascal, using the mercury-column barometer of Evangelista Torricelli, demonstrated the decrease in atmospheric pressure with increasing height and determined that atmospheric force at a point exerted equal pressure in all directions. More exactly, in a static fluid, force is transmitted at the velocity of sound throughout the fluid. The force acts normal to any surface. This natural phenomenon is the basis of the pneumatic tire, balloon, hydraulic jack, and related devices. SEE HYDROSTATICS; TORRICELLI'S THEOREM.

BUOYANCY
VICTOR L. STREETER

The resultant vertical force exerted on a body by a static fluid in which it is submerged or floating. The buoyant force F_B acts vertically upward, in opposition to the gravitational force that causes it. Its magnitude is equal to the weight of fluid displaced, and its line of action is through the centroid of the displaced volume, which is known as the center of buoyancy. With V the displaced volume of fluid and γ the specific weight of fluid (weight per unit volume), the buoyant force equation becomes $F_B = \gamma V$. The magnitude of the buoyant force must also be given by the difference of vertical components of fluid force on the lower and upper sides of the body. SEE AEROSTATICS; FLUID STATICS; HYDROSTATICS.

By weighing an object when it is suspended in two different fluids of known specific weight, the volume and weight of the solid may be determined. SEE ARCHIMEDES' PRINCIPLE.

Horizontal buoyancy. Another form of buoyancy, called horizontal buoyancy, is experienced by models tested in wind or water tunnels. Horizontal buoyancy results from variations in static pressure along the test section, producing a drag in closed test sections and a thrust force in open sections. These extraneous forces must be subtracted from data as a boundary correction. Wind tunnel test sections usually diverge slightly in a downstream direction to provide some correction for horizontal buoyancy. SEE WATER TUNNEL; WIND TUNNEL.

Stability. A body floating on a static fluid has vertical stability. A small upward displacement decreases the volume of fluid displaced, hence decreasing the buoyant force and leaving an unbalanced force tending to return the body to its original position. Similarly, a small downward displacement results in a greater buoyant force, which causes an unbalanced upward force.

A body has rotational stability when a small angular displacement sets up a restoring couple that tends to return the body to its orignal position. When the center of gravity of the floating body is lower than its center of buoyancy, it will always have rotational stability. Many a floating body, such as a ship, has its center of gravity above its center of buoyancy. Whether such an object is rotationally stable depends upon the shape of the body. When it floats in equilibrium, its center of buoyancy and center of gravity are in the same vertical line. When the body is tipped, its center of buoyancy shifts to the new centroid of the displaced fluid and exerts its force vertically upward, intersecting the original line through the center of gravity and center of buoyancy at a point called the metacenter. A floating body is rotationally stable if the metacenter lies above the center of gravity. The distance of the metacenter above the center of gravity is the metacentric height and is a direct measure of the stability of the object.

ARCHIMEDES' PRINCIPLE
VICTOR L. STREETER

A body immersed in static fluid is acted upon by a vertical force equal to the weight of fluid displaced, and a body floating in the fluid displaces its own weight of fluid. For example, a balloon ascends because it displaces a volume of air which weighs more than the weight of the balloon. This principle was first stated by Archimedes (about 287–212 B.C.) and was used by him to determine the relative amounts of gold and silver in a crown. The principle can be proved by determining the difference in vertical components of fluid force acting on the lower and upper curved surfaces of the body. This force, called the buoyant force, acts vertically upward through the centroid of the displaced volume of fluid. SEE BUOYANCY; FLUID STATICS; HYDROSTATICS.

To find the specific gravity of a body, it is weighed separately in two fluids of specific weights γ_1 and γ_2 (see **illus**.). If its volume is V and its weight W, and it weighs F_1 in the fluid of specific weight γ_1 and F_2 in the fluid of specific weight γ_2, then Eq. (1) and (2) hold. Its specific weight is then given by Eq. (3), and its specific gravity is the value of γ divided by the specific

$$V = \frac{F_1 - F_2}{\gamma_2 - \gamma_1} \quad (1) \qquad W = \frac{F_1\gamma_2 - F_2\gamma_1}{\gamma_2 - \gamma_1} \quad (2) \qquad \gamma = \frac{W}{V} = \frac{F_1\gamma_2 - F_2\gamma_1}{F_1 - F_2} \quad (3)$$

weight of water at standard conditions. That is, specific gravity is the ratio of the density of the substance to the density of water, or it can be given as the ratio of specific weight of the substance to specific weight of water, at standard conditions.

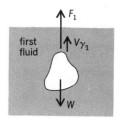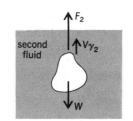

Free-body diagram for body suspended first in one fluid and then in a second fluid.

Specific gravity of a liquid may be related directly to its significant property. Thus, charge condition of an electrolyte in a storage battery, freezing temperature of a coolant, and energy density of a fuel such as kerosine are proportional to the specific gravities of these liquids. This correspondence provides a convenient and rapid means for measurement. The hydrometer uses Archimedes' principle to determine the specific gravity of liquids. It is a weighted body with a thin stem arranged so that it floats vertically with the liquid surface at some position along the stem, depending upon the specific gravity of the liquid. With a liquid of less specific gravity than water, more of the hydrometer is submerged, so that the weight of displaced liquid is the same in each case. By graduating the stem, specific gravities may be read directly from its depth of immersion in the liquid. Because specific gravity may vary rapidly with temperature, a temperature correction or measurement at a stated temperature is necessary. SEE AEROSTATICS.

HYDROMECHANICS
WILLIAM ALLAN

That branch of physics which deals with liquids, traditionally water, as a medium for the transmission of forces. A body immersed in a liquid experiences a vertical force proportional to the volume of liquid that it displaces. A fluid whose velocity changes produces forces as a result of changing pressure and of changing momentum. Such phenomena are applied in various machines. SEE ARCHIMEDES' PRINCIPLE; BERNOULLI'S THEOREM; BUOYANCY; HYDRAULICS; HYDROSTATICS.

HYDROSTATICS
WILLIAM ALLAN

The study of liquids at rest. In the absence of motion, there are no shear stresses; the internal state of stress at any point is determined by pressure alone. Hence, the pressure at a point is the same in all directions. Pressure acts normally to all boundary surfaces. For equilibrium under gravity, regardless of the shape of the containing vessel, the pressure is uniform over any horizontal cross section. Pressure varies with height or depth in accordance with the relation $dp = wdz$, where w is the specific weight of the liquid (pounds per cubic foot), and z is the height or depth in feet measured positive downward. Two different reference levels are used in measuring pressure. For many engineering purposes, gage pressure is used with pressure measured relative to atmospheric pressure as zero. For most scientific purposes, pressure is referred to true zero. Normal atmospheric pressure at sea level caused by the weight of the air above is approximately 14.7 psia (101.3 kilopascals). *See* PRESSURE MEASUREMENT.

Applications of hydrostatics. Storage tanks, underwater tunnels, gates for hydraulic structures, walls, dams, sheetpiling and bulkheads, pressure-measuring devices, hydraulic presses, and other pressure-actuated systems are applications of hydrostatics.

Hydrostatic forces on immersed surfaces. Force F is the pressure p multiplied by the area A on which it acts. Its magnitude is in pounds and its direction is normal to the area. It is a vector quantity and may be broken into components usually taken horizontally and vertically. On all surfaces, plane or curved, forces acting on elementary areas are evaluated as $dF = pdA$.

On a plane surface all elementary force vectors are parallel and the total force is the sum of the elementary forces. Hence, the total force is the product of the total area and the average pressure acting on it. The average pressure is that at the centroid of the area. The total force is independent of the orientation of the plane surface on which it acts as long as the depth to its centroid is the same.

Pressure volume. In calculating the total hydrostatic force acting on a plane surface, a solid is imagined with the plane surface area as its base and the fluid pressure at each point on the base erected there as an altitude. The total force is the volume of the solid. If the total force were imagined to act at a single point, the center of pressure of the surface, it would pass through the center of gravity of the solid and be normal to the surface.

On a nonplane surface, the elementary force vectors are not parallel. The summation of their horizontal and vertical components gives respectively the horizontal and vertical components of the total force. The total force is determined from these components by vector addition. The horizontal component of the total force is that which would be exerted on the vertical plane projected area of the nonplane surface. The vertical component of the total force is the weight of the liquid volume extending vertically from the nonplane surface up to the free surface of the liquid. As single forces, these components would pass through the centers of gravity of their respective volumes. When combined, they give the magnitude, line of action, and point of application of the total force on the nonplane surface.

Buoyant force. This is the force exerted vertically upward by a fluid on a body wholly or partly immersed in it. Its magnitude is equal to the weight of the fluid displaced by the body. This value is also the vertical component of the fluid pressure force acting upward against the bottom of the body minus the fluid pressure force component (if any) acting vertically downward against the top of the body. If this buoyant force equals the weight of the body, the body will remain at the given level. If it exceeds the weight of the body, the latter will rise, and vice versa. The buoyant force as a single magnitude acts vertically upward through the center of buoyancy which is the center of gravity of the displaced fluid. *See* ARCHIMEDES' PRINCIPLE; BUOYANCY.

Stability. The stability of a wholly or partly immersed body is determined by the relative positions of its center of gravity G and center of buoyancy B. The position of G depends upon the distribution of the mass within the body; the position of B depends upon the shape of the submerged portion of the body. If G lies directly below B, the body will be stable; under an angular displacement, a righting moment will tend to restore the body to its original position.

A floating body, depending upon its shape, may be stable even if G lies above B. An angular rotation or heel will not change the volume of the displaced fluid but may change its shape and the lateral position of B so that a righting moment through B and G may exist to restore the body to its original position. The point of intersection of the vertical line through the displaced position of B with the line drawn through G and the original position of B is M. If M lies above G for a given angle of heel, a righting moment will exist; if M lies below G, an overturning moment will arise to capsize the body. In most ships, for angles of heel up to 10–15°, M remains in a practically constant position—the metacenter.

Pressure-measuring instruments. Two types of instruments measure pressure—gages and manometers.

Gages. A metallic element such as a curved tube or a flexible diaphragm which deforms under liquid pressure is the usual sensing element of a gage. The deformation is changed mechanically or electrically into a calibrated dial reading. The bourdon gage, used for measurement of static or slowly changing pressure, converts mechanically the deformation of a curved metal tube into a reading in pressure units. Each gage is designed to be accurate for a selected range of pressures but does not give accurate readings of short-time pressure fluctuations.

A pressure transducer converts deformation of a metallic diaphragm into an electric current differential, which is calibrated to read in pressure units. Each transducer is designed to be accurate for a selected range of pressures and reacts to fluctuations of microsecond duration. The cost is high and accessory instrumentation elaborate; application is generally to the measurement of rapidly fluctuating pressures.

Manometers. Glass tubes in which the height of a liquid is a measure of the pressure being sought are called manometers. A simple manometer is open to the atmosphere at one end and connected to the pressure source at the other; it is called a piezometer if its liquid is the same as that in the tank, pipe, or other device to which it is connected. Measured above a selected datum, the height to the top of the liquid in the open-ended tube is the piezometer head. If the pressure source is a pipe, the height of the piezometer liquid level above the pipe center is the pressure head in the pipe. The pressure in pounds per square foot is obtained by multiplying this height in feet by the specific weight of the liquid.

A differential manometer measures the differences in pressure between two sources. The glass tube is usually an erect or inverted U partly filled with a liquid other than those liquids (or gases) of which the pressure difference is desired. One leg of the U is connected to one pressure source, the other leg to the other sources. If the pressure sources are at the same elevation and contain the same liquid, the pressure difference is the vertical distance (displacement) between the tops of the manometer liquid in the two legs of the U multiplied by the difference between the specific weight of the manometer liquid and that of the displacing liquid. Measurement of small differences can be magnified by using a manometer liquid of low specific gravity; for large pressure differences mercury is commonly used. SEE MANOMETER.

Pressure transmission. Pressure applied to a confined liquid is transmitted with equal intensity throughout the liquid and by it to all surfaces of the confining vessel or piping. Hence, a small force applied to a small area of a confined liquid can create a large force against a large area. If the small and large areas are pistons the device may be a hydraulic press or jack. Because the transmitting liquid is practically incompressible and its volume virtually constant, the linear movement of the large piston will be to that of the small piston in inverse proportion to their areas. The principle of multiplying a force by means of liquid pressure applies also to hydraulic brakes, power steering, control systems, and the like; the actuating force may be a pump instead of a small piston. SEE HYDRAULICS; HYDROMECHANICS.

GAS MECHANICS
FRANK H. ROCKETT

The action of forces on gases. One of the simplest forces on a gas is illustrated by the gravitational attraction of the Earth for its atmosphere. From this and other observations, the weight of an ideal gas is found to be proportional to its molecular weight. Another mechanical property of a gas is its ability to transmit pressure in any direction. In this respect a gas and a liquid behave similarly. SEE PASCAL'S LAW.

AEROMECHANICS
William C. Walter

The science of air and other gases in motion or equilibrium. Aeromechanics has two branches, aerostatics and aerodynamics. Aeromechanics is a special case of the more general field of fluid mechanics, the science of fluids in motion or equilibrium.

Aerostatics is the branch of aeromechanics dealing with the equilibrium of air or other gases, and also with the equilibrium of bodies immersed in a gaseous medium. Examples of aerostatic phenomena are air being compressed in a closed container and the behavior of a dirigible or balloon. See Aerostatics.

Aerodynamics is the branch of aeromechanics dealing with the properties and characteristics of, and the forces exerted by, air and other gases in motion. The resistance and pressure of air flowing through a duct such as a wind tunnel, and the forces exerted by airflow over an airfoil-shaped compressor blade in a turbojet engine are aerodynamic in nature. See Aerodynamics; Wind tunnel.

AEROSTATICS
John R. Sellars

The science of the equilibrium of gases and of solid bodies immersed in them when under the influence only of natural gravitational forces. Aerostatics is concerned with the balance between the weight of the gases and the weight of any object within them. Archimedes' law that an immersed body experiences a buoyance force equal to the weight of the fluid displaced is the principal law of aerostatics, if the fluid is air, or of hydrostatics, if the fluid is water. Some phases of meteorology and the flight of balloons and dirigibles are based on aerostatics. In meteorology cloud and fog subsidence and simple pressure and temperature relations with altitude are predicted from aerostatic principles. See Archimedes' principle; Buoyancy.

Strictly speaking, the air and the immersed body must be at rest for aerostatic principles to apply, but there are many problems where aerostatic forces essentially govern despite some movement. A convenient example of this is given by the motion of a dirigible through the air. Aerodynamic force (drag) limits the speed which the dirigible can achieve, yet the aerostatic forces essentially support the vehicle. This contrasts with the airplane, where aerodynamic forces provide both the lift and the drag. Another example is given by the atmosphere, where the pressure and density relations are determined to a first order by aerostatics, although some motion of the atmosphere takes place through winds and turbulence. See Fluid statics; Hydrostatics.

2

FLUID FLOW

Fluid flow	14
Laminar flow	21
Streamline flow	21
Turbulent flow	22
Wake flow	26
Boundary-layer flow	26
Skin friction	32
Pipe flow	33
Friction factor	35
Open channel	36
Hydraulic gradient	38
Jet flow	38
Ducted flow	39
Nozzle	40
Throttled flow	42
Incompressible flow	42
Compressible flow	43
Isentropic flow	44
Prandtl-Meyer expansion fan	45
Vortex	46
Karman vortex street	47
Source flow	48
Sink flow	50
Doublet flow	50
Stokes stream function	51

FLUID FLOW
ROBERT L. DAUGHTERY

Motion of a fluid as a continuum. Fluids flow whereas solids move as bodies. In flow, the individual particles of a substance move relative to each other as well as to their surroundings. A fluid is a substance that flows under the slightest shear stress. Thus, nonuniform pressure of a gas is sufficient to cause it to flow throughout a container to which it is admitted. The weight of a liquid is sufficient to cause it to flow throughout a container to which it is admitted. The weight of a liquid is sufficient to cause it to flow inside a container but without significant change in volume. Because chemical bonds have to be broken, considerable external force is required to cause a solid to flow.

By analogy, electricity, heat, and other forms of energy are said to flow. This article discusses only flow of fluids.

Fluids. A fluid may be liquid, vapor, or gas. A liquid will fill the container which holds it, but it may have a free surface, that is, a surface at which pressure is that of its own vapor over the surface. All liquids are relatively incompressible. SEE FLUIDS.

A vapor is a gas whose temperature and pressure are such that is very near the liquid phase. Thus, steam is considered to be a vapor because its state is not far from that of water.

A gas may be defined as a highly superheated vapor; that is, its state is far removed from the liquid phase. Thus, air is considered to be a gas because its state is normally very far from that of liquid air. A gas is very compressible, and when all external pressure is removed, it tends to expand indefinitely. A gas is therefore in equilibrium only when it is completely enclosed.

The volume of a liquid is altered only slightly by changes in either pressure or temperature unless the temperature change is considerable or unless the initial temperature is near the critical temperature, but the volume of a gas or a vapor is greatly affected by changes in pressure or temperature. SEE GAS; LIQUID.

Ideal and real flow. When there is no friction between adjacent moving particles, flow is termed ideal; that is, viscosity is zero. In ideal flow, internal forces at any section are always normal to the section. Forces are purely pressure forces. Such flow is approached but never achieved in reality.

In a real fluid, tangential or shearing forces always cause flow and fluid friction, because these viscous forces oppose the sliding of one particle past another. These friction forces are due to a property called viscosity.

Viscosity. The viscosity of a fluid is a measure of its resistance to shear or to angular deformation. Consider two parallel plates large enough that edge conditions may be neglected, and placed a small distance apart, and assume that the space between is filled with the fluid (**Fig. 1**). Assume that the lower surface is stationary while the upper one is moved relative to it with a

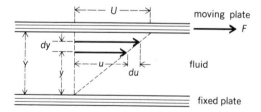

Fig. 1. Diagrammatic representation of viscous flow.

velocity U by the application of a force F corresponding to some area A of the moving plate. Particles of the fluid in contact with each plate will adhere to it, and if the distance Y is not too great and the velocity U not too high, the velocity gradient will be a straight line. The action of the fluid may be likened to a series of thin sheets, each of which slips a little relative to the next. Experiment has shown that for a large class of fluids the shear stress τ between any two thin sheets of fluid may be expressed as Eq. (1), where μ is the coefficient of viscosity. This is also called the absolute or dynamic viscosity. SEE NEWTONIAN FLUID.

$$\tau = \frac{F}{A} = \mu\frac{U}{Y} = \mu\frac{du}{dy} \qquad (1)$$

In a newtonian fluid, the viscosity does not change with the rate of deformation and is represented by the straight line (**Fig. 2**). The slope of this line determines the magnitude of the viscosity. In a non-newtonian fluid, μ varies with the rate of deformation. Printer's ink, pastes,

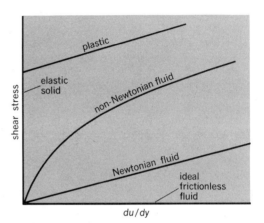

Fig. 2. Types of frictional flow.

paints, greases, and most suspensions in water, such as coal slurries and drilling muds, are non-newtonian fluids. SEE NON-NEWTONIAN FLUID; NON-NEWTONIAN FLUID FLOW.

Absolute viscosity. In the metric system, the unit of absolute viscosity is kg/m·s; in the cgs system, it is the poise (1 poise = 1 dyne-s /cm^2). Because most fluids have low viscosities, the centipoise (0.01 poise) is frequently a more convenient unit. It has further advantage that the viscosity of water at 68.4°F or 20.2°C (1 centipoise) provides a convenient reference.

In the English gravitational, or engineers', system, the viscosity unit in pounds (force) is 1 lb-s/ft^2 or 1 slug/ft-s. In the English absolute system, viscosity is expressed in terms of poundal-s/ft^2 or lb/ft-s. There are no names for the English units.

The conversion factors are 10^{-1} kg/m·s ≡ 1 poise ≡ 100 centipoises ≡ 0.00209 lb-s/ft^2 ≡ 0.0672 poundal-s/ft^2.

The viscosities of all liquids decrease with an increase in temperature, whereas those of all gases and vapors increase. The absolute viscosity of both liquids and gases is practically independent of pressure except for extremely high pressures, at which there is a small increase.

Kinematic viscosity. In many fluid-flow problems there occurs the value of viscosity divided by density. This is called kinematic viscosity, and is $\nu = \mu/\rho$. In the metric system the unit of kinematic viscosity is the m^2/s; in the cgs system it is the stoke (1 stoke = 1 poise divided by density, g/cm^3). A convenient unit is the centistoke (0.01 stoke). The dimensions of kinematic viscosity are cm^2/s in the cgs system.

There is no name for the unit of kinematic viscosity in the English system, but the dimensions are ft^2/s. It is necessary to be consistent in the units employed. Thus, viscosity in lb(force)-s/ft^2 should be divided by density in slugs/ft^3, or viscosity in poundal-s/ft^2 should be divided by density in lb/ft^3, which is numerically the same as specific weight.

The conversion factors are 10^{-4} m^2/s ≡ 1 stoke ≡ 100 centistokes ≡ 0.001076 ft^2/s and 1 ft^2/s ≡ 929 cm^2/s. SEE VISCOSITY.

Density and specific weight. Density is mass per unit volume and is normally indicated by ρ. In the metric system the dimensions of ρ are kg/m^3. In the English systems they are either slugs/ft^3 or lb/ft^3 according to whether the engineers' system or the absolute system is used. The value of density is, however, the same for any location.

Specific weight is weight or gravitational force per unit volume and is commonly designated by either w or γ. In the English engineers' system dimensions are lb(force)/ft^3. In the absolute system they are slugs/ft^3 where 1 slug $\equiv$ 32.174 lb(mass). Specific weight in the engineers' system varies with the value of gravity and hence varies with location.

The relation between density and specific weight is $\rho = w/g$, where g is the acceleration of gravity. The standard sea-level values of g are 980.66 cm/s^2 or 32.174 ft/s^2.

Compressibility. All liquids are relatively incompressible, and for most purposes water in particular can be treated as incompressible; yet it is 10 times as compressible as steel. The passage through water of a sound wave, which is really a pressure wave, is a result of the water's compressibility. *See* Water hammer.

In dealing with the flow of air or other gases when the change in pressure is small, so that the change in density is negligible, even gases can be treated as incompressible. For an airplane flying at speeds of less than 250 mi/h (110 m/s) the air can be considered to be of constant density. However, as the speed approaches that of sound in air, which is of the order of 700 mi/h (300 m/s), the pressure of the air adjacent to the body becomes materially different from that at some distance away, and air must then be considered to be compressible. *See* Compressible flow.

The change in volume from v to v_1 of a fluid as the result of a change in pressure Δp can be determined by $\Delta v/v_1 = -\Delta p/E_v$, where E_v is the mean value of the volume modulus of elasticity for the pressure range. For water the value of the volume modulus varies with both temperature and pressure, but a typical value for most conditions is 300,000 psi or 2 gigapascals (Δp and E_v must both be in the same units). For isothermal compression of a perfect gas, $E_v = p$ and for an adiabatic compression $E_v = kp$, where k is the ratio of the specific heat at constant pressure to that at constant volume. For air the value of k is normally about 1.4, and p is the average pressure range. For low pressures air is many thousand times as compressible as water.

For practical use the change in volume of a gas is better determined by means of the ideal gas equation of state.

Ideal gas. An ideal gas is one whose equation of state is $\rho = pM/RT$, where ρ is density in metric units of kg/m^3 or engineering units of lb(mass)/ft^3, p is absolute pressure in metric units of N/m^2 or engineering units of lb(force)/ft^2, M is molecular weight of the gas in metric units of g/mol or engineering units of lb(mass)/lb mole, T is absolute temperature in metric units of kelvins [$T(K) = T(°C) + 273.15$] or engineering units of degrees Rankine [$T(°R) = T(°F) + 459.7$], and R is a universal constant with value in metric units of 8.314 N · m/mol · K or in engineering units 1545 ft · lb(force)/lb mole · °R. The average value of M for dry air is 29 g/mol in metric units and 29 lb(mass)/lb mole in engineering units.

There is no ideal gas, but air and other so-called permanent gases (gases at conditions far from their liquid states) may usually be so considered. For normal steam pressures, the equation for ideal gas density does not hold, and vapor tables must be used. Also, for real gases when high pressure or low temperatures are involved the equation of state is more complicated.

Real gases. For real gases, in which the conditions are such that the ideal gas law is not sufficiently accurate, a compressibility factor Z may be used in the equation of state $pv = ZRT$, where Z may range from 0.2 to 3 in extreme cases.

The van der Waals equation of state is sometimes useful at high pressures.

Ideal fluid flow. The flow of an ideal fluid is also termed inviscid flow, because it assumes an imaginary fluid of zero viscosity which therefore has no fluid friction. In such flow all particles of the fluid would move along individual streamlines and with equal velocities (**Fig. 3**).

Because there is no fluid friction, the total energy is the same throughout. For an incom-

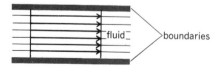

Fig. 3. Diagrammatic representation of ideal fluid flow.

pressible liquid Bernoulli's equation with units in terms of "head" (m or ft), Eq. (2), can be used

$$\frac{p}{w} + z + \frac{V^2}{2g} = \text{constant} \qquad (2)$$

from one streamline to the next and also along any streamline. In Eq. (2) p/w is pressure head in height of the fluid. If p is pressure in lb(force)/ft^2 and w is specific weight in lb(force)/ft^3, p/w is a linear quantity in feet of the fluid. Potential energy of a unit weight of fluid is represented by z, which is a height above any arbitrary datum plane. The kinetic energy per unit weight is represented by $V^2/2g$ with V the velocity in ft/s and g the acceleration of gravity in ft/s^2.

Each term in Bernoulli's equation is a quantity that also represents energy per unit weight. Strictly speaking, the energy possessed by a particle of fluid is the sum of its potential and kinetic energy; that is, $z + V^2/2g$, whereas p/w represents the potential work which can be done as a result of pressure and motion.

Bernoulli's equation can also be expressed with terms in units of energy per unit mass [N · m/kg or ft · lb(force)/lb(mass)], as in Eq. (3).

$$\frac{p}{\rho} + gz + \frac{V^2}{2} = \text{constant} \qquad (3)$$

In many real cases fluid friction is so small that results obtained by Bernoulli's theorem are sufficiently accurate for practical purposes or in many cases need only be modified slightly. Thus, the velocity of a jet issuing from a tank under a head h can be obtained by introducing a velocity coefficient into the equation such that $V = C_v\sqrt{2gh}$, where C_v may have a value as high as 0.98 or 0.99. Thus, the actual value of V is little less than the ideal value. *See* BORDA MOUTHPIECE; JET FLOW.

However, caution must be used in neglecting the effect of viscous friction; in many cases, the fluid friction is large, and the actual result may be very different from the ideal frictionless value.

The effect of viscous friction really originates where the fluid is in contact with a solid surface, and at a considerable distance from such surface the effect of friction is much less. Thus, fluid friction resulting from viscosity produces significant results near a body, such as an airplane, but at some distance away the relative velocity of the air may be considered as uniform (Fig. 3). *See* BOUNDARY-LAYER FLOW.

Real fluid flow. Real fluid flow and viscous flow are synonymous terms; all real fluids are more or less viscous. Because of fluid friction, Bernoulli's equation must be modified by the introduction of a term h_f, which represents the loss of head, or energy. Thus, an energy equation between two points in a one-dimensional steady flow is written as Eq. (4), which shows that the

$$\frac{p_1}{w} + z_1 + \frac{V_1^2}{2g} = \frac{p_2}{w} + z_2 + \frac{V_2^2}{2g} + h_f \qquad (4)$$

total head or mechanical energy per unit mass always decreases in the direction of flow. This equation applies to an incompressible fluid. *See* BERNOULLI'S THEOREM.

In some real cases the loss of head may be small and can be disregarded with slight error, but in other cases it may be large. Values of h_f are obtained for specific cases by equations that are more or less empirical and are based upon experimental data.

Laminar flow. Laminar flow is also a kind of streamline flow, because all particles of the fluid move in distinct and separate lines. It is called laminar flow because the action is as if layers or lamina of fluid slide relative to each other. In the case of a laminar flow in a circular pipe, the velocity adjacent to the wall is zero and increases to a maximum in the center of the pipe (**Fig. 4**). The velocity profile in a circular pipe is a parabola, and the average velocity for volume flow is 0.5 times the maximum velocity in the center. *See* PIPE FLOW; STREAMLINE FLOW.

In the case of laminar flow in a circular pipe, the loss of head due to fluid friction is not given by any empirical equation but is given by an equation known as the Hagen-Poiseuille law, Eq. (5), where h_f is linear head in feet of the fluid, p is pressure in lb(force)/ft^2, μ is absolute

$$h_f = \frac{p_1 - p_2}{w} = 32\frac{\mu}{w}\frac{L}{D^2}V = 32\nu\frac{L}{gD^2}V \qquad (5)$$

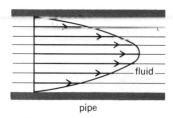

Fig. 4. Laminar flow in a circular pipe.

viscosity in lb(force)/ft-s, w is specific weight in lb (force)/ft^3, ν is kinematic viscosity in ft^2/s, L is flow distance between measured pressure levels 1 and 2 in ft, D is diameter in ft, and V is velocity in ft/s.

Laminar flow in a circular pipe will be found when the Reynolds number, $DV\rho/\mu = DV/\nu$, is less than 2000. SEE LAMINAR FLOW.

Turbulent flow. In turbulent flow no distinct streamlines are found. Instead, the fluid consists of a mass of eddies (**Fig. 5**a). No two particles can follow the same or similar paths (Fig. 5b). The velocity profile shows a maximum velocity in the center, while near the wall the velocity is about one-half the center velocity. The profile is flatter for a smooth pipe than it is for a rough one, and the ratio of average velocity to maximum velocity in the center ranges from about 0.74 for a very rough pipe to about 0.88 for a very smooth one.

In a circular pipe the loss of head due to fluid friction is given by $h_f = f(L/D)V^2/2g$, where f is a function of both Reynolds number and the wall roughness, L/D is the ratio of pipe length to pipe diameter, V is the average velocity in ft/s, and g is the acceleration of gravity, normally taken as 32.2 ft/s^2.

Turbulent flow usually occurs in a circular pipe when Reynolds number has values greater than 2000.

In the case of a solid body immersed in a stream, there is a turbulent wake in the rear (**Fig. 6**a and b). This produces a pressure difference which results in a force on the body known as form or pressure drag. SEE TURBULENT FLOW; WAKE FLOW.

Uniform flow. If at a given instant the velocity is the same in both magnitude and direction at every point in space, the flow is uniform. This strict definition can have little meaning for the flow of a real fluid when the velocity varies across a section. However, when the size and shape of cross section are the same in any given length, the flow is said to be uniform. Specifically, the flow in a pipe of constant diameter is uniform; the flow in a pipe of varying size is not. Also, the flow in an open canal is uniform if the size and shape of the cross section of the stream are the same at different locations along the canal. SEE OPEN CHANNEL.

Steady flow. By steady flow is meant that all conditions at any one point are constant with respect to time. True steady flow is found only with laminar flow. In turbulent flow there are continual fluctuations in velocity and pressure at every point. However, if the values fluctuate on both sides of an average value that is constant, then the flow is mean steady flow.

In steady flow, conditions are usually constant in time from one section to another, although not necessarily the same at different sections. Thus, along the line of flow the equation of continuity applies, which is written as Eq. (6), or for a fluid of constant specific weight, as Eq. (7).

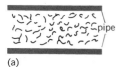
(a)

(b)

Fig. 5. Turbulent flow in a circular pipe. (a) General turbulence. (b) Path of a single particle.

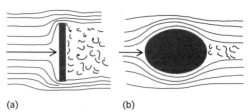
(a)　　　　　(b)

Fig. 6. Turbulent wake of solid body immersed in a stream. (a) Rectangular body. (b) Oval body.

$$W = \rho_1 A_1 V_1 = \rho_2 A_2 V_2 = \text{constant} \quad (6) \qquad Q = A_1 V_1 = A_2 V_2 = \text{constant} \quad (7)$$

Here W is mass flow rate in lb(mass)/s, w is density in lb(mass)/ft^3, A is cross-sectional area in ft^2 normal to the velocity, V is average velocity across the section in ft/s, and Q is flow in ft^3/s.

Unsteady flow. This means that conditions are changing with respect to time and ultimately may become either steady flow or zero flow. This changing rate of flow may take place slowly, as when the action of a valve in a channel produces a gradual change in the rate, or it may take place rapidly as a result of a sudden closure, which produces a phenomenon known as water hammer. It is also found in such a case as the flow from one reservoir to another, in which equilibrium is approached as the two levels approach each other.

Unsteady flow also includes periodic motion such as that of waves on beaches, tidal motion in estuaries, and other oscillations. The difference between such cases and mean steady flow in turbulent flow is that the deviations from the mean are much greater and the time scale is also much longer. *See Wave motion in liquids.*

Compressible flow. For compressible fluids such as gases and vapors it is necessary to add thermal terms to the Bernoulli equation to obtain an equation accounting for total energy, not just mechanical energy. The energy equation between station 1 and station 2 then becomes Eq. (8), where q is Btu/lb of fluid flowing, which may be transferred from the fluid to the surroundings.

$$\frac{p_1}{\rho_1} + JI_1 + gz_1 + \frac{V_1^2}{2} - Jq = \frac{p_2}{\rho_2} + JI_2 + gz_2 + \frac{V_2^2}{2} \quad (8)$$

If a turbine or pump occurs between the two sections, a work input term must also be added. If the heat flow is into the fluid, then q is negative, as the equation is written. Internal energy is thermal energy and is represented by I in Btu/lb(mass), and J is 778 ft-lb(force)/Btu. For a perfect gas $\Delta = c_v \Delta T$ for each unit of mass. For a gas or a vapor the quantity p/ρ is usually large relative to $g(z_1 - z_2)$, because of the small value of ρ in general. Therefore the z terms are usually, but not always, neglible.

Because p/ρ, which equals pv, and I both describe the state of gases and vapors, it is customary to replace them by a single term called enthalpy H, which is defined as $H = I + pv/J$ in Btu/lb(mass). Therefore the energy equation becomes Eq. (9). If metric units are used, J has a value of unity.

$$JH_1 + \frac{V_1^2}{2} - Jq = JH_2 + \frac{V_2^2}{2} \quad (9)$$

Isothermal flow of a gas. For a perfect gas $I_1 = I_2$ if the temperature is unchanged, and $p_1/\rho_1 = p_2/\rho_2$. Hence for this special case Eq. (10) holds, which shows that an isothermal flow has

$$V_2^2 - V_1^2 = -2gJq \quad (10)$$

to be accompanied by an absorption of heat when there is an increase in kinetic energy. This equation is true either with or without friction, because friction supplies some of energy necessary to maintain constant temperature; however, less heat is absorbed from the surrounding medium. Consequently q is less, when the increase in kinetic energy is the same.

Adiabatic flow. In adiabatic flow, heat transfer is zero; hence the energy equation becomes Eq. (11).

$$JH_1 + \frac{V_1^2}{2} = JH_2 + \frac{V_2^2}{2} \quad (11)$$

For a perfect gas $\Delta H = c_p \Delta T$, where c_p is specific heat per pound at constant pressure. For such a gas undergoing adiabatic, frictionless compression or expansion, Eq. (12) holds, where

$$\frac{V_2^2 - V_1^2}{2g} = \frac{k}{k-1}(p_1 v_1 - p_2 v_2) \quad (12)$$

$k = c_p/c_v$ (ratio of specific heats at constant pressure and at constant volume), and where $v = 1/\rho$.

The preceding equations are valid with or without fluid friction. For a frictionless flow the values of H_2, T_2, and v_2 are all determined as the result of an isentropic expansion from p_1 to p_2. SEE ISENTROPIC FLOW.

Rotational and irrotational flow. In rotational flow, each minute particle of fluid appears to a stationary observer to rotate about its own axis. A specific case is a forced vortex in which a fluid rotates like a solid body. Such a case might be obtained with a fluid in a cylindrical vessel rotating about its central axis, assuming no motion of the fluid relative to the container. A more common case is that in which the container is fitted with vanes, as the impeller of a centrifugal pump. As a small element rotates about the central axis, it also rotates about its own axis (**Fig. 7***a*).

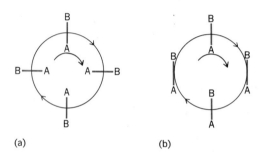

(a)　　　(b)

Fig. 7. Rotational and irrotational flow. (*a*) Particles rotate on their axes as they move around. (*b*) Particles retain absolute orientation during motion.

In irrotational flow, each infinitesimal particle or element of the fluid preserves its original orientation (Fig. 7*b*). Because an element of fluid can be caused to rotate about its axis only by the application of viscous forces, rotational flow is possible only with a real or viscous fluid, whereas irrotational flow is possible only for an ideal or nonviscous fluid. For fluids of low viscosity, such as air or water, irrotational flow may be approached in a free vortex. In a free vortex, a body of fluid rotates without the application of external torque because of angular momentum previously imparted to it. Examples are the rotation of fluid after leaving the impeller of a centrifugal pump, a tornado, or the rotation of water entering the drain of a tub. SEE VORTEX.

Boundary-layer flow. For an ideal frictionless fluid the velocity of flow adjacent to a surface would be the same as that at a distance. In actuality, adhesion between fluid and boundary surface tends to make the velocity of the fluid at the surface equal to the velocity of the surface body. For a very small distance from the surface, the velocity increases with distance at a very rapid rate because of viscosity within the fluid. The flow in this thin layer is laminar in character. This thin layer is known as the laminar boundary layer.

There is then a transition zone, the boundaries of which are indefinite, and beyond this the flow is fully turbulent. Much farther from the surface, the effect of the surface vanishes and the flow is undisturbed. The layer between the laminar one and the undisturbed field is known as the turbulent boundary layer. It is thick and there is no sharp line of demarcation between it and the undisturbed uniform flow. Such a case may be found with an airplane at a considerable height above the ground or a submarine deeply submerged beneath the surface. In the case of flow in a pipe or other conduit the turbulent boundary layers from opposite sides may meet so that there is no zone of undisturbed flow. Viscosity effects are most pronounced at or near a solid boundary and diminish rapidly with distance from the boundary. SEE FLOW MEASUREMENT; FLUID-FLOW PRINCIPLES; GAS DYNAMICS; HYDRODYNAMICS; LOW-PRESSURE GAS FLOW.

Bibliography. J. W. Daily and D. R. F. Harleman, *Fluid Dynamics*, 1966; R. L. Daugherty and J. B. Franzini, *Fluid Mechanics*, 7th ed., 1977; I. H. Shames, *Mechanics of Fluids*, 2d ed., 1982; V. L. Streeter, *Handbook of Fluid Dynamics*, 1961; V. L. Streeter and E. B. Wylie, *Fluid Mechanics*, 8th ed., 1985; J. K. Vennard and R. L. Street, *Elementary Fluid Mechanics*, 6th ed., 1982.

LAMINAR FLOW
Frank M. White

A smooth, streamline type of viscous fluid motion characteristic of flow at low-to-moderate deformation rates. The name derives from the fluid's moving in orderly layers or laminae without the formation of small eddies or irregular fluctuations.

The chief criterion for laminar flow is a relatively small value—less than a few thousand—for the Reynolds number, Re = $\rho VL/\mu$, where ρ is fluid density, V is flow velocity, L is body size, and μ is fluid viscosity. Thus laminar flow may be achieved in many ways: low-density flows as in rarefied gases; low-velocity or "creeping" motions; small-size bodies such as microorganisms swimming in the ocean; or high-viscosity fluids such as lubricating oils. At higher values of the Reynolds number, the flow becomes disorderly or turbulent, with many small eddies, random fluctuations, and streamlines intertwining like spaghetti. SEE CREEPING FLOW; REYNOLDS NUMBER; TURBULENT FLOW; VISCOSITY.

Nearly all of the many known exact solutions of the equations of motion of a viscous fluid are for the case of laminar flow. These mathematically accurate descriptions can be used to give insight into the more complex turbulent and transitional flow patterns for which no exact analyses are known. SEE NAVIER-STOKES EQUATIONS.

The theory of viscous lubricating fluids in bearings is a highly developed area of laminar flow analysis. Even large Reynolds number flows, such as aircraft in flight, have regions of laminar flow near their leading edges, so that laminar flow analysis can be useful in a variety of practical and scientifically relevant flows. SEE BOUNDARY-LAYER FLOW; FLUID FLOW.

Bibliography. W. F. Hughes, *An Introduction to Viscous Flow*, 1979; H. Schlichting, *Boundary Layer Theory*, 7th ed., 1979; R. K. Shah and A. L. London, *Laminar Flow Forced Convection in Ducts*, 1978; F. M. White, *Viscous Fluid Flow*, 1974.

STREAMLINE FLOW
Arthur G. Hansen

A condition of fluid flow characterized by the absence of turbulence. Other designations employed are laminar flow or viscous flow.

Fluid-flow particles in streamline flow follow well-defined continuous paths or streamlines. At a fixed point in streamline flow, the flow velocity either remains constant (steady flow) or varies in a regular fashion with time (unsteady flow). In turbulent flow the velocity at a given point exhibits irregular, high-frequency fluctuations with time. In some instances, streamline flow can best be depicted as formed from thin layers of fluid which slip past each other (lamellar flow). As an illustration, streamline flow in a straight pipe might be considered as formed from layers in the shape of concentric annuli. If the flow can properly be represented by thin, plane layers (laminae) sliding past each other, the flow is commonly referred to as laminar flow, although this term is often used to designate streamline flow in general.

The persistence of streamline flow in a given system is largely dependent on the value of a nondimensional parameter called the Reynolds number, defined as $\rho LU/\mu$, where ρ is the fluid density, U is a reference flow velocity, L is a reference length, and μ is the coefficient of viscosity. When the Reynolds number of a particular flow exceeds a certain value (the critical Reynolds number), transition from streamline flow to turbulent flow generally takes place.

Under special circumstances flow may alternate between streamline flow and turbulent flow. This phenomenon is readily observed in pipe flow when the Reynolds number approaches the critical value. The pipe flow may become locally turbulent, the turbulent region passing downstream as a plug followed by streamline flow. As the plug leaves the pipe, the entire process is repeated.

It is possible for both streamline flow and turbulent flow regimes to exist simultaneously in fluids with low values of viscosity, such as air. In such fluids, frictional effects are often confined to thin layers of fluid, known as boundary layers, in the immediate neighborhood of the bounding

surfaces. The boundary layer may be locally turbulent, whereas the flow in the mainstream away from the surfaces may be streamline flow. SEE BOUNDARY-LAYER FLOW; LAMINAR FLOW; STREAMLINING; TURBULENT FLOW.
Bibliography. S. Pai, *Viscous Flow Theory*, vol. 1, 1965; L. Prandtl and O. G. Tietjens, *Applied Hydro- and Aeromechanics*, 1934, reprint 1957; V. L. Streeter and E. B. Wylie, *Fluid Mechanics*, 8th ed., 1985.

TURBULENT FLOW
SHIH I. PAI

Motion of fluids in which local velocities and pressures fluctuate irregularly. Most flows observed in nature, such as rivers and winds, are turbulent. Such flows occur at high Reynolds numbers. In turbulent flow, motion of the fluids is steady only insofar as the temporal mean values of velocities and pressures are concerned. The velocity and the pressure distributions in turbulent flows as well as the energy losses are determined mainly by the turbulent fluctuations. SEE REYNOLDS NUMBER.

Random nature. The essential characteristic of turbulent flow is that the fluctuations are random. Hence, the solution of the turbulence problem requires the application of methods of statistical mechanics. In turbulent flow, the most important phenomenon is the transfer of forces by such random motion. The rate of turbulent transfer, such as heat transfer, shearing stress, and diffusion, is much higher than that due to molecular mechanism in laminar flow. SEE VISCOSITY.

In turbulent motion, even though the fluid is regarded as a continuum with an average overall molecular motion, turbulent velocity fluctuations must be superimposed on the mean motion. The separation of mean motion and turbulent fluctuation depends mainly on the scale of turbulence. Different scales give different descriptions of turbulent flow. Once the scale of turbulence is chosen, the instantaneous velocity component u_i for instance, is given by Eq. (1), where

$$u_i = \bar{u}_i + u'_i \qquad (1)$$

u_i is the ith component of the total fluid velocity, $\bar{u}_i$ is the ith mean velocity component, and u'_i is the ith component of the turbulent fluctuating velocity.

Turbulent stresses. The instantaneous velocity component u_i satisfies the Navier-Stokes equations of motion of a viscous fluid. The substitution of the expression for the instantaneous velocity components into the Navier-Stokes equations and the use of the mean values of the equations give the Reynolds equations for turbulent flow. The difference of the Reynolds equation from the Navier-Stokes equations is due to the additional terms of turbulent stresses. The turbulent normal stresses are $-\rho \overline{u'^2_i}$, and the turbulent shearing stresses are $-\rho \overline{u'_i u'_j}$, where $i \neq j$ and ρ is the density of the fluid. SEE NAVIER-STOKES EQUATIONS.

These stresses represent the rate of transfer of momentum across the corresponding surfaces because of turbulent velocity fluctuations.

Semiempirical theories of turbulence. To illustrate various semiempirical theories of turbulent flow, consider a simple parallel mean flow $\bar{u} = \bar{u}(y)$, $\bar{v} = 0$, $\bar{w} = 0$, with fluctuating velocity components u', v', and w', where u, v, and w are the x, y, and z components of velocity, respectively. Semiempirical theories of turbulence are formulated based on various hypotheses about the turbulent stresses.

J. Boussinesq introduced the turbulent exchange coefficient ϵ such that Eq. (2) holds. In

$$\overline{u'v'} = -\epsilon \frac{d\bar{u}}{dy} \qquad (2)$$

actual analysis, further hypotheses are necessary about the variations of ϵ, which are different for different flow.

L. Prandtl originated the mixing-length theory, in which the fluctuating value of a transferable quantity q of the fluid may be written as Eq. (3), where l is the mixing length. For simple parallel flow, the exchange coefficient becomes Eq. (4).

$$|q'| = l\frac{d\bar{q}}{dy} \quad (3) \qquad \epsilon = -\overline{lv'} \quad (4)$$

There are many different mixing-length theories based on different quantities being transferred. In Prandtl's momentum transfer theory, the momentum of the fluid elements is assumed to be preserved in the mixing process. Prandtl obtained a formula for shearing stress τ of nearly parallel turbulent flow as defined by Eq. (5). This formula has been used successfully in the cal-

$$\tau = \rho l^2 \left|\frac{d\bar{u}}{dy}\right|^2 \quad (5)$$

culation of many turbulent flow problems such as flow along a flat plate and in jet and wakes.

In the vorticity transfer theory of G. I. Taylor, the transferable quantity is vorticity. T. von Kármán made a similarity hypothesis to determine the mixing length so that no special model for a transferable quantity is required.

Logarithmic velocity profile. The most important deduction from von Kármán's similarity hypothesis is the universal velocity distribution for the flow in circular pipes or between parallel plane walls. Von Kármán first pointed out that the ratio between the velocity defect $U_m - \bar{u}$ and the quantity $\sqrt{\tau_0/\rho}$ is a universal function of the ratio $(y_0 - y)/y_0$, as in Eq. (6), where y_0 is

$$\frac{U_m - \bar{u}}{\sqrt{\tau_0/\rho}} = f\left(\frac{y_0 - y}{y_0}\right) \quad (6)$$

the radius of the circular pipe or the half-width between the two plates, y is the distance from the wall, $\bar{u}$ is axial velocity, U_m is the maximum axial velocity occurring at $y = y_0$, which is the center of the channel, and τ_0 is the shearing stress at the wall $y = 0$.

The universal velocity distribution is given by Eq. (7), where Δ is of the same order of

$$\bar{u} = 2.5 \sqrt{\frac{\tau_0}{\rho}} \log \frac{y + \Delta}{\delta} \quad (7)$$

magnitude as the thickness of the laminar sublayer, which is negligible. For smooth wall, length δ is determined by a physical parameter such as density and viscosity of the fluid; for rough surface, it is determined by the roughness of the wall.

For engineering application, a nondimensional pipe-resistance coefficient λ is used such that Eq. (8) holds. Here $p_1 - p_2$ is the pressure drop along a pipe of length L and diameter d, and

$$\frac{dp}{dx} = \frac{p_1 - p_2}{L} = \frac{\lambda}{d}\frac{\rho}{2} u_0^2 \quad (8)$$

u_0 is the average mean velocity over a section of the pipe. *See Dimensionless groups.*

For smooth pipe, Eq. (9) holds.

$$\frac{1}{\sqrt{\lambda}} = 2.0 \log_{10}\left(\frac{u_0 d}{\nu}\sqrt{\lambda}\right) - 0.80 \quad (9)$$

For pipe of rough surface, the resistance depends on size, shape, and spacing of the roughness elements. Only for closely packed roughness can linear dimension h of the roughness alone be used to describe the roughness. For a completely rough pipe in which $(\sqrt{\tau_0/\rho})h/\nu > 100$, where ν is the coefficient of kinematic viscosity, the resistance law is given by Eq. (10).

$$\frac{1}{\sqrt{\lambda}} = 2.00 \log_{10}\left(\frac{y_0}{h}\right) + 1.74 \quad (10)$$

In the noncircular pipe, the characteristic length is often represented by the hydraulic mean length L_h, which is $L_h = 2A/L_w$, where A is the cross-sectional area of the pipe and L_w is the wetted circumferential length. If the hydraulic mean length is used instead of the radius of the circular pipe, resistance law (9) may be used for noncircular pipes with an accuracy within a few percent.

Turbulent jet mixing. Another type of turbulent flow without a solid wall in the flow field is known as the free-turbulence problem. Jet mixing and wakes are in this class. *See* JET FLOW; WAKE FLOW.

In free-turbulence problems, the application of Prandtl's mixing length theory is more successful than that for the turbulent boundary-layer flow along a solid wall. In the free-turbulence problem, simple and plausible assumptions on the variation of mixing length in the flow field are possible. The mean velocity distribution calculated on the basis of these assumptions agrees well with the experimental results over a major portion of the flow field. *See* BOUNDARY-LAYER FLOW.

For a turbulent jet in a medium at rest, the jet spreads linearly. For a wake of a body of revolution, the width of the wake increase with $(C_D S_b x)^{1/3}$, where C_D is the drag coefficient of the body, S_b is the reference area of the body, and x is the distance from the body.

For a first approximation, velocity distributions in a jet mixing region and in a wake may be represented by error functions.

For turbulent jet mixing of fluids of different temperatures or of different densities, the spread of temperature and of concentration are about the same, and they are usually wider than the spread of velocity profile.

Statistical theory of turbulence. Even though the semiempirical theory has had successfully predicted mean velocity distributions in many practical problems, it has serious limitations and inconsistencies. For an understanding of turbulent flow in general, a study of the mean velocity distribution is insufficient. The fields of turbulent fluctuations must be studied in detail. Because turbulent-velocity fluctuations of a fluid are much too complicated, changing too rapidly in time and location to be known in all their details, a study of only some mean values is feasible. These mean quantities include the intensity of turbulent fluctuations, the correlation functions, and the spectrum of turbulence.

Modern statistical theory of turbulence was developed by Taylor, who introduced the correlation function, the spectrum, and the concept of statistically isotropic turbulence. Great simplification can be obtained by the isotropic property. Hence, most results from statistical theory of turbulence are concerned with isotropic turbulence.

Correlation function. Consider the fluctuating variables u_A and u_B between stations A and B, and assume that there exists a certain correlation between them. The correlation function is then given by Eq. (11), where the bar means taking the average. The correlation coefficient R_{AB} is given by Eq. (12). The correlation coefficent lies within the limits of -1 and $+1$.

$$\rho_{AB} = \overline{u_A u_B} \quad (11) \qquad R_{AB} = \frac{\overline{u_A u_B}}{\sqrt{\overline{u_A^2}} \sqrt{\overline{u_B^2}}} \quad (12)$$

Von Kármán first pointed out the tensor character of the correlation function. Both von Kármán and Taylor studied extensively the correlation functions between the components of fluctuating velocity at the same time at two different points of the fluid for isotropic turbulence. These correlations functions had been measured by hot-wire anemometer. Experimental results check well with theory.

For isotropic turbulence, the correlation function is a function of time t and distance r between two points. The curvature of the double correlation curve at $r = 0$ determines a microscale of turbulence, which is a measure of the size of the smallest eddies in the turbulent flow, these eddies being responsible for the dissipation of turbulent energy. The integration of the correlation coefficient over $0 < r < \infty$ gives the scale of turbulence; the scale of turbulence is a measure of the large eddies in the turbulent flow.

Spectrum. A more detailed description of turbulence can be obtained by considering the distribution of energy among eddies of different sizes. This description can be put into precise mathematical form by considering the distribution of energy with frequency or with wave number, which is known as the spectrum of turbulence. Spectral density is a Fourier transform of correlation coefficient. Spectrum of turbulence can be measured by hot-wire anemometer.

Local isotropy. The most significant idea contributed to the problem of turbulent shear flow in recent years is the hypothesis of local isotropy proposed by A. N. Kolmogoroff. He suggested that the fine structure in turbulent shear flow may be isotropic. Turbulent motion is considered to be a mixture of eddies of all sizes from the largest, whose dimensions are comparable with those of the main flow or of the turbulence-producing mechanism such as a grid of bars in a wind tunnel, down to the smallest eddies. When turbulent motion starts, the mean flow breaks up, or the eddies produced by the grid break up into smaller eddies, their motions being unstable: these in turn break up into smaller eddies and so on, until eddies are produced of a small enough size to be stable; this gives a lower bound to the eddy size. Kolmogoroff's idea may be expressed by saying that there is something universal about small eddies; below a certain eddy size the nature of the motion is unaffected by the origin of the turbulence, and it is expected that eddies, small compared with the dimensions of the mean flow, will be statistically isotropic. Kolmogoroff's idea of local isotropy has been verified experimentally by many research workers.

Except for the concept of local isotropy, little has been accomplished for the statistical theory of maintained shear turbulence. However, the statistical theory of isotropic turbulence shows which quantities are important in describing the fluctuating field; they include turbulent intensities, correlation function, spectrum, and probability distribution. In the experimental investigations of shear flow, such as flow in circular pipe, channel, boundary layer over a flat plate, jets, and wakes, these are the quantities to be measured.

Turbulent diffusion. Diffusion is a fundamental process of turbulence. There is an essential difference between molecular and turbulent diffusion. In molecular diffusion, the medium consists of discrete particles, while in turbulence diffusion, the medium is continuous. The old method of investigating turbulent diffusion is semiempirical; it uses Boussinesq's turbulent exchange coefficient. The new method for solution for turbulent diffusion uses the statistical theory of turbulence. In statistical theory, two different approaches have been used. One is the continuous stochastic process in which the diffusion equation is obtained from a probabilistic integral equation. The other approach is the random walk method.

Compressible fluid flow. For the turbulent flow of an incompressible fluid, the effect of variation of density in the expression of turbulent stresses is neglected. This effect is no longer negligible for the turbulent flow of a compressible fluid and cannot be neglected for high-speed flow, flow with large variation of temperature, or both. The study of the turbulent flow of a compressible fluid requires the correlation of velocity components, of velocity and density, and of pressure and velocity. To obtain these three correlations is a complicated procedure.

Mixing-length theories may be extended to compressible fluid. For two-dimensional parallel flow with mean flow field Eqs. (13) hold. The fluctuations of velocity component u, density ρ, and

$$\bar{u} = \bar{u}(y) \qquad \bar{v} = 0$$
$$\bar{\rho} = \bar{\rho}(y) \qquad \bar{T} = \bar{T}(y) \qquad (13)$$

temperature T of the fluid may be written as Eqs. (14), where l, l_ρ, and l_T are the corresponding

$$|u'| = l\frac{d\bar{u}}{dy} \qquad |\rho'| = l_\rho \frac{d\bar{\rho}}{dy} \qquad |T'| = l_T \frac{d\bar{T}}{dy} \qquad (14)$$

mixing lengths for the velocity, density, and temperature. It is customary to assume that these mixing lengths are equal to simplify the analysis and to aid in solving practical problems. Experimental evidences indicate, however, that they are not equal. For instance, in jet mixing of a compressible fluid, the spread of temperature is wider than that of velocity. One way to explain this phenomenon is to assume that l_T is larger than l.

The statistical theory of isotropic turbulence has been extended to the case of compressible fluids. SEE COMPRESSIBLE FLOW.

Electrically conducting fluid flow. Magnetohydrodynamics deals with flow in electrically conducting fluids in which electromagnetic forces are of the same order of magnitude as gas-dynamic forces such as pressure and viscosity. Magnetohydrodynamics is important in problems of astrophysics, geophysics, and the behavior of interstellar gas masses, as well as in such engineering problems as reentry of intercontinental ballistic missiles, controlled fusion, and plasma jets. Because of the large dimensions, it seems probable that the normal state of motion in the cosmos should be turbulent. The high speed of intercontinental ballistic missiles also causes the flow to be turbulent. Controlled fusion research shows that the turbulent dissipation in magnetohydrodynamics is a main difficulty to be overcome before controlled fusion becomes successful.

In the study of turbulence in magnetohydrodynamics, correlations between magnetic field and velocity components are important. SEE FLUID FLOW, FLUID-FLOW PRINCIPLES.

Bibliography. P. Bradshaw, *An Introduction to Turbulence and Its Measurements*, 1971; W. Frost and T. H. Moulden (eds.), *Handbook of Turbulence: Fundamentals and Applications*, 1977; J. O. Hinze, *Turbulence*, 2d ed., 1975; W. Kollman, *Prediction Methods for Turbulent Flows*, 1980; D. C. Leslie, *Developments in the Theory of Turbulence*, 1983; A. J. Reynolds, *Turbulent Flows in Engineering*, 1974; H. Tennekes and J. L. Lumley, *First Course in Turbulence*, 1972.

WAKE FLOW
VICTOR L. STREETER

Turbulent eddying flow that occurs downstream from bluff bodies. When fluid flows along the boundary of a solid body, the fluid near the boundary is slowed down by viscous shear stresses exerted at the boundary. This action is progressive along the body and, under certain conditions, fluid near the boundary is brought to rest, which causes the fluid moving near the body to separate from the body (see **illus.**). A wake develops and produces additional form drag on the body.

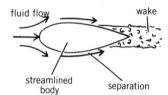

Wake formed downstream from a streamlined body.

When the flow does not separate from the boundary, as in ideal fluid flow situations, the high fluid velocity at the largest cross section of the body is reduced along the downstream portion of the body with recovery of pressure. Formation of a wake stops the pressure recovery process, leaving a low pressure intensity in the wake and a resultant pressure force on the body acting in the downstream direction. The fluid in the wake is highly turbulent, containing vortices that are shed from the body. The wake continues downstream with the flow. SEE KARMAN VORTEX STREET.

A wake is formed downstream from bluff bodies such as bridge piers, smoke stacks, buildings, or trees. For unsteady flow cases, such as the motion of a train or a ship, the fluid behind the moving object has great turbulence remaining in it and, in the case of a ship, is easily discerned for a great distance. SEE FLUID-FLOW PRINCIPLES; TURBULENT FLOW.

BOUNDARY-LAYER FLOW
JOSEPH J. CORNISH, III

The flow of that portion of a viscous fluid which is in the neighborhood of a body in contact with the fluid and in motion relative to the fluid. Wherever a viscous fluid flows past a boundary, the layers of the fluid nearest the boundary are subjected to shearing forces, which cause the velocity of these layers to be reduced. As the boundary is approached, the velocity continuously decreases until, immediately at the boundary, the fluid particles are at rest relative to the body. This region of retarded velocity is called the boundary layer, and a graph of the variation of velocity with distance from the wall or boundary describes a boundary-layer profile (**Fig. 1**). The primary effects of the viscosity of the fluid are concentrated in this boundary layer, whereas in the outer or free-stream flow the viscous forces are negligible. L. Prandtl founded modern boundary-layer research in 1904, when he recognized that the flow in the boundary layer could be treated separately from the free-stream flow and simplified the Navier-Stokes equations for use in the boundary layer.

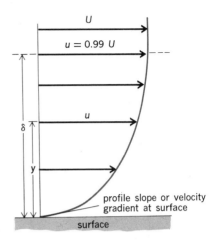

Fig. 1. Boundary-layer profile.

Because the friction forces due to the viscosity of the fluid are restricted to the boundary layer, this region is also often called the friction layer. SEE FLUID FLOW; NAVIER-STOKES EQUATIONS; VISCOSITY.

The velocity of the fluid within the boundary layer approaches the local free-stream velocity asymptotically as the distance from the wall is increased; thus the actual thickness of the boundary layer is difficult to determine. An arbitrary boundary-layer thickness δ is usually defined as that distance from the wall or boundary where the velocity of the flow in the boundary layer reaches 99% of the local free-stream velocity (Fig. 1). The thickness of a boundary layer depends upon the viscosity of the flowing fluid, the free-stream flow conditions, the roughness of the immersed surface or boundary, and the extent of the surface over which the fluid has passed. Each body moving through a viscous fluid is sheathed in a mantle of boundary-layer flow. The thicknesses of boundary layers surrounding vehicles moving through the air range from less than 0.1 in. (2.5 mm) near the front of a high-speed aircraft to more than 10 ft (3 m) at the rear of a dirigible or an airship.

The viscous shearing force created within the boundary layer constitutes a major portion of the resistance or drag experienced by an airplane in flight or by fluids passing through pipes and channels; this shearing force opposes the relative motion of any viscous fluid past a bounding or immersed surface. The boundary-layer thickness generally increases with the extent of the surface over which the fluid has passed. In the flow of fluid through pipes, the boundary layer formed at the walls thickens until the pipe is entirely filled with boundary-layer flow (**Fig. 2**). When such a condition is reached, fully developed pipe flow is said to exist. In cases of unbounded flows, such as a thin, flat plate placed parallel to the flow direction, the boundary layer continues to increase in thickness as the fluid passes downstream (**Fig. 3**). This growth of the boundary layer results in a deflection or displacement of the free-stream flow away from the surface, thus apparently increasing the thickness of the plate.

Boundary-layer separation. In flowing over a surface of changing contour, the velocity and pressure of a fluid vary. As the flow velocity increases, the pressure must decrease. In the free stream, where there are negligible losses to viscosity, the flow is always able to proceed into

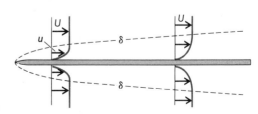

Fig. 2. Boundary-layer development in a pipe.

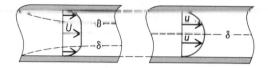

Fig. 3. Boundary-layer development on flat plate.

regions of increasing pressure at the expense of its velocity, the opposing pressure forces being balanced by the forces that result from the change in momentum of the flow. However, within the boundary layer, where the momentum of the fluid has been reduced in overcoming viscous forces, the remaining momentum may be insufficient to allow the flow to proceed into regions of increasing pressure. At the position on the body where this condition exists, the boundary layer no longer continues to follow the contour of the boundary but instead leaves, or separates from, the surface. This phenomenon is known as boundary-layer separation (**Fig. 4**). Beyond the point of separation, between the surface and the separated boundary layer, a region of reversed or upstream flow exists. This reversed flow and the separated boundary layer subsequently coalesce to form a wake of low-velocity fluid behind the body. The separation of the flow in the boundary layer limits the lifting capabilities of wings and causes the stalling of airplanes. The efficiencies of fluid machines and mechanisms, such as turbines, compressors, pumps, and propellers, are decreased by flow separation originating in the boundary layer. S*ee* B*ernoulli's* *theorem*.

Laminar and turbulent layer. The flow in the boundary layer formed on a body exists in one of two characteristic states: laminar or turbulent. If the flow is laminar, strata or laminae of fluid pass smoothly across the surface, whereas if it is turbulent, the flow is characterized by a rapid churning or mixing between layers of flow having different velocities. These two flow states are familiarly illustrated by the smoke rising from a cigarette in a still room. At first the smoke rises in smooth, continuous filaments, which after some distance begin to oscillate and finally erupt into boiling turbulent flow. This same phenomenon of transition from laminar to turbulent flow may occur within the boundary layer after the flow has passed over a surface for some distance. The flow on the forward or upstream portion of a body is generally laminar. However, as the boundary layer develops on the after or downstream part of the body, the flow may become turbulent. The flow in the boundary layer, at first smooth or laminar, eventually begins to fluctuate and finally bursts into random turbulence (**Fig. 5**). Turbulent flow may occur within the boundary layer even when, as is generally the case, the free-stream flow outside the boundary layer is laminar. Turbulent flow is readily distinguished from laminar by experiment; however, a complete and precise theoretical explanation of the processes which cause a laminar boundary layer to become turbulent has not yet been formulated. S*ee* L*aminar flow*; T*urbulent flow*.

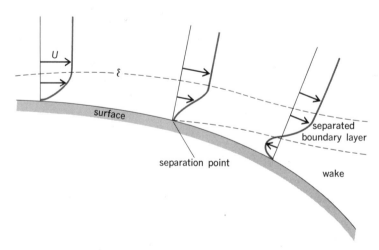

Fig. 4. Boundary-layer separation.

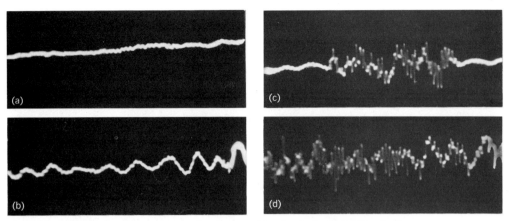

Fig. 5. Photographs of oscilloscope traces showing transitional velocity fluctuations in the boundary layer. (a) Laminar flow with small oscillations. (b) Oscillations amplified to form Tollmien-Schlichting waves. (c) Turbulent bursts indicating onset of transition. (d) Fully developed turbulent fluctuations.

Critical Reynolds number. Among the first to investigate the transition from laminar to turbulent flow, O. Reynolds suggested a dimensionless ratio, the Reynolds number, as a flow parameter useful in determining the extent of laminar boundary-layer flow over a surface. It has since been determined that for many specific shapes there exists a Reynolds number above which the flow in the boundary layer no longer remains laminar but becomes turbulent. This Reynolds number is known as the critical Reynolds number of the boundary layer. SEE REYNOLDS NUMBER.

In dealing with boundary-layer flows, the Reynolds number is usually written as in Eq. (1), where U is the local free-stream velocity, ν is the kinematic viscosity of the fluid, and δ^* is the displacement thickness of the boundary layer, a characteristic parameter defined as Eq. (2), where

$$R = U\delta^*/\nu \qquad (1) \qquad \delta^* = \int_0^\delta \left(1 - \frac{u}{U}\right) dy \qquad (2)$$

u is the velocity at any distance y away from the surface. Transition generally will not occur below a Reynolds number $R = 575$.

The term critical Reynolds number is often used in a slightly different sense when bluff or blunt bodies, such as spheres or circular cylinders, placed perpendicular to the flow are being considered. In these cases, the critical Reynolds number is that value at which there occurs a sudden drop or decrease in the drag coefficient of the body. Here the Reynolds number is defined using the diameter D of the sphere or cylinder in place of the displacement thickness δ^* of the boundary layer. The critical Reynolds number of a sphere is about 325,000 and that for a circular cylinder about 450,000.

Stability of laminar layer. To predice the theoretical value of the critical Reynolds number of an arbitrary boundary layer, it has been found convenient to assume that the laminar boundary-layer flow is not absolutely smooth and laminar, but that it contains velocity fluctuations of infinitesimal amplitude. It is further assumed that these fluctuations cover a wide range of frequency of oscillation. Based upon these assumptions, the Tollmien-Schlichting stability theory has been developed. This theory shows that under certain conditions in the boundary layer, velocity fluctuations of particular frequencies are amplified, whereas fluctuations of all other frequencies are damped. The amplified fluctuations increase in magnitude and cause the flow within the boundary layer to oscillate at their particular frequency. These large oscillations within the boundary layer are known as Tollmien-Schlichting waves (Fig. 5b). The oscillations increase in amplitude and eventually cause turbulent flow. When the flow conditions are such that no fluctuations, regardless of frequency, are amplified, the boundary layer is said to be stable. If, however, the

Reynolds number of the boundary layer is large enough, fluctuations of some frequency are amplified, the boundary layer is unstable, and the flow eventually becomes turbulent.

Transition to turbulent flow may also be caused by a finite disturbance to the boundary layer, such as surface roughness. A roughness element on the surface may result in localized separation of the boundary layer, causing turbulence to be shed from the element; frequently, the wake of the roughness element may cause a wedge of turbulent flow to be formed downstream, just as turbulent flow may be formed behind an object dipped into a smooth stream of water. It has been further postulated that such a transition can be excited even on a smooth surface by large velocity fluctuations within the boundary layer. Thus, amplified Tollmien-Schlichting waves can have the same effect in causing transition as disturbance or roughness elements on the surface. Both the Tollmien-Schlichting stability theory and Taylor's finite disturbance theory have been confirmed conclusively by experimentation.

Skin friction. The viscous shearing forces which retard the flow within the boundary layer stem from the friction developed between the moving fluid and the surface or skin of the immersed body. The accurate calculation and prediction of the surface shear or skin friction which will exist on a given body is the primary goal of boundary-layer theory. Because skin friction produces a direct shearing resistance or drag and also may result in separation of the boundary layer, a knowledge of its distribution and magnitude is useful in determining the performance of bodies moving through viscous fluids.

The shearing stresses existing within a fluid are proportional to the viscosity of the fluid and to the rate of change of strain within the fluid. Thus, for laminar flows, a relatively simple relation for the shearing stress may be written as in Eq. (3), where τ is the shearing or frictional

$$\tau = \mu \frac{du}{dy} \qquad (3)$$

stress, μ is the viscosity of the fluid, and du/dy is the gradient or rate of change of velocity normal to the surface. The skin friction or shearing stress at the surface is found by evaluating this relation at the surface where $y = 0$, so that Eq. (4) holds. Thus the skin friction of a laminar bound-

$$\tau_0 = \mu \frac{du}{dy}\bigg|_0 \qquad (4)$$

ary layer can be determined directly from its boundary-layer profile.

For turbulent flows, however, the shearing stress is further increased by turbulent mixing, which results in exchange of mass and momentum within the boundary layer. Laminar shear is caused only by molecular interaction within the fluid, whereas turbulent shear is determined mainly by the interaction of macroscopic portions or particles of fluid. In a turbulent flow, finite particles of fluid having different velocities are churned or mixed together by the turbulent fluctuations. In this manner, the exchange of momentum among the particles produces an increased resultant shearing stress. A particle moving from a low-velocity region into a region of higher velocity must be accelerated to the velocity of its new environment, and the force required to produce this acceleration reduces the momentum of the surrounding particles. The turbulent mixing and consequent momentum exchange within the boundary layer tend to produce a profile in which there is a more uniform distribution of velocity than in the case of laminar flow (**Fig. 6**). Prandtl's mixing-length theory treats the additional shearing stresses caused by turbulence as an apparent increase in the viscosity of the fluid, and gives the friction in the boundary layer as in Eq. (5), where ρ is the density of the fluid and ϵ is the apparent kinematic viscosity, or eddy

$$\tau = \rho \epsilon \frac{du}{dy} \qquad (5)$$

viscosity, of the fluid. The determination of ϵ is dependent on the scale of the turbulence and is indicated by the size of the turbulent eddies.

In the part of a turbulent boundary layer which is in the immediate proximity of the surface, the turbulent fluctuations are almost completely damped by the wall. This region is called the laminar sublayer, and the shear within this region, including the skin friction, follows the laws of

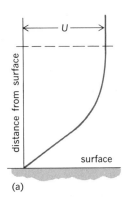
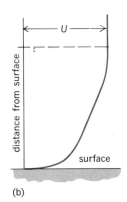

Fig. 6. Typical boundary-layer velocity profile. (*a*) Laminar. (*b*) Turbulent.

laminar flow. This sublayer is, however, extremely thin and the velocity gradients are much larger; hence, the skin friction is much greater than for completely laminar boundary layers.

Layers in compressible flow. As the free-stream flow approaches sonic velocity, additional considerations complicate the behavior of the boundary layer. For example, the energy lost to skin friction produces sufficient heat to invalidate the assumptions of constant fluid density and viscosity. Thus the heat produced by skin friction results in a thermal boundary layer of varying temperature near the surface. At a Mach number $M=3$, the surface temperature is raised about 50% above the temperature of the ambient fluid. The interaction of this variation of temperature with the variation of velocity in the boundary layer makes a rigorous theoretical analysis of the flow intractable. However, several parameters have been shown to influence the behavior of boundary layers in compressible flows. Among these is the Mach number $M=U/c$, where U is the velocity of the fluid and c is the sonic velocity. Also of importance is the Prandtl number $P=\nu/a$, where ν is the kinematic viscosity of the fluid and a is the thermal diffusivity. SEE MACH NUMBER.

The shock waves which may develop on a body also have a considerable and generally adverse effect upon the boundary layer. A shock wave almost inevitably results in premature transition from laminar to turbulent flow in the boundary layer and, if strong enough, will cause an early separation of the flow. Although theoretical treatments have been developed, the greater part of compressible boundary-layer technology is based upon empirical relations obtained from experimental results. SEE COMPRESSIBLE FLOW; SHOCK WAVE.

Boundary-layer control. The natural development of a boundary layer is affected by the contour and surface roughness of the body on which it is formed. Therefore, by the reduction of surface roughness and the choice of surface contours, some amount of control may be exerted upon the development of the boundary layer. Attention to surface roughness, waviness, and continuity is particularly important in the preservation of laminar boundary layers. Even the protuberance on a surface which is apparently most insignificant may produce localized transition to turbulent flow. Transition on aircraft wings is often produced by the remains of insects which have impinged upon the wings at high speeds. Rivet heads, skin laps and joints, and other surface discontinuities readily produce turbulent flow on airplanes.

The overall contour of the body is also an important consideration if laminar boundary-layer flow is to be maintained. Many laminar airfoil sections have been developed for high-speed aircraft. In fact, the term streamlining has come to describe the technique of designing shapes upon which the boundary-layer growth is minimized. The body shape also has a large influence upon the separation of the flow, and, to give low resistance or drag, shapes much be designed which prevent or delay boundary-layer separation. These techniques of influencing the development of the boundary layer by prescribing the geometry of bodies may be called geometric boundary-layer control. SEE STREAMLINING.

Because the amount of control available by purely geometric means is limited, additional methods for influencing the development of the boundary layer have evolved. These methods in general are intended either to remove the low-momentum flow from the boundary layer or to restore the lost momentum. In the latter method, the retarded flow in the boundary layer is re-

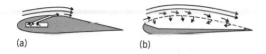

(a) (b)

Fig. 7. Airfoils (a) with flowing boundary-layer control and (b) with suction boundary-layer control.

energized by supplying high-velocity flow through slots or jets in the surface of the body (**Fig. 7a**). This blowing boundary-layer control requires suitable pumps and ducts within the body to provide the necessary quantities of blown or ejected fluid. This technique has, as yet, not proven to be an economical method of control, although several aircraft utilizing variations of the principle have been flown. Blowing boundary-layer control is intended to suppress or delay the separation of the boundary layer rather than to preserve laminar flow and, as a result, is limited in its applications.

The method of controlling the boundary layer by the removal of the retarded flow, called suction boundary-layer control, may be applied both to maintain laminar flow and to prevent boundary-layer separation. By sucking away the flow in the lower region of the boundary layer through slots or perforations in the surface, the development of the boundary layer can be readily influenced (Fig. 7b). The proper distribution of this suction, as determined by the requirements of the boundary-layer equations, allows a designer to tailor the development of the boundary layer to his particular demands. By maintaining the boundary-layer Reynolds number below its critical value, the flow can be kept in a stable condition and thus remain laminar. By thinning the boundary layer properly, it can be made to resist separation. Practical suction sytems for maintaining laminar flow and for delaying flow separation have been demonstrated in flight.

Bibliography. S. Goldstein (ed.), *Modern Developments in Fluid Dynamics*, 2 vols., 1938; H. Schlichting, *Boundary Layer Theory*, 7th ed., 1979.

SKIN FRICTION
Bernard M. Leadon

A type of friction force which exists at the surface, or skin, of a solid body immersed in a much larger volume of fluid which is in motion with velocity u_1 relative to the body (see **illus.**). The magnitude of skin friction per unit surface area, the shear stress τ_w, was equated by Isaac Newton to the rate of deformation of an adjacent fluid element $(\partial u/\partial y)_w$ times a transport property of the fluid called the absolute or dynamic viscosity coefficient μ. Such flow distortion is significant only in a thin boundary layer, which may be laminar or turbulent and outside of which, with y greater than $\delta(x)$, the motion is essentially inviscid. SEE VISCOSITY.

Although many fluids such as air and water behave in this way, there are non-newtonian fluids of importance in technology which exhibit more complicated skin friction laws because in effect the viscosity coefficient itself may depend upon the rate of fluid element distortion or upon the time duration of the state of stress. SEE NON-NEWTONIAN FLUID; RHEOLOGY.

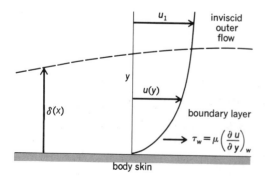

Boundary-layer velocity profile.

The skin friction force contributes directly to the drag of the body; it also contributes indirectly because, by its action, the inviscid outer flow may be modified with effect upon the pressure distribution. Conveniently, although far from completely, it has been possible to relate the skin friction quantitatively to conditions of the outer flow, such as its velocity, viscosity, density, and Mach number, and to body size and shape.

Because skin friction does work upon the fluid, its action in gases at high speeds tends to produce high temperatures in the boundary layer, which lead to the problem of cooling the skin. SEE AEROTHERMODYNAMICS; BOUNDARY-LAYER FLOW.

Bibliography. H. Schlichting, *Boundary Layer Theory*, 7th ed., 1979; V. L. Streeter and E. B. Wylie, *Fluid Mechanics*, 8th ed., 1985.

PIPE FLOW
Victor L. Streeter

Conveyance of fluids in closed conduits. Pipes have been used for thousands of years, but the detailed understanding of velocity distributions and of energy losses has come about during the twentieth century. Satisfactory equations for predicting the losses in flow of water have been in use since about the middle of the nineteenth century.

Laminar flow. The fluid moves parallel to the pipe axis in a straight, round pipe when the Reynolds number VD/ν is less than 2000, in which V is the average velocity, D the pipe diameter, and ν the kinematic viscosity. The velocity distribution in laminar flow is parabolic (**Fig. 1**) and the equation for velocity v is Eq. (1), in which Δp is the pressure drop in length L, μ is the absolute viscosity, and r_0 and r are as shown in Fig. 1. Discharge Q in laminar flow is given by Eq. (2), and is independent of the roughness of the interior pipe wall surface. SEE REYNOLDS NUMBER.

$$v = \frac{\Delta p}{4\mu L}(r_0^2 - r^2) \quad (1) \qquad Q = \frac{\Delta p \pi D^4}{128 \mu L} \quad (2)$$

In laminar flow the fluid particles move in straight lines parallel to the pipe axis, in telescoping layers with each inner layer moving more rapidly than its adjacent outer layer. Energy losses vary as the first power of the velocity, so that by doubling the discharge (or average velocity) the pressure drop is doubled as well. SEE LAMINAR FLOW.

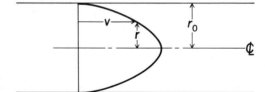

Fig. 1. Laminar velocity distribution.

Turbulent flow. In a pipe when the Reynolds number is greater than 2000–4000, normally the fluid particles no longer move parallel to the pipe axis. The exact transition velocity depends upon the nature of the piping system. For high Reynolds numbers, the orderly motion of laminar flow becomes unstable, with fluid particles moving in random paths with large transverse velocity components. Energy is dissipated in the turbulent motion, with the loss varying as the velocity to the 1.7–2.0 power. The Darcy-Weisbach equation for head loss h_f due to turbulent flow in a pipe is Eq. (3), in which V is the average velocity, L is the length, D the diameter, and f a dimensionless

$$h_f = f \frac{L}{D} \frac{V^2}{2g} \quad (3)$$

factor dependent on the wall roughness, the fluid properties, and on the velocity and pipe diameter $f = f(V,D,\rho,\mu,\epsilon)$ with ρ the fluid density and ϵ a measure of the absolute roughness of the pipe wall, having the dimensions of a length. SEE TURBULENT FLOW.

For turbulent flow in smooth pipe, $\epsilon = 0$ and the expression for f becomes $f = f(VD\rho/\mu)$, in which $VD\rho/\mu$ is Reynolds number R. The form of the functional relation between f and R must be determined by experiment, and is shown as the lowest curved line on the Moody diagram (**Fig. 2**).

Laminar flow may also be shown on the Moody diagram, because its equation may be written as (4).

$$h_f = \frac{64}{R} \frac{L}{D} \frac{V^2}{2g} \tag{4}$$

For rough pipes $f = f(R,\epsilon/D)$, in which ϵ/D is known as the relative roughness. An empirical equation, worked out by C. F. Colebrook, is the basis for the Moody diagram. It gives good results for new commercial pipes, with valves of ϵ.

To find the head loss for flow of a given amount of liquid per unit time through a pipe of known size, length, and type of manufacture, the Reynolds number and the relative roughness are computed and then used in the Moody diagram to determine f. With f known, all quantities in the Darcy-Weisbach equation are known except h_f, so it can be determined.

When the amount of head loss is known but the discharge (volume per unit time flowing) is desired, a trial solution is required. An f is assumed from the Moody diagram for the known ϵ/D, and by its use a trial value of V is found from the Darcy-Weisbach equation. With this V, a trial Reynolds number is computed which permits a better value of f to be found from the Moody diagram.

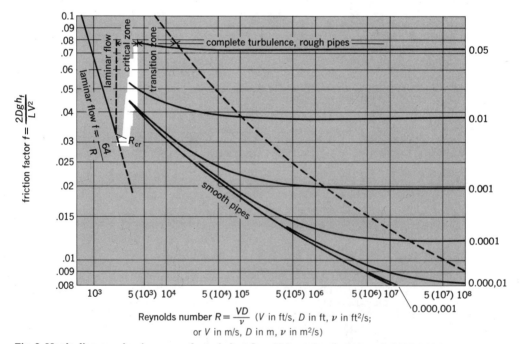

Fig. 2. Moody diagram, showing curves for turbulent flow. Values of ϵ: riveted steel, 0.003–0.03 ft (900–9000 μm); concrete, 0.001–0.01 ft (300–3000 μm); wood stave, 0.0006–0.003 ft (180–900 μm); cast iron, 0.00085 ft (260 μm); galvanized iron, 0.0005 ft (150 μm); asphalted cast iron, 0.0004 ft (120 μm); commercial steel or wrought iron, 0.00015 ft (45 μm); drawn tubing, 0.00005 ft (1.5 μm).

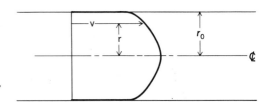

Fig. 3. Turbulent velocity distribution.

With flow of a gas, the same methods may be used as with a liquid if the change of density is smaller (less than 10%). For large density changes the equation of state relating density and pressure intensity is required, as well as special methods for obtaining head loss or weight per unit time flowing.

Head losses due to changes in direction and in size of pipe and those due to valving are grouped as minor losses, and tend to vary as the square of velocity. They may be expressed as an equivalent length of pipe L_e which is added to the actual length of pipe in using the Darcy-Weisbach equation.

With old pipe, wall roughness ϵ' tends to increase linearly with time so that $\epsilon' = \epsilon + \alpha t$ in which α is a constant determined by test on the particular pipeline and fluid, and t is time.

Velocity distribution in turbulent flow in a pipe is more uniform than for laminar flow, due to the large transfer of momentum radially across the flow (**Fig. 3**). A simple equation that gives reasonably good results is Prandtl's one-seventh power law, Eq. (5), in which y is the distance from the pipe wall, and r_0 is the pipe radius.

$$v = v_{\max}\left(\frac{y}{r_0}\right)^{1/7} \quad (5)$$

Bibliography. R. P. Benedict, *Fundamentals of Pipe Flow*, 1980; J. O. Hinze, *Turbulence*, 2d ed., 1975; V. L. Streeter and E. B. Wylie, *Fluid Mechanics*, 8th ed., 1985.

FRICTION FACTOR
Charles E. Lapple

The name given to the proportionality factor in Eq. (1) relating the friction or head loss encoun-

$$E_f = 4\tau_w L/\rho D = af(L/D)(u^2/2g_c) \quad (1)$$

tered when a fluid flows in a pipe to the kinetic energy of the flowing fluid and the length and diameter of the pipe. Here E_f is the energy converted into heat as the result of friction [(ft-lb force)/(lb fluid flowing) or J/kg of fluid flowing], τ_w, is the shear stress at the wall (lb force/ft^2 or N/m^2), ρ is the fluid density (lb mass/ft^2 or kg/m^3), f is a dimensionless friction factor, L is pipe length (ft or m), D is pipe diameter (ft or m), u is average fluid velocity in the pipe (ft/s or m/s), g_c is the conversion factor [32.17 (lb mass/lb force)(ft/s^2) or 1 (kg/N)(m/s^2)], and a is an arbitrary dimensionless constant.

There are a variety of friction factors in use, differing from each other only by the value assigned to the term a. If a is assigned the value 4, f value 1, f is the Darcy-Weisbach resistance coefficient. Thus, the Darcy-Weisbach resistance coefficient is four times as large as the Fanning friction factor. A friction factor defined by setting a equal to 2 has also been employed. *See* Pipe Flow.

The friction factor is related to the Chezy coefficient C, commonly used in civil engineering, by Eq. (2), where C has the dimensions {[(lb mass/lb force)(ft/s^2)]$^{1/2}$ or [(kg/N)(m/s^2]$^{1/2}$}, and f is the Fanning friction factor.

$$C = \sqrt{2g_c/f} \quad (2)$$

OPEN CHANNEL
William Allan

A covered or uncovered conduit in which liquid (usually water) flows with its top surface bounded by the atmosphere. Typical open channels are rivers, streams, canals, flumes, reclamation or drainage ditches, sewers, and water-supply or hydropower aqueducts.

Open-channel flow is classified according to steadiness, a condition in relation to time, and to uniformity, a condition in relation to distance. Flow is steady when the velocity at any point of observation does not change with time; if it changes from instant to instant, flow is unsteady. At every instant, if the velocity is the same at all points along the channel, flow is uniform; if it is not the same, flow is nonuniform. Nonuniform flow which is steady is called varied; nonuniform flow which is unsteady is called variable (**Fig. 1**).

Flow occurs from a higher to a lower elevation by action of gravity. If the phenomenon is short, wall friction is small or negligible, and gravity shapes the flow behavior. Gravity phenomena are local; they include the hydraulic jump, flow over weirs, spillways, or sills, flow under sluices, and flow into culvert entrances.

If the phenomenon is long, friction shapes the flow behavior. Friction phenomena include flows in rivers, streams, canals, flumes, and sewers.

All analyses and designs require the use of the fundamental laws of continuity and of conservation of energy or force. Friction data are usually obtained from experience tables or from rational or empirical equations which reflect observed facts. Models are sometimes used to show flow patterns and to measure forces for projection to full scale. By the laws of dynamic similarity, friction phenomena are reproduced to a chosen scale if the model and prototype have the same Reynolds number; gravity phenomena, if the same Froude number. Where gravity and friction play commensurate roles, a combination of a Froude number model study and an analytical friction study may be desirable. *See Dynamic similarity*.

From the laws of dynamics, the instantaneous dynamic relation in all open-flow hydraulics is expressed per pound for a given length in a direction, such as the direction of flow, by Eq. (1). The specific energy per pound at a point is the depth of flow plus the kinetic energy there. In

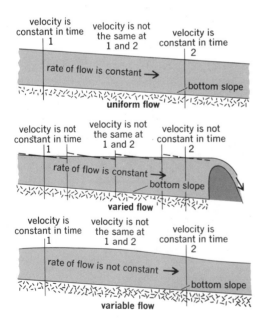

Fig. 1. Three types of flow in open channels.

$$\begin{pmatrix}\text{Gravity}\\ \text{work}\end{pmatrix} - \begin{pmatrix}\text{resistance}\\ \text{work}\end{pmatrix} = \begin{pmatrix}\text{change in}\\ \text{specific energy}\end{pmatrix} \quad (1)$$

uniform flow there is no change in depth or velocity from point to point and therefore no change in specific energy per pound; hence Eq. (2) holds.

$$\text{Gravity work} - \text{resistance work} = 0 \quad (2)$$

In varied flow, where the specific energy is constant in time but changes from point to point, the channel is assumed to be composed of a succession of short lengths in each of which flow is uniform; the length-depth rate-of-flow computation goes stepwise from one short length to the next until the desired length has been traversed. Depending upon the channel characteristics and the precision required, the stepwise computation may be made arithmetically or by methods of the calculus.

In variable flow the change in specific energy occurs in both time and distance. Hence, in each short length of the whole, consideration is required of continuously changing resistance and of storage because the rate of flow in at one end will be more than that out at the other. Computers facilitate the calculations; otherwise, even with simplifying assumptions, the solution requires much time. An example is flood flow in a river.

Uniform flow at a given rate in a cross section of given shape and wall roughness occurs at a greater depth and lesser velocity if the bottom slope is small (mild) and at a lesser depth and greater velocity if the bottom slope is large (steep). At one intermediate slope (critical) the given flow occurs at a depth (critical) at which the specific energy content per pound is minimum; this flow is critical; its Froude number is unity. Flow at greater than critical depth occurs with more than the minimum specific energy content; this flow is tranquil; its Froude number is less than unity. In flow at less than critical depth the specific energy content is also more than the minimum; this flow is rapid; its Froude number is greater than unity.

Hydraulic jump. If design conditions or natural circumstances require flow to change from rapid to tranquil state, the transition will involve a hydraulic jump. This is a local phenomenon in which the specific energy content changes abruptly from low potential and high kinetic to high potential and low kinetic (**Fig. 2**). Usually the wall friction is negligible, but in the transition some energy goes into the creation of eddies and rollers which break up and dissipate their mechanical energy as heat. Thus, a jump usually converts energy from a high-kinetic to a less eroding, high-potential form. SEE HYDRAULIC JUMP.

Stream gaging. Methods of gaging depend upon the size of the channel and the conditions at the site. Flow in small streams is measured by means of weirs or similar calibrated de-

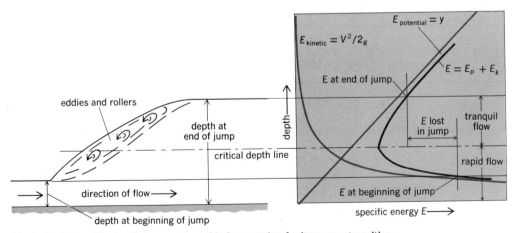

Fig. 2. Depiction of hydraulic jump and graphical accounting for its energy transition.

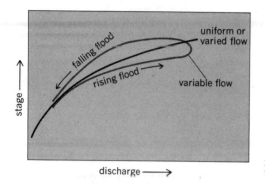

Fig. 3. Relation between flow or discharge and water level or stage. This depends on nature of flow.

vices. Large streams are gaged by means of velocity measurements taken at many points laterally and vertically in a measured cross section. A calibrated current meter is generally used. The relation between discharge and stage (water level on a fixed gage) depends upon the steadiness of flow. If flow is uniform or varied, a single curve exists; if variable, as in time of flood, the curve is hairpin-shaped (**Fig. 3**). Complete study of a stream requires simultaneous determination of the slope of the water surface, dimensions of cross sections, and distribution of velocities. SEE FLOW MEASUREMENT; HYDRAULICS.

Bibliography. E. F. Brater, *Handbook of Hydraulics: For the Solution of Hydrostatic and Fluid Flow Problems*, 6th ed., 1976; N. P. Chermisinoff, *Fluid Flow: Pumps, Pipes and Channels*, 1981; Ven Te Chow, *Open-Channel Hydraulics*, 1959; C. V. Davis and K. E. Sorenson (eds.), *Handbook of Applied Hydraulics*, 3d ed., 1968; R. H. French, *Open-Channel Hydraulics*, 1985; F. M. Henderson, *Open-Channel Flow*, 1966; K. Mahmood et al. (eds.), *Unsteady Flow in Open Channels*, 3 vols., 1975; H. Rouse (ed.), *Engineering Hydraulics*, 1950.

HYDRAULIC GRADIENT
DONALD R. F. HARLEMAN

The slope along a closed water conduit that measures the sum of elevation and pressure heads. Total head consists of these heads plus the velocity head. The elevation and pressure heads at a point can be measured directly by connecting a tube to a small hole in the conduit wall and observing the vertical height to which the water rises in the tube, called a piezometer. SEE BERNOULLI'S THEOREM.

JET FLOW
VICTOR L. STREETER

A local high-velocity stream in a relatively stationary surrounding fluid. Fluid may be caused to flow in a jet for a variety of reasons. Propulsion of a body through a fluid may be by jet propulsion, by a propeller, or by a pump or compressor. The impulse wheel used in hydroelectric power plants takes energy from a jet of water and converts it into torque applied to a rotating shaft. In fighting fires, a smooth jet of water is produced by a nozzle to carry water to the fire without separating it into droplets. Water jets are also used to move earth in gold mining, and as a means of dissipating fluid energy.

A jet may be formed by flow out of a closed conduit, or downstream from a propeller. Flow through an orifice or nozzle causes a jet to form. In the case of power development, a needle nozzle is used to convert the pressure and kinetic energy in the penstock (pipe leading from the reservoir to the turbine) into a smooth jet of variable diameter and discharge but practically constant velocity (**Fig. 1**). The change in size of jet is accomplished by movement of the needle

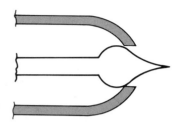

Fig. 1. Flow through needle nozzle.

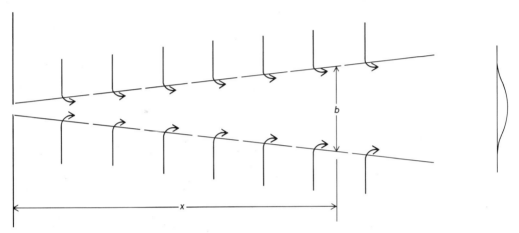

Fig. 2. Fluid jet issuing into some fluid medium.

forward or backward. With flow through a nozzle of fixed diameter a change in discharge causes a corresponding change in velocity of the jet. *See Nozzle.*

The formation of a jet requires that a force be exerted on the fluid in the direction of the jet. An equal and opposite force is exerted on the machine causing the jet, and this is the propulsive force in the case of an airplane or a ship.

When a jet of liquid issues into a gas, such as a water jet entering into air, turbulence within the liquid and friction between gas and liquid cause the liquid jet eventually to pull gas along with it and to allow penetration of the gas into the stream. At length the jet breaks into a spray.

When a jet issues into fluid of the same density, fluid is brought or induced into the jet (**Fig. 2**). The jet spreads at a linear rate with $b = x/8$. Turbulent shear forces reduce the jet velocity within the central cone, and equal turbulent shear forces act to increase velocity in the outer portions of the jet. The momentum of the jet must remain substantially constant in the axial direction. because no forces act external to the jet.

DUCTED FLOW
Victor L. Streeter

Fluid flow with zero velocity at the boundary relative to the boundary. This condition is distinguished from jet flow, in which the boundary is a fluid, either liquid or gas, and does not remain stationary.

Under certain conditions a propeller has a housing or shrouding around it, as in the **illustration**, to control the fluid approaching it, as well as to control the induction of fluid into the jet

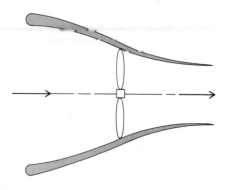

Ducted flow past a propeller.

downstream from the propeller. Shrouding a propeller may be used on a ship to decrease interference of propeller and hull, or it may be used to protect the propeller. Additional frictional losses are incurred by the high-velocity flow near the surface of the ducts, which in some instances may be offset by improved guidance and control of the jet. SEE JET FLOW.

NOZZLE
Fred D. Kochendorfer

A projecting opening that directs the flow of fluid into an open space. Some nozzles maintain the fluid in a jet; an example is the needle nozzle that directs water against the buckets of an impulse turbine. Other nozzles disperse the fluid in an atomized mist; an example is the cone nozzle that sprays liquid fuel into a combustion chamber. The nozzle may be an integral part of a machine, as the nozzle of a steam turbine, or it may be a separate interchangeable piece as on a fire truck. SEE ATOMIZATION; JET FLOW.

Energy exchanges. The quantity Q of incompressible liquid such as water discharged from a smooth-walled nozzle supplied by liquid at head h at the entrance to the nozzle convergence is given by the equation below, where A_1 is the entrance area at which head h is measured,

$$Q = CA_2 \sqrt{2gh} \sqrt{\frac{1}{1 - (A_2/A_1)^2}}$$

A_2 is the discharge area, g is constant of gravity, and C is the discharge coefficient for the particular nozzle structure. A smooth, tapered nozzle has a coefficient near 0.98; rough-walled nozzles and nozzles with abrupt changes in diameter have smaller coefficients, the coefficient being an indication of portion of the pressure head converted into discharge velocity.

In an atomizing nozzle, some of the pressure energy is expended in separating the liquid into droplets.

In a nozzle for compressible vapor or gas, as a steam nozzle, the energy changes are best determined by following the action through the nozzle on an enthalpy chart. Experience indicates that actual velocity will be 0.98–0.96 the ideal velocity because of friction losses. For jet propulsion, the nozzle converts chamber pressure to exhaust velocity. High-temperature combustion products may undergo dissociation, introducing a further energy exchange within the nozzle.

Wind-tunnel nozzle. As used in a wind tunnel, a nozzle increases fluid velocity but with the added requirement that the higher-velocity stream be uniform and parallel. Physically the nozzle consists of a contracting section. If the final fluid velocity is to be supersonic, a divergent portion downstream of the contraction is also required (**Fig. 1**). The region at the minimum section is called the throat. The shape of the cross section is arbitrary; however, most tunnel nozzles are either circular (axisymmetric) or rectangular (two-dimensional). SEE WIND TUNNEL.

Because the fluid, in passing through the nozzle, neither produces work nor gives up heat

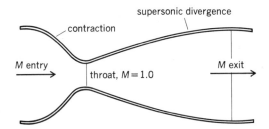

Fig. 1. Diagram of supersonic nozzle.

and because, in addition, area changes are usually gradual, the one-dimensional isentropic relations between fluid properties and velocity are useful approximations. Local pressure, temperature, flow area, and velocity relative to their values at the throat plotted as functions of local Mach number, M, for air show that during acceleration the pressure and temperature decrease continuously; the area, however, must first decrease, then increase (**Fig. 2**). SEE MACH NUMBER.

Design considerations. In the contracting section the velocity should increase fairly uniformly and there should be no local regions of rising pressure. The amount of area contraction is arbitrary, but, to keep pressure loss in the upstream ducting at a minimum, the entry Mach number should be low. Contractions of 10 or greater are common, in which case the resulting entry Mach number is 0.06 (Fig. 2). Another advantage of large contractions is that they permit more effective use of screens to produce low turbulence. SEE ISENTROPIC FLOW.

The ratio of exit to throat area of the supersonic section is fixed by the desired exit Mach number. To obtain a uniform, parallel exit flow free from shock disturbances, the divergent contour must be carefully designed. The usual procedure requires two steps: first, the theoretical wall shape for nonviscous flow is obtained. The boundary layer corrections are added to obtain the final shape.

Because the utility of a nozzle is greatly increased if its exit Mach number can be varied, the area ratio may be made adjustable by flexing or translating the divergent walls.

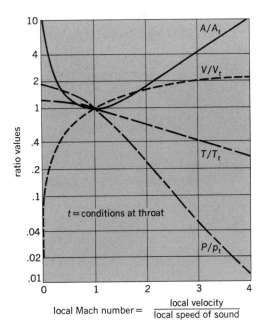

Fig. 2. Isentropic flow of air. Variation of pressure, temperature, velocity, and area with Mach number.

Condensation. The large temperature drop in nozzles for high Mach numbers can result in condensation of one or more of the constituents of the working fluid. Upstream dryers or heaters or both are usually employed to ensure a minimum temperature not more than 50°F (28°C) below the condensation temperature. SEE GAS DYNAMICS.

THROTTLED FLOW
VICTOR L. STREETER

Flow which is forced to pass through a restricted area, where the velocity must be increased. This is also known as choked flow. Most of the kinetic energy produced by reduction of pressure in passing through the constriction is generally converted into thermal energy by turbulent eddying. The net result is a loss in mechanical energy in the system. When a gas is throttled, as by a globe valve, the velocity a short distance downstream from the valve is only a little higher than before the throttling section in most cases, the process being one of constant enthalpy. By introducing mechanical energy losses into a flow system by a throttling valve, the amount of flow may be controlled.

A special throttling effect is produced when gas flows through a constriction, such as a nozzle, at sonic velocity. When this occurs, further reduction of downstream pressure does not alter upstream conditions and the flow remains constant. SEE NOZZLE.

INCOMPRESSIBLE FLOW
FRANK M. WHITE

Fluid motion with negligible changes in density. No fluid is truly incompressible, since even liquids can have their density increased through application of sufficient pressure. But density changes in a flow will be negligible if the Mach number, Ma, of the flow is small. This condition for incompressible flow is given by Eq. (1), where V is the fluid velocity and a is the speed of

$$Ma = V/a < 0.3 \tag{1}$$

sound of the fluid. It is nearly impossible to attain Ma = 0.3 in liquid flow because of the very high pressures required. Thus liquid flow is incompressible. SEE MACH NUMBER.

Speed ranges in gas dynamics. Gases may easily move at compressible speeds. Doubling the pressure of air—from, say, 1 to 2 atm—may accelerate it to supersonic velocity. In principle, practically any large Mach number may be achieved in gas flow, as shown in the **illustration**. As Mach number increases above 0.3, the four compressible speed ranges occur: subsonic, transonic, supersonic, and hypersonic flow. Each of these has special characteristics and methods of analysis. SEE COMPRESSIBLE FLOW.

Air at 68°F (20°C) has a speed of sound of 760 mi/h (340 m/s). Thus inequality (1) indicates that air flow will be incompressible at velocities up to 228 mi/h (102 m/s). This includes a wide variety of practical air flows: ventilation ducts, fans, automobiles, baseball pitches, light aircraft, and wind forces. The result is a wide variety of useful incompressible flow relations applicable to both liquids and gases.

Mass, momentum, and energy relations. The incompressible flow assumption yields very useful simplifications in analysis. For example, a mass balance reduces to a velocity-space relation, independent of fluid density. In steady flow through a variable-area duct, the volume flow Q (m³/s or ft³/s) is constant as given in Eq. (2), where V is the average velocity and A the cross-

$$Q = VA = \text{constant} \tag{2}$$

section area. Constricted areas (nozzles) yield high velocities; for example, in a hypodermic syringe the thumb moves slowly but the jet emerges rapidly from the needle.

The momentum equation for steady, inviscid, incompressible flow reduces to the Bernoulli theorem, given by Eq. (3), where p is pressure, ρ is density (assumed constant), g is the acceler-

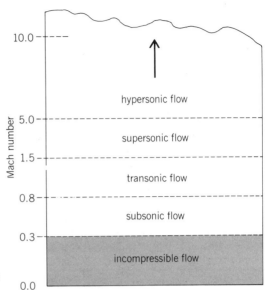

Schematic of various speed ranges of fluid motion. Incompressible flow means small but not zero Mach number.

$$\frac{p}{\rho} + \frac{V^2}{2} + gz = \text{constant} \qquad (3)$$

ation of gravity, and z is the elevation. Combining Eqs. (2) and (3) gives, for example, the formula for flow rate of a Bernoulli-type flowmeter. SEE BERNOULLI'S THEOREM; FLOW MEASUREMENT.

Similarly, the general energy equation reduces for incompressible flow to a form of the heat conduction equation, for which many practical solutions are known.

The incompressible flow assumption also greatly simplifies analysis of viscous flow in boundary layers. Pressure and density are eliminated as variables, and velocity becomes the only unknown, with many accurate solution techniques available. SEE BOUNDARY-LAYER FLOW.

Any disturbance in either a gas or liquid causes the propagation of pressure pulses or "sound waves" in the fluid. This is not a flow but rather a problem in wave propagation and particle oscillation in an otherwise nearly still fluid. SEE FLUID FLOW; WAVE MOTION IN FLUIDS.

Bibliography. I. G. Currie, *Fundamental Mechanics of Fluids*, 1974; R. L. Panton, *Incompressible Flow*, 1984; V. L. Streeter and E. B. Wylie, *Fluid Mechanics*, 8th ed., 1985; B. Thwaites, *Incompressible Aerodynamics*, 1960; F. M. White, *Fluid Mechanics*, 2d ed., 1986; C. S. Yih, *Fluid Mechanics*, rev. ed., 1979.

COMPRESSIBLE FLOW
JOHN E. SCOTT, JR.

Flow in which the fluid density varies. In aerodynamic phenomena, when the flow velocity is large, it is necessary to consider that the fluid is compressible rather than to carry over from classical aerodynamics the assumption that the fluid has a constant density. Under this condition the speed of sound becomes an important factor. At relatively low speeds the changes of temperature and density of a fluid caused by the motion of a body in the fluid are almost negligible. However, if the body moves at high speed through the fluid, the motion can cause pronounced changes in density and temperature of the fluid. Hence, consideration of phenomena of this type involves not only classical fluid mechanics but thermodynamics as well. SEE AEROTHERMODYNAMICS.

The essential difference between an incompressible fluid and a compressible fluid is in the speed of sound. In an incompressible fluid the propagation of pressure change is essentially instantaneous, in a compressible fluid the propagation takes place with finite velocity. For example, if one strikes the surface of an incompressible fluid, the effect observed at great distances is, of course, less than at a smaller distance, but it reaches even infinite distance in essentially zero time. In a compressible fluid the effect propagates at finite velocity. A small disturbance propagates at the velocity of sound. SEE INCOMPRESSIBLE FLOW.

In aerodynamic phenomena the effects of the compressibility of the fluid can be important if the variation of fluid volume caused by the variation of pressure occurring in the flow is of the same order of magnitude as the variation in velocity which corresponds to the variation in pressure. The ratio of the fractional change in volume $\Delta v/v$ to the fractional change in velocity $\Delta u/u$ is equal to the ratio of the square of the velocity to the square of the velocity of sound $(u/a)^2$, where a is sonic velocity. Hence, when the velocity of the flow is the same order of magnitude as the velocity of sound in the flow, the variation of volume (or density) is the same order of magnitude as the variation in velocity. The large velocity variations occurring in high-speed flows therefore cause large changes in the fluid density.

The ratio of the local fluid velocity to the local sound velocity is the local Mach number. Mach number is an index of compressibility, serving as a measure of the relative importance of density changes in a fluid-flow field. In aerodynamic forces the error which results from the assumption of incompressibility in flow problems amounts to roughly $M_0^2/2\%$, when M_0 is the flight Mach number. SEE MACH NUMBER.

ISENTROPIC FLOW
PHILIP E. BLOOMFIELD

The flow of a fluid is isentropic when its entropy is identical at all point in the flow. Isentropic flow can be approached for fluids flowing either in a duct or over the outside of a body. Because the entropy of the fluid is a thermodynamic property, similar to the enthalpy or energy of a fluid, the value of the entropy is fixed by the state of the fluid. For a pure substance, in the absence of external forces, entropy is a function of two independent properties. For example, in the absence of gravity, capillarity, electricity, and magnetism, the entropy of a single-phase fluid is a function of pressure and temperature.

One of the simplest examples of isentropic flow is the flow of a fluid through a nozzle wherein the fluid is accelerated by means of a pressure gradient. This flow can be easily computed for the situation shown in the **illustration** from the conservation of energy (first law of thermodynamics). Per unit mass of fluid Eq. (1) holds, where v_1 and v_2 are the entering and exiting

$$\tfrac{1}{2}v_1^2 + h_1 = \tfrac{1}{2}v_2^2 + h_2 \qquad (1)$$

velocities of the fluid and h_1 and h_2 are the corresponding fluid enthalpies. [Equation (1) in this form is applicable in SI units or any other coherent system of units.] Here the cross-sectional areas and pressures are assumed constant in each half of the nozzle. Thus Eq. (2) holds, where P_1 is the

$$h_1 = P_1 V_1 + U_1 \qquad (2)$$

fluid pressure, V_1 is the volume per unit mass, and U_1 is the internal energy per unit mass. SEE NOZZLE.

In an actual nozzle, the fluid flow is not completely isentropic because (1) the fluid shear

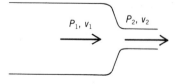

Flow through a nozzle. Entering and exiting velocities (v_1 and v_2) and pressures (P_1 and P_2) obey conservation of energy.

stress at the walls is not zero, thereby introducing some friction; (2) a significant rate of heat transfer can occur between fluid and walls, as in a rocket nozzle; (3) a significant rate of mass transfer or diffusion may occur normal to the streamlines, thus producing local changes of entropy in the real flow; and (4) chemical reactions can occur in the flow, thus causing local changes of entropy.

Isentropic flow is often used as a basis of comparison of the real flow with the ideal flow. The figure of merit for flow in a nozzle is defined by Eq. (3), where v_a is the actual measured

$$\text{Nozzle efficiency} = v_a/v_s \qquad (3)$$

velocity issuing from the nozzle, and v_s is a hypothetical velocity for isentropic flow of the same fluid from the same initial state to the same exit pressure as the real flow. The concept of isentropic flow is useful for fluid flow inside ducts and outside of variously shaped bodies. Isentropic flow is also used for predicting such flows as those of perfect gases; real gases; dissociating and chemically reacting systems; liquids; two-phase, single-, and multicomponent systems; and plasmas.

Isentropic fluid flow can be obtained in irreversible processes by selecting a process in which the local entropy could increase and then providing sufficient heat transfer to maintain the entropy constant at all points. *See* FLUID-FLOW PRINCIPLES; GAS DYNAMICS.

Bibliography. F. W. Sears and G. L. Salinger, *Thermodynamics, the Kinetic Theory of Gases and Statistical Mechanics*, 3d ed., 1975; V. L. Streeter (ed.), *Handbook of Fluid Dynamics*, 1961; V. L. Streeter and E. B. Wylie, *Fluid Mechanics*, 8th ed., 1985.

PRANDTL-MEYER EXPANSION FAN
ARTHUR E. BRYSON, JR.

A steady planar compressible fluid flow that occurs only at supersonic speeds. The Prandtl-Meyer expansion fan is essentially the isentropic flow around a corner from a uniform supersonic flow. The **illustration** shows a typical expansion fan, starting from the uniform flow at Mach number

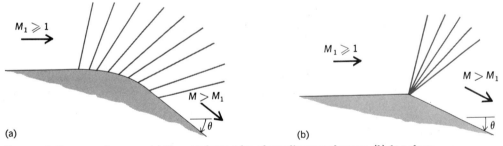

Supersonic flow around a curve. (*a*) Flow produces a fan of standing sound waves. (*b*) At a sharp corner, expansion fan is centered.

M_1 along a flat wall, and turning a corner through an angle θ. The straight lines in the fan are Mach waves, which are standing sound waves in the fluid. The flow behind the fan is also uniform, but the velocity is higher and the pressure is lower. A centered Prandtl-Meyer expansion fan develops at a sharp corner in a supersonic flow. All flow quantities are constant on a given Mach wave in the fan and depend only on M_1 and the angle through which the flow has turned. Using $M_1 = 1$ as a convenient reference, the relationship between flow angle θ and local Mach number M is given by Eq. (1). The local pressure p is obtained from Eq. (2), where p_0 is the

$$\theta = \sqrt{\frac{\gamma + 1}{\gamma - 1}} \arctan \sqrt{\frac{\gamma - 1}{\gamma + 1}} \sqrt{M^2 - 1} - \arctan \sqrt{M^2 - 1} \qquad (1)$$

$$\frac{p_0}{p} = \left(1 + \frac{\gamma - 1}{2} M^2\right)^{\gamma/(\gamma - 1)} \qquad (2)$$

upstream total pressure, and $\gamma = C_p/C_v$, where C_p = specific heat at constant pressure, and C_v = specific heat at constant volume. SEE FLUID-FLOW PRINCIPLES; MACH NUMBER.

VORTEX
VICTOR L. STREETER

In a general sense, any flow that possesses vorticity. The term is frequently used to refer to a flow with closed streamlines or to the idealized case of a line vortex. A line vortex in two-dimensional fluid flow produces a flow or circulation around the line.

Free vortex. Consider the effect of rotating a right-circular cylinder of radius r_0 about its axis with a peripheral velocity v_0 in a fluid otherwise at rest. The fluid in contact with the surface of the cylinder rotates with the cylinder. Fluid at greater radius is also set in motion in concentric circles with velocity diminishing as the radius increases. This type of fluid motion, in which the velocity varies inversely as the radius, is referred to as a free vortex.

If the cylinder is reduced to zero radius so that $v_0 r_0$ remains constant in the limit as r_0 approaches zero, a line vortex results. The velocity at the line is infinite, so the line itself must be considered as a singular line, to be excluded from the actual fluid.

Examples of vortices occur frequently in nature. The tornado is an example of a free vortex, with high velocities near its center, and correspondingly low pressure intensities. The waterspout is its counterpart over water.

The fluid motion in the case of a line vortex in an ideal (frictionless and incompressible) fluid is irrotational; that is, its motion may be described in terms of a velocity potential. SEE LAPLACE'S IRROTATIONAL MOTION.

Vortex tube. If a small spherical particle of a frictionless fluid could be considered as suddenly solidified, its resulting rotation could be expressed by a vector parallel to the axis of rotation, with its length proportional to the angular velocity, and with its direction indicating the sense of rotation by the right-hand rule. When the rotation vector is everywhere zero throughout a region of fluid, the motion of that fluid is irrotational. When some finite fluid regions have nonzero values of the rotation vector, then this fluid has vorticity. A vortex line is a line drawn through the fluid such that it is everywhere tangent to the rotation vector. A collection of vortex lines through a small closed curve defines a vortex tube, which has certain special properties.

1. The circulation about a vortex tube is everywhere the same along its length. Circulation is defined as the line integral of the velocity vector around a closed path.

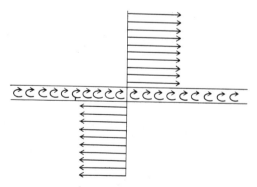

Fig. 1. Vortex sheet which is formed between oppositely directed streams.

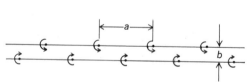

Fig. 2. Kármán vortex street which consists of alternate vortices of opposite rotational sense.

2. The vortex tube cannot end in the fluid. It must either extend to a boundary or close upon itself.

3. Vortex lines move with the fluid. Vorticity of a fluid is a property of the fluid itself and not the space it occupies.

A smoke ring is a practical example of a closed vortex tube. A circular vortex tube in otherwise still fluid will translate perpendicular to the plane of the ring without change in size.

Vortex arrays. A discontinuity in fluid velocity along a surface, such as slippage of one layer of fluid over another, may be handled as a vortex sheet in an otherwise continuous flow. In this case all vortex lines are in the surface (**Fig. 1**). A practical case of a vortex sheet is the flow downstream from an airfoil when the velocity leaving the upper surface is higher than the velocity leaving the under surface.

When real fluid flows around a body, such as wind blowing across a cable, the fluid rotation in the boundary layer causes vortices to form along the downstream side of the body. For certain conditions they remain near the body and are referred to as bound vortices.

For higher velocities the vortices form, grow, and are then systematically shed from the downstream side of the body, forming a vortex street (**Fig. 2**). The unsymmetrical case shown is called the Kármán vortex street after T. von Kármán, who first identified them and showed that the motion is stable when $b/a = 0.281$. SEE FLUID-FLOW PRINCIPLES; KARMAN VORTEX STREET.

KARMAN VORTEX STREET
ARTHUR E. BRYSON, JR.

A double row of line vortices in a fluid. Under certain conditions a Kármán vortex street is shed in the wake of bluff cylindrical bodies when the relative fluid velocity is perpendicular to the generators of the cylinder (see **illus.**). This periodic shedding of eddies occurs first from one side of the body and then from the other, an unusual phenomenon because the oncoming flow may be perfectly steady. Vortex streets can often be seen, for example, in rivers downstream of the col-

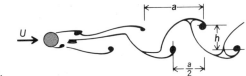

Kármán vortex street.

umns supporting a bridge. The streets have been studied most completely for circular cylinders at low subsonic flow speeds. Regular, perfectly periodic, eddy shedding occurs in the range of Reynolds number (Re) of 50–300, based on cylinder diameter. Above a Re of 300, a degree of randomness begins to occur in the shedding and becomes progressively greater as Re increases, until finally the wake is completely turbulent. The highest Re at which some slight periodicity is still present in the turbulent wake is about 10^6. SEE REYNOLDS NUMBER.

Vortex streets can be created by steady winds blowing past smokestacks, transmission lines, bridges, missiles about to be launched vertically, and pipelines aboveground in the desert. The streets give rise to oscillating lateral forces on the shedding body. If the vortex shedding frequency is near a natural vibration frequency of the body, the resonant response may cause structural damage. The aeolian tones, or singing of wires in a wind, is an example of forced oscillation due to a vortex street. T. von Kármán showed that an idealized infinitely long vortex street is stable to small disturbances if the spacing of the vortices is such that $h/a = 0.281$; actual spacings are close to this value. A complete and satisfying explanation of the formation of vortex streets has, however, not yet been given. For $10^3 < \text{Re} < 10^5$ the shedding frequency f for a circular cylinder in low subsonic speed flow is given closely by $fd/U = .21$, where d is the cylinder diameter and U is stream speed; h/a is approximately 1.7. A. Roshko discovered a spanwise periodicity of vortex shedding on a circular cylinder at $\text{Re} = 80$ of about 18 diameters; thus, it appears that the line vortices are not quite parallel to the cylinder axis. SEE FLUID-FLOW PRINCIPLES; VORTEX.

Bibliography. V. L. Streeter, *Handbook of Fluid Dynamics*, 1961; C. S. Yih, *Fluid Mechanics*, rev. ed., 1979.

SOURCE FLOW
VICTOR L. STREETER

A source, in three-dimensional flow, is a point from which fluid issues at a uniform rate in all directions (**Fig. 1a**). The arrows in Fig. 1 indicate flow direction along the streamlines. The circles are the lines of equal potential.

Characteristics of a source. The strength of a source is defined as the volume per unit time issuing from the point. Because the flow is outward and is uniform in all directions, velocity v_r at distance r from the source is the strength m divided by the area of sphere through the point with center at the source, or $v_r = m/4\pi r^2$. By defining velocity potential ϕ as a scalar function of space and time, whose negative derivative with respect to any direction is the velocity component in that direction, Eq. (1) can be written. Because the velocity vector is entirely in the r direction, $\phi = m/4\pi r$ by integration.

$$v_r = \frac{\partial \phi}{\partial r} = \frac{m}{4\pi r^2} \tag{1}$$

Streamlines are radial lines through the point of the source, and equipotential surfaces are given by letting ϕ = constant in the equation. A sink is a negative source, or a point into which fluid flows uniformly in all directions.

A well point is the physical equivalent of a sink. It is a relatively small region in a porous medium into which fluid flows uniformly from all directions. Similarly, if fluid were injected into the porous medium through the well point it would be equivalent to a source flow.

When the radial lines (streamlines) (Fig. 1) are rotated about the horizontal axis of symmetry, they generate surfaces known as stream surfaces. There is no flow normal to a stream surface. The lines are drawn so that the amount of flow between adjacent stream surfaces is the same everywhere. Because of the three-dimensional nature of the flow, the radial lines are not uniformly located, as they are in the two-dimensional case in Fig. 1b. SEE STREAMLINE FLOW.

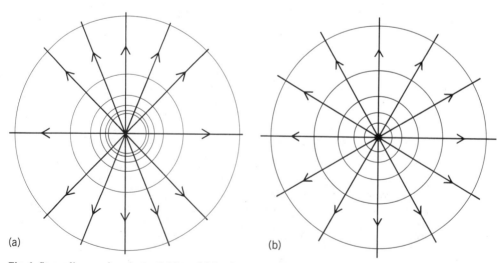

Fig. 1. Streamlines and equipotential lines (a) for the source of a three-dimensional flow and (b) for the source of a two-dimensional flow.

In two-dimensional flow, all flow occurs in a set of parallel planes, with no flow normal to them. The flow is identical in each of these parallel planes.

A source, in two-dimensional flow, is a line normal to the planes of flow, from which fluid is imagined to flow uniformly in all directions at right angles to the line (Fig. 1b). The source appears as a point on the customary two-dimensional flow diagram.

The total flow per unit time per unit length is the strength of the source. By calling the strength $2\pi\mu$, the velocity at distance r from the source is $2\pi\mu/2\pi r = \mu/r$. Then Eqs. (2) hold.

$$-\frac{\partial \phi}{\partial r} = \frac{\mu}{r} \qquad \frac{\partial \phi}{\partial \theta} = 0 \qquad (2)$$

Here θ is the angle in polar coordinates associated with r. By integrating, $\phi = -\mu \ln r$, in which ln is the natural logarithm. A negative source is a sink, into which fluid is imagined to flow uniformly from all directions at right angles to its line.

Application to flow about a body. Sources and sinks are used as flow elements in conjunction with doublets, vortices, and uniform flow to develop complex flow situations. The combination of a source, an equal sink, and a uniform flow, properly placed, results in flow about a closed body in three-dimensional flow and about a cylinder in two-dimensional flow.

In the three-dimensional case of a Rankine body, a source of strength m is located at $(a,0)$, a sink of strength m at $(-a,0)$, and a uniform flow U in the $-x$ direction (**Fig. 2**). Equation (3)

$$\phi = Ux + \frac{m}{4\pi}\left(\frac{1}{r_1} - \frac{1}{r_2}\right) \qquad (3)$$

defines the relationship using auxiliary coordinates (**Fig. 3**). This flow case has axial symmetry; that is, all streamlines are in planes that pass through the x axis, and all such planes have identical

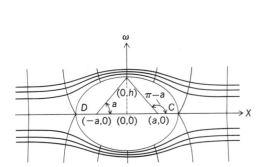

Fig. 2. Three-dimensional flow net about Rankine body.

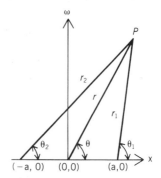

Fig. 3. Auxiliary coordinate systems used for Rankine body.

streamlines. The length of the Rankine body is the distance between stagnation points D and C. With $(x_0,0)$ the coordinate of the upstream stagnation point C, the half-length x_0 is given by Eq. (4), which is most conveniently solved by trial. The value of (x_0/a) depends on the value of Ua^2/m.

$$0 = \frac{x_0/a}{[(x_0/a)^2 - 1]^2} - \frac{\pi U a^2}{m} \qquad (4)$$

The half-breadth h, which occurs at the midsection of the body, is given by Eq. (5), which also is solved most conveniently by trial.

$$\left(\frac{h}{a}\right)^2 \sqrt{\left(\frac{h}{a}\right)^2 + 1} = \frac{m}{\pi U a^2} \qquad (5)$$

By expressing dynamic pressure p as zero at a great distance from the body, where the velocity is U, from Bernoulli's equation, Eq. (6) is obtained. Here ρ is the mass density of fluid and

$$p = \frac{\rho}{2}(U^2 - q^2) \qquad (6)$$

q is the velocity at any point (x,ω) where the pressure is p. Speed q is determined from the components q_x and q_ω by Eq. (7), in which Eqs. (8) apply. Frequently it is easier to compute the

$$q = \sqrt{q_x^2 + q_\omega^2} \qquad (7) \qquad q_x = -\frac{\partial \phi}{\partial x} \qquad q_w = -\frac{\partial \phi}{\partial \omega} \qquad (8)$$

velocity due to each flow element at a point, and then to add them vectorially to find the speed q at a point. Other solutions to two- and three-dimensional flow are made in an analogous manner.

The source and sink concepts are generally useful for frictionless flow situations, and have little meaning in frictional situations, such as flow in a pipe. Pipe flow is said to be one-dimensional, in that calculations of pressure and velocity are made as if the flow were constant over each cross section of the pipe. SEE PIPE FLOW; SINK FLOW.

Bibliography. V. L. Streeter, *Fluid Dynamics*, 1948; V. L. Streeter and E. B. Wylie, *Fluid Mechanics*, 8th ed., 1985.

SINK FLOW
VICTOR L. STREETER

A point in three-dimensional flow, into which fluid is presumed to flow uniformly from all directions. The strength of a sink is defined as the volume per unit time flowing into the point. A sink may also be defined as a negative source. SEE SOURCE FLOW.

In two-dimensional flow, in which all flow occurs in parallel planes that have identical flow patterns, a sink is a straight line into which fluid flows uniformly from all directions at right angles to the line. It appears as a point on the customary two-dimensional flow diagram.

By analogy, flow of groundwater into a well point closely approximates three-dimensional sink flow. The concept of the sink is useful in building up complex flow patterns when used in conjunction with sources, doublets, and uniform flow. SEE FLUID FLOW.

DOUBLET FLOW
VICTOR L. STREETER

In hydrodynamics, a doublet is the combination of a source and a sink of equal strength which are allowed to approach each other in such a manner that the product of their strength and the distance between them remains constant in the limit. Doublets have directional properties, the line drawn from the sink toward the source being the axis of the doublet. The strength of a doublet is proportional to the product of strength of source and distance between source and sink before the limit is taken. SEE SINK FLOW; SOURCE FLOW.

The doublet is a flow element that is used in combination with other elements to build up special flow cases. For example, a uniform flow superposed on a two-dimensional doublet so that the axis of the doublet is directed upstream yields the flow case of uniform flow around a circular cylinder. In three-dimensional flow, uniform flow and a doublet directed upstream result in the case of uniform flow around a sphere.

The flow net consists of equipotential lines and streamlines such that the change in value between adjacent lines is the same. In two-dimensional flow the streamlines are circles with centers on the y axis and the equipotential lines are circles with centers on the x axis (**Fig. 1**). In three-dimensional flow the flow net is somewhat similar, except that the closed loops are not circles (**Fig. 2**). SEE STREAMLINE FLOW.

Three-dimensional doublets may be distributed along lines, over surfaces, or through volumes in such a way that the strength over the periphery of a circle, with axes normal to the plane of the circle, yields flow around a torus-shaped body when a uniform flow is superposed in the direction of the negative axes of the doublets.

Steady flow around a sphere is a result of doublet flow and flow at uniform velocity (**Fig.**

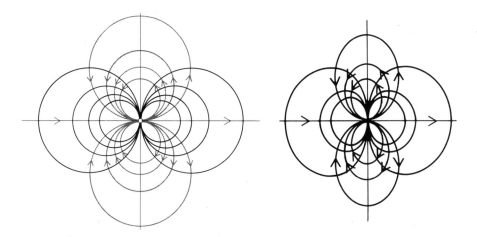

Fig. 1. Equipotential lines and streamlines for the two-dimensional doublet.

Fig. 2. Streamlines and equipotential lines for a three-dimensional doublet.

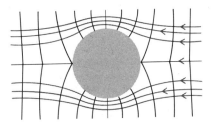

Fig. 3. Streamlines and equipotential lines for uniform flow about a sphere at rest.

3). The equation for velocity potential ϕ is given below. The first term on the right is the velocity

$$\phi = \frac{Ua^3}{2r^2} \cos\theta + Ur\cos\theta$$

potential for a three-dimensional doublet with axis in the $+x$ direction, and the second term is for uniform velocity U in the $-x$ direction. The spherical polar coordinates are r and θ, and a is the radius of sphere.

STOKES STREAM FUNCTION
ARTHUR E. BRYSON, JR.

A degenerate (one-component) vector potential used in analyzing and describing axially symmetric fluid-flow fields. In a steady axially symmetric flow, the rotation of a streamline about the axis of symmetry generates a stream surface. A certain mass rate of flow exists inside this stream surface which is the same at every axial station because, by definition, there is no flow through the stream surface. The value of the Stokes stream function ψ at a point in the flow is equal to

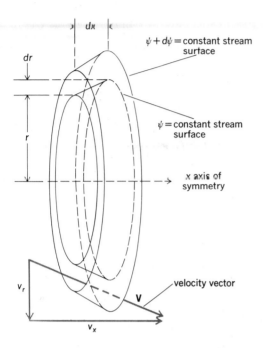

Section of axially symmetric fluid-flow field.

$1/2\pi$ times mass rate of flow inside the stream surface passing through that point. If r is the radial coordinate of a point, and x the axial coordinate, then $\psi = \psi(x,r) = $ constant is the equation of a stream surface as shown in the **illustration**. At a station where $x = $ constant, the differential amount of mass flow between two stream surfaces with radial distance dr between them and of density ρ is given by Eq. (1). Similarly, on an $r = $ constant cylinder, the mass flow between two stream surfaces with axial distance dx between them is given by Eq. (2). Thus the radial and axial

$$2\pi\, d\psi = 2\pi \frac{\partial \psi}{\partial r} dr = (2\pi r\, dr)\,\rho v_x \quad (1) \qquad 2\pi\, d\psi = 2\pi \frac{\partial \psi}{\partial x} dx = -(2\pi r\, dx)\,\rho v_r \quad (2)$$

velocity components at any point are given by Eqs. (3), where ρ is fluid density. If the fluid motion

$$v_r = -\frac{1}{\rho}\frac{1}{r}\frac{\partial \psi}{\partial x} \qquad\qquad v_x = \frac{1}{\rho}\frac{1}{r}\frac{\partial \psi}{\partial r} \quad (3)$$

is slow enough to neglect compressibility, $\rho = $ constant and may be dropped from the above relations; ψ then measures the volume rate of flow inside a stream surface. *See* F<small>LUID-FLOW PRINCIPLES</small>.

FLUID DYNAMICS

Fluid dynamics	54
Fluid-flow principles	54
Navier-Stokes equations	66
Hydrokinematics	67
Hydrodynamics	67
Hydrokinetics	74
Hydraulics	74
Gas kinematics	75
Gas dynamics	76
Low-pressure gas flow	89
Knudsen number	98
Gas kinetics	99
Aerodynamics	99
Aerothermodynamics	101

FLUID DYNAMICS
Victor L. Streeter

The science of fluids in motion. Fluid dynamics attempts to describe the motion of a fluid as it is displaced and deformed by the action of moving or fixed boundaries. Fluid dynamics may be divided into two parts, hydrodynamics and aerodynamics. For low Mach numbers (ratio of velocity of fluid to local acoustic velocity) both hydro- and aerodynamics may be treated in the same manner, but for Mach numbers over about 1/2, compressibility must be taken into account.

Fluid dynamics makes use of both theoretical developments and experimental results. Simplifying assumptions are generally made in theoretical studies in order to make the equations manageable. The extent to which the results differ from the flow of real fluids must be determined by experiment, and corrections applied to the theoretical treatment to obtain practical results.

Theoretical hydrodynamics has been made a useful tool in solution of flow problems by use of the Prandtl hypothesis, which states that with fluids of low viscosity the effects of viscosity are limited to a narrow region along the boundaries. The problem may be solved as if the fluid were frictionless (nonviscous) to determine the velocity and pressure intensity throughout the fluid except at the boundaries. The flow near the boundaries is called boundary-layer flow, and takes into account viscous effects and the fact that the velocity at the boundary relative to the boundary is zero. From a study of the boundary layer, its growth may be computed, as well as the tangential shear force it exerts on the boundary. Under certain conditions of adverse pressure distribution, the boundary-layer film immediately adjacent to the wall comes to rest, and the bounding streamline separates from the wall. This phenomenon is known as separation and results in turbulence and formation of a wake downstream from the separation point, which increases the energy losses. Beyond the separation point, the fluid does not follow the boundary and does not regain pressure as velocity is reduced. There remains an additional drag or pressure force on a body immersed in a flowing fluid, known as form drag. SEE BOUNDARY-LAYER FLOW; HYDRODYNAMICS.

At very high altitude or low pressure instead of the Prandtl hypothesis, it is assumed that individual molecules slip along the surface of a flight vehicle in slip flow. If the fluid is ionized and reacts with electric or magnetic fields as well as with thermal and mechanical boundary conditions, additional behavior is obtained.

Bibliography. I. G. Currie, *Fundamental Mechanics of Fluids*, 1974; H. Lamb, *Hydrodynamics*, 6th ed., 1945; H. Rouse (ed.), *Advanced Mechanics of Fluids*, 1959, reprint 1976; I. H. Shames, *Mechanics of Fluids*, 2d ed., 1982; V. L. Streeter, *Fluid Dynamics*, 1948.

FLUID-FLOW PRINCIPLES
Stephen Whitaker

The fundamental principles that govern the motion of fluids. This article is concerned with these principles and the manner in which they are applied to problems of practical interest. It is assumed that a fluid can be modeled as a continuum and that the details of the molecular structure will appear only in the coefficients associated with constitutive equations or equations of state.

FUNDAMENTALS

Understanding the motion of a continuum and the laws that govern that motion is not easy; thus, it should not be surprising that the understanding of motion has its origins in the study of particles, or rigid bodies. While Newton's laws provided a basis for the study of the motion of particles, or more precisely mass points, it was L. Euler (1707–1783) who organized newtonian mechanics into a form which allows a variety of mechanical problems to be solved with relative ease.

Euler's laws of mechanics. Euler represented Newton's ideas in terms of two laws of mechanics, which can be stated as follows: (1) The time rate of change of the linear momentum of a body equals the force acting on the body. (2) The time rate of change of the angular momentum of a body equals the torque acting on the body.

It is understood that the distances and velocities associated with these two laws are measured relative to an inertial frame and that the torque and angular momentum are measured relative to the same fixed point. Euler's first law contains all of Newton's three laws except for the part of the third law requiring that the actions of two bodies be directed to contrary parts. This is most often referred to as the central force law, and it is valid only for nonrelativistic systems.

In addition to providing a concise statement of the laws of mechanics for continua, Euler postulated that these two laws hold not only for distinct bodies but also for any arbitrary body one might wish to consider. This idea is referred to as the Euler cut principle, and it provides the key to the derivation of governing differential equations from axiomatic statements.

Conservation of mass. The body referred to in Euler's laws consists of some fixed quantity of material that may move and deform in an arbitrary but continuous manner with time t. The region occupied by a body is designated by $V_m(t)$ [**Fig. 1**], and the principle of conservation of mass for a body is stated as follows: The time rate of change of the mass of a body is zero.

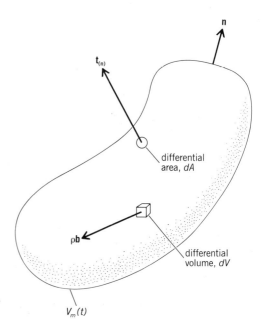

Fig. 1. A material volume.

Density. If the mass density of the continuum is designated by ρ, the mass of the body is expressed by Eq. (1), where the right-hand side is a volume integral over $V_m(t)$, dV being a differ-

$$\text{Mass of a body} = \int_{V_m(t)} \rho dV \qquad (1)$$

ential volume element. The principle of conservation of mass takes the form of Eq. (2), where d/dt

$$\frac{d}{dt} \int_{V_m(t)} \rho dV = 0 \qquad (2)$$

represents differentiation with respect to time, or rate of change. The density has units of mass/volume, and is nearly constant for liquids and solids. For gases, ρ depends on the temperature and pressure, and an equation of state is required in order to predict the density. SEE GAS; LIQUID.

To obtain a governing differential equation from the axiomatic statement given by Eq. (2), it is necessary to interchange differentiation and integration. G. W. Leibnitz (1646–1716) derived a theorem for accomplishing this, and when this theorem is applied to the time-dependent volume

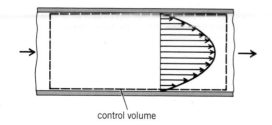

Fig. 2. Steady laminar flow in a tube.

$V_m(t)$, it is usually known as the Reynolds transport theorem. Application of this theorem to Eq. (2) leads to Eq. (3), and since the volume $V_m(t)$ is arbitrary, the integrand must be zero and continuity equation (4) results, where the first term on the left-hand side represents the time rate

$$\int_{V_m(t)} \left[\frac{\partial \rho}{\partial t} + \nabla \cdot (\rho \mathbf{v})\right] dV = 0 \quad (3) \qquad \frac{\partial \rho}{\partial t} + \nabla \cdot (\rho \mathbf{v}) = 0 \quad (4)$$

of change of mass per unit volume, and the second term represents the net flux of mass, leaving a differential volume. In going from Eq. (3) to Eq. (4), it is assumed that the density ρ and the velocity $\mathbf{v}$ are continuously differentiable with respect to both time and space and that $V_m(t)$ is arbitrary. The velocity vector can be expressed in terms of base vectors and components as Eq. (5), and the gradient operator can be expressed in similar form as Eq. (6). Continuity equation (4) will be used repeatedly below.

$$\mathbf{v} = \mathbf{i}v_x + \mathbf{j}v_y + \mathbf{k}v_z \quad (5) \qquad \nabla = \mathbf{i}\frac{\partial}{\partial x} + \mathbf{j}\frac{\partial}{\partial y} + \mathbf{k}\frac{\partial}{\partial z} \quad (6)$$

The principle of conservation of mass, as given by Eq. (2), should have obvious intuitive appeal; however, when this concept is applied to a vanishingly small volume (that is, when dV in Fig. 1 is allowed to approach zero), (4) is obtained and the physical meaning is less clear. When Eq. (4) is integrated over the control volume shown in **Fig. 2**, the macroscopic mass balance is obtained, and for steady flow in a tube it leads to the statement (7).

$$\begin{matrix} \text{Rate at which} \\ \text{mass flows into} \\ \text{the control volume} \end{matrix} = \begin{matrix} \text{Rate at which} \\ \text{mass flows out of} \\ \text{the control volume} \end{matrix} \quad (7)$$

Linear momentum principle. Euler's laws may be expressed in precise mathematical form. The linear momentum of the differential volume element shown in Fig. 1 is the mass ρdV times the velocity $\mathbf{v}$, and this leads to Eq. (8). The total force acting on a body consists of body

$$\begin{matrix} \text{Linear momentum} \\ \text{of a body} \end{matrix} = \int_{V_m(t)} \rho \mathbf{v} dV \quad (8)$$

forces (gravitational, electrostatic, and electromagnetic) that act at a distance upon the mass, and surface forces that act on the bounding surface of the body. As indicated in Fig. 1, the body force per unit volume is represented by $\rho \mathbf{b}$, and the body force acting on the entire body is given by Eq. (9).

$$\begin{matrix} \text{Body force acting} \\ \text{on the body} \end{matrix} = \int_{V_m(t)} \rho \mathbf{b} dV \quad (9)$$

The surface force per unit area is expressed by the stress vector, which is indicated in Fig. 1 by $\mathbf{t}_{(n)}$. The subscript $(\mathbf{n})$ is used as a reminder that the stress vector is a function of the unit outwardly directed normal vector for the surface under consideration. In terms of $\mathbf{t}_{(n)}$, the surface force acting on the body becomes Eq. (10), in which $A_m(t)$ represents the surface of the body illustrated in Fig. 1.

FLUID DYNAMICS   57

$$\text{Surface force acting on the body} = \int_{A_{m(t)}} \mathbf{t}_{(\mathbf{n})} dA \tag{10}$$

Equations (8) through (10) can be used to express Euler's first law as Eq. (11), where the

$$\frac{d}{dt}\int_{V_{m(t)}} \rho \mathbf{v} dV = \int_{V_{m(t)}} \rho \mathbf{b} dV + \int_{A_{m(t)}} \mathbf{t}_{(\mathbf{n})} dA \tag{11}$$

left-hand side is the time rate of change of momentum, and the terms on the right-hand side represent the body force and the surface force respectively.

The Euler cut principle forced the scientists of the eighteenth century to contemplate the internal state of stress that appeared when an arbitrary body was cut out of a distinct body. It was nearly a century after Euler presented his ideas on the mechanics of continua that A. L. Cauchy (1789–1857) determined the functional dependence of the stress vector on the unit normal vector $\mathbf{n}$. This is given in terms of two lemmas that can be expressed as Eqs. (12) and (13), and

$$\mathbf{t}_{(\mathbf{n})} = -\mathbf{t}_{(-\mathbf{n})} \tag{12} \qquad \mathbf{t}_{(\mathbf{n})} = \mathbf{T} \cdot \mathbf{n} \tag{13}$$

both of these relations are a natural consequence of the form of Eq. (11). The first lemma is reminiscent of Newton's third law of action and reaction. The second lemma is not so obvious, since it introduces the doubly directed quantity known as the stress tensor. It is best to think of $\mathbf{T}$ as a function that maps the unit normal vector $\mathbf{n}$ into the stress vector $\mathbf{t}_{(\mathbf{n})}$. This idea can be expressed in matrix form as Eq. (14).

$$\begin{bmatrix} t_{(\mathbf{n})x} \\ t_{(\mathbf{n})y} \\ t_{(\mathbf{n})z} \end{bmatrix} = \begin{bmatrix} T_{xx} & T_{xy} & T_{xz} \\ T_{yx} & T_{yy} & T_{yz} \\ T_{zx} & T_{zy} & T_{zz} \end{bmatrix} \begin{bmatrix} n_x \\ n_y \\ n_z \end{bmatrix} \tag{14}$$

Angular momentum principle. A precise mathematical statement of Euler's second law is obtained by following the procedure given by Eqs. (8) through (11) to obtain Eq. (15), where

$$\frac{d}{dt}\int_{V_{m(t)}} \mathbf{r} \times \rho \mathbf{v} dV = \int_{V_{m(t)}} \mathbf{r} \times \rho \mathbf{b} dV + \int_{A_{m(t)}} \mathbf{r} \times \mathbf{t}_{(\mathbf{n})} dA \tag{15}$$

the left-hand side represents the time rate of change of angular momentum, and the two terms on the right-hand side represent the torque owing to the body force and the torque owing to the surface force. The mathematical analysis of this equation is extremely complex, and after a great deal of effort it can be proved that Eq. (15) yields only the result that the stress tensor is symmetric. This result can be expressed as Eqs. (16) and indicates that the stress matrix represented in

$$T_{xy} = T_{yx} \qquad T_{xz} = T_{zx} \qquad T_{yz} = T_{zy} \tag{16}$$

Eq. (14) is symmetric. With this result from Euler's second law, (11) can be used to obtain the differential form of Euler's first law. This is given by Eq. (17), where the two terms on the left-

$$\frac{\partial}{\partial t}(\rho \mathbf{v}) + \nabla \cdot (\rho \mathbf{v} \mathbf{v}) = \rho \mathbf{b} + \nabla \cdot \mathbf{T} \tag{17}$$

hand side represent the time rate of change of momentum per unit volume, and the net flux of momentum leaving a differential volume; and the two terms on the right-hand side represent the body force per unit volume and the surface force per unit volume. Equation (17) is referred to as Cauchy's first equation or the stress equation of motion, since the surface forces are represented in terms of the stress tensor $\mathbf{T}$. The continuity equation in the form of Eq. (18) can be used in order to simplify Eq. (17) to Eq. (19). It is important to think of Eq. (4) as the governing differential

$$\mathbf{v}\left[\frac{\partial \rho}{\partial t} + \nabla \cdot (\rho \mathbf{v})\right] = 0 \quad (18) \qquad \rho\left(\frac{\partial \mathbf{v}}{\partial t} + \mathbf{v} \cdot \nabla \mathbf{v}\right) = \rho \mathbf{b} + \nabla \cdot \mathbf{T} \quad (19)$$

equation for the density ρ and Eq. (19) as the governing differential equation for the velocity $\mathbf{v}$. Euler's second law, given by Eq. (15), provides only a constraint on the stress field, so that there are only six distinct components of the stress tensor $\mathbf{T}$. At this point there are more unknowns than equations, and in order to solve fluid-mechanical problems a constitutive equation for **T** is needed.

Fluid characteristics. A fluid is a material that deforms continuously under the action of a shear stress. Air and water are obvious examples of fluids, and their mechanical characteristics are clearly much simpler than catsup, meringue, or toothpaste. It is convenient to separate fluids into those which possess elastic characteristics (often called viscoelastic fluids) and those which do not (often called viscous fluids). In this article, only viscous fluids will be discussed, and for such fluids it is convenient to decompose the stress tensor according to Eq. (20). Here p is the thermodynamic pressure, **I** is the unit tensor, and τ is the symmetric viscous stress tensor. Use of Eq. (20) in Eq. (19) leads to Eq. (21). Here a constitutive equation for τ is needed in order to obtain

$$\mathbf{T} = -p\mathbf{I} + \boldsymbol{\tau} \qquad (20)$$

$$\rho\left(\frac{\partial \mathbf{v}}{\partial t} + \mathbf{v} \cdot \nabla \mathbf{v}\right) = -\nabla p + \rho \mathbf{b} + \nabla \cdot \boldsymbol{\tau} \qquad (21)$$

an equation that can be used to determine the velocity field. *See* FLUIDS.

Viscosity. The viscosity of a fluid is a measure of its resistance to deformation, and a convenient method of deforming a fluid is illustrated in **Fig. 3**. The lower plate is fixed while the

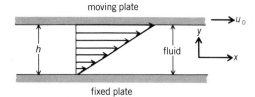

Fig. 3. Velocity distribution between two parallel plates.

upper plate is moving at a constant velocity u_0, giving rise to a simple shear flow. Experiments of this type have shown that for a large class of fluids the force required to maintain the motion of the upper plate is given by Eq. (22). Here F represents the force, A is the area of the surface of

$$F = \mu \left(\frac{u_0}{h}\right) A \qquad (22)$$

the plate in contact with the fluid, h is the perpendicular distance between the plates, and μ is the coefficient of shear viscosity. More commonly, μ is referred to simply as the viscosity. If the force per unit area acting in the x direction on a surface having a normal vector parallel to the y direction is designated τ_{yx}, the constitutive equation that is consistent with Eq. (22) is given by Eq. (23). Fluids which obey this relation between the shear stress τ_{yx} and the rate of deformation

$$\tau_{yx} = \left(\frac{F}{A}\right) = \mu \left(\frac{\partial v_x}{\partial y}\right) \qquad (23)$$

$\partial v_x/\partial y$ are called newtonian fluids. Many fluids exhibit this type of behavior; however, other fluids do not. These other fluids can be separated into two groups: viscoelastic fluids and non-newtonian fluids. The behavior of non-newtonian fluids is illustrated in **Fig. 4**. The Bingham plastic is representative of fluids which remain rigid until a certain critical shear stress is applied. Such fluids tend to slip at solid surfaces, and in tubes they flow as a plug of toothpastelike material. Shear-thinning fluids tend to become less viscous as they are sheared more vigorously, while shear-thickening fluids become more viscous. A concentrated suspension of cornstarch in water is an example of a shear-thickening fluid. *See* NEWTONIAN FLUID; NON-NEWTONIAN FLUID; NON-NEWTONIAN FLUID FLOW; VISCOSITY.

Newton's law of viscosity. For complex flows, all six of the distinct components of the viscous stress tensor must be represented in terms of the components of the velocity gradient tensor. For linear, isotropic, viscous fluids, this representation takes the form of Eq. (24). Known

$$\boldsymbol{\tau} = \mu(\nabla \mathbf{v} + \nabla \mathbf{V}^T) + [(\kappa - 2/3\mu) \nabla \cdot \mathbf{v}]\mathbf{I} \qquad (24)$$

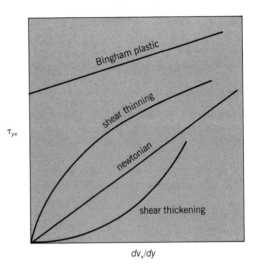

Fig. 4. Behavior of non-newtonian fluids.

as Newton's law of viscosity, it is one of a series of constitutive equations used to describe the behavior of linear, isotropic materials. Others are Ohm's law of electrical conductivity, Hooke's law of elasticity, and Fourier's law of heat conduction. In Eq. (24), κ is known as the bulk coefficient of viscosity, and it is of importance only for a special class of compressible flows, such as the damping of sonic disturbances. For a very wide class of flows, the second term in Eq. (24) is negligible, and use of this result in Eq. (21) leads to Eq. (25). Here ∇^2 represents the laplacian

$$\rho\left(\frac{\partial \mathbf{v}}{\partial t} + \mathbf{v} \cdot \nabla \mathbf{v}\right) = -\nabla p + \rho \mathbf{b} + \mu \nabla^2 \mathbf{v} + \mu \nabla(\nabla \cdot \mathbf{v}) + \nabla \mu \cdot (\nabla \mathbf{v} + \nabla \mathbf{v}^T) \tag{25}$$

operator, and in rectangular coordinates it takes the form of Eq. (26).

$$\nabla^2 \mathbf{v} = \frac{\partial^2 \mathbf{v}}{\partial x^2} + \frac{\partial^2 \mathbf{v}}{\partial y^2} + \frac{\partial^2 \mathbf{v}}{\partial z^2} \tag{26}$$

Navier-Stokes equations. Since the viscosity is a function of temperature and pressure, gradients in the viscosity do exist, and $\nabla \mu$ is nonzero. However, the last term in Eq. (25) is usually small compared to the dominant viscous term $\mu \nabla^2 \mathbf{v}$, and can be discarded for all but the most sophisticated analyses. In addition, the term $\mu \nabla(\nabla \cdot \mathbf{v})$ is comparable in magnitude to those terms which result from the second term on the right-hand side of Eq. (24) and can therefore be discarded. This leads to Eq. (27), where the two terms on the left-hand side represent the local

$$\rho\left(\frac{\partial \mathbf{v}}{\partial t} + \mathbf{v} \cdot \nabla \mathbf{v}\right) = -\nabla p + \rho \mathbf{b} + \mu \nabla^2 \mathbf{v} \tag{27}$$

acceleration and the convective acceleration, and the three terms on the right-hand side represent the pressure force, body force, and viscous force. These equations are known as the Navier-Stokes equations, and serve as the starting point for the analysis of a wide variety of flows. SEE NAVIER-STOKES EQUATIONS.

Incompressible flow. Since the density is a function of temperature and pressure, there are gradients in ρ for all flows; however, in many flows the variation in ρ can be neglected. These flows are called incompressible flows, and for such flows, a suitable approximation for the continuity equation given by Eq. (4) is Eq. (28). While there are many flows that can be treated as

$$\nabla \cdot \mathbf{v} = 0 \tag{28}$$

incompressible, that is, the density can be assigned a constant value, there are no incompressible fluids. Both air and water, for example, are compressible fluids; however, there are many situations for which the flow of both these fluids can be accurately described by Eqs. (27) and (28). In general, this situation occurs when the velocity is small compared to the speed of sound (about 1150 ft/s or 350 m/s for air and about 5200 ft/s or 1600 m/s for water), but there are other situations in which density variations must be considered. The flow that occurs when a pan of water is heated on a stove is an example in which the variation of the density must be taken into account, as is the process of pumping up a bicycle tire. SEE INCOMPRESSIBLE FLOW.

Laminar and turbulent flow. Laminar flow is characterized by the smooth motion of one lamina (or layer) of fluid past another, while turbulent flow is characterized by an irregular and nearly random motion superimposed on the main (or time-averaged) motion of the fluid. The two types of flow can be observed in the heated plume of smoke shown in **Fig. 5**. The smoke originally rises in a smooth laminar stream; however, this buoyancy-driven flow is unstable, and a transition to turbulent flow takes place. This is characterized by an irregular motion as the smoke rises. The transition from laminar to turbulent flow can often be characterized by the Reynolds number, which is a dimensionless group defined by Eq. (29). Here v^* and L^* represent a

$$\text{Re} = \frac{\rho v^* L^*}{\mu} \tag{29}$$

characteristic velocity and characteristic length respectively. The Reynolds number can be thought of as the ratio of inertial effects to viscous effects, but this is a simplistic interpretation, and a more detailed discussion is below. SEE LAMINAR FLOW; REYNOLDS NUMBER; TURBULENT FLOW.

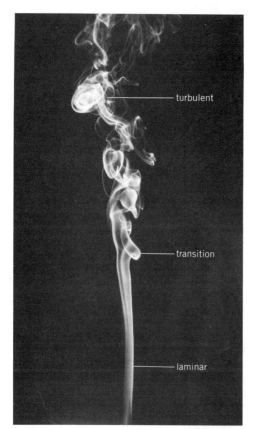

Fig. 5. Laminar and turbulent flow in a smoke plume. (From S. Whitaker, Introduction to Fluid Mechanics, Krieger Publishing, 1981)

APPLICATIONS

Some applications of the Navier-Stokes equations to incompressible flows will be considered in order to illustrate the role played by the individual terms in Eq. (27).

Static fluids. When the velocity **v** is zero, Eq. (27) reduces to Eq. (30). If gravity is the only body force under consideration, and the coordinate system shown in **Fig. 6** is used, Eq. (30)

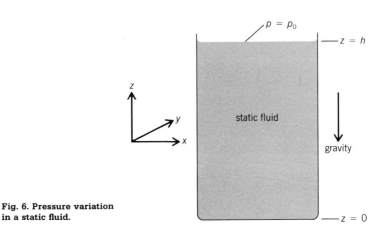

Fig. 6. Pressure variation in a static fluid.

reduces to Eq. (31), where g is the acceleration of gravity. At the surface of the liquid the pressure

$$0 = -\nabla p + \rho \mathbf{b} \qquad (30) \qquad\qquad 0 = -\frac{\partial p}{\partial z} - \rho g \qquad (31)$$

must be equal to the ambient atmospheric pressure p_0, and under these circumstances Eq. (31) requires that the pressure be given by Eq. (32). Here it has been assumed that the density is

$$p = p_0 + \rho g(h - z) \qquad (32)$$

constant, and this leads to a linear variation of pressure with position. The situation is not so simple in the atmosphere, since the density of the air depends on both the pressure and the temperature. The hydrostatic variation of the pressure that occurs in the atmosphere is not trivial, and at 10,000 ft (3000 m) above sea level it is responsible for a shortness of breath and a lowering of the boiling point of water from 212°F to 193°F (from 100 to 89°C).

Archimedes' principle. The force exerted on an object immersed in a static fluid takes an especially simple form, which is generally attributed to Archimedes: A body is buoyed up by a force equal to the weight of the displaced fluid. The buoyancy force for a constant density fluid is given by Eq. (33), where V is the volume of the object and ρ is the density of the fluid.

$$F_{\text{buoy}} = \rho g V \qquad (33)$$

See ARCHIMEDES' PRINCIPLE; BUOYANCY; FLUID STATICS; HYDROSTATICS.

Uniform, laminar flow. When steady, laminar flow occurs in a conduit of uniform cross section, the velocity becomes independent of the position along the conduit after a certain distance, referred to as the entrance length. Under these circumstances, Eq. (27) reduces to Eq. (34).

$$0 = -\nabla p + \rho \mathbf{b} + \mu \nabla^2 \mathbf{v} \qquad (34)$$

For flow in the cylindrical conduit shown in **Fig. 7**, there is only one nonzero component of the velocity vector, and Eq. (34) can be solved to determine the velocity profile, yielding Eq. (35). Here

$$v_z = \left(\frac{\Delta p}{L} + \rho b_z\right) \frac{r_0^2}{4\mu} \left[1 - \left(\frac{r}{r_0}\right)^2\right] \qquad (35)$$

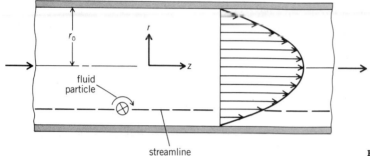

Fig. 7. Uniform flow in a tube.

Δp represents the pressure drop that occurs over a distance L, and Eq. (35) indicates that the velocity profile is parabolic, as is illustrated in Fig. 7. If a small amount of dye were injected into a flow of this type, it would trace out a streamline. For this flow, all the streamlines are parallel, and one of these is indicated in Fig. 7. An arbitrarily small element of fluid that moves along this streamline rotates in the manner indicated in Fig. 7. This rotation is caused by the viscous forces, and when viscous effects are small, an irrotational flow is often encountered. When the Reynolds number is greater than 2100, the laminar-flow field becomes unstable and turbulent flow results. The details of the flow are extremely complex, and the irregular motion that accompanies turbulent flow causes the velocity profile to be nearly flat over the central portion of the tube, with very steep velocity gradients near the wall of the tube. *See* Pipe flow.

For turbulent flow, the relation between the volumetric flow rate Q and the pressure drop Δp must be determined experimentally; however, for laminar flow Eq. (35) can be used to obtain Eq. (36). This result is known as the Hagen-Poiseuille law in honor of G. H. L. Hagen and J. L. M.

$$Q = \frac{\pi r_0^4}{8\mu}\left(\frac{\Delta p}{L} + \rho b_z\right) \tag{36}$$

Poiseuille, who experimentally discovered this result in the early 1800s.

Boundary-layer flow. Boundary-layer flow is characterized by a large gradient of the velocity in a direction perpendicular to the direction of the main flow, and a small gradient of the velocity in a direction parallel to the direction of the main flow. This type of flow is traditionally discussed in terms of the flow past a thin flat plate, illustrated in **Fig. 8**, where u_∞ is the uniform fluid velocity (relative to the plate) at large distances from the plate, and $\delta(x)$ is the thickness of the boundary layer. The practical applications of boundary-layer theory have historical roots in the design of airfoils such as that in **Fig. 9**. The gradients of v_x in a boundary-layer flow are characterized by the inequality (37), and this inequality is not valid in the region near the leading edge

$$\frac{\partial v_x}{\partial x} \ll \frac{\partial v_x}{\partial y} \tag{37}$$

of the plate, where v_x undergoes rapid changes in the x direction. *See* Airfoil.

When the length Reynolds number, given by Eq. (38), is large compared to 1 and the flow is steady, it is found that the Navier-Stokes equations are characterized by the restrictions of Eq. (39), which indicates that the pressure field is hydrostatic, and inequality (40), which indicates

$$\text{Re},x = \frac{\rho u_\infty x}{\mu} \tag{38} \qquad 0 = -\nabla p + \rho\mathbf{b} \tag{39} \qquad v_x \gg v_y \tag{40}$$

that the flow is quasi–one-dimensional. Under these circumstances, the Navier-Stokes equations and the continuity equation take the form of Eqs. (41) and (42). These two equations are usually

$$\rho\left(v_x\frac{\partial v_x}{\partial x} + v_y\frac{\partial v_x}{\partial y}\right) = \mu\left(\frac{\partial^2 v_x}{\partial y^2}\right) \tag{41} \qquad \frac{\partial v_x}{\partial x} + \frac{\partial v_y}{\partial y} = 0 \tag{42}$$

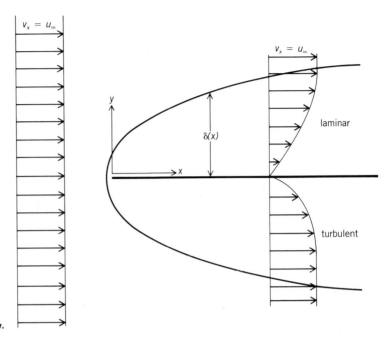

Fig. 8. Boundary-layer flow.

referred to as the Prandtl boundary-layer equations in the analysis of flow around airfoils. Equations (41) and (42) can be solved to provide values for the velocity field that are in excellent agreement with experimental measurements. In addition, the theoretical values of the velocity can be used to determine the shear stress exerted by the fluid on the flat plate, according to Eq. (43).

$$\tau_{yx}\Big|_{y=0} = \mu \frac{\partial v_x}{\partial y}\Big|_{y=0} \qquad (43)$$

The drag force acting on both sides of the flat plate can then be computed from Eq. (44), in which

$$\text{Drag force} = 2w \int_{x=0}^{x=L} \tau_{yx}\Big|_{y=0} dx \qquad (44)$$

w is the width of the plate. The drag force determined by Eq. (44) is also in excellent agreement with experimental values, provided the length Reynolds number is constrained by $10^3 \leq \text{Re},x \leq 10^5$. The reason for this constraint is that Re,x must be larger than 10^3 in order for Eq. (41) to be valid, and for values of Re,x larger than 2×10^5 the flow becomes turbulent. This gives rise to

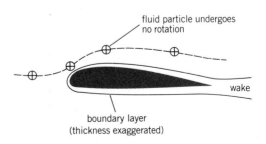

Fig. 9. Flow past an airfoil.

the turbulent velocity profile shown in Fig. 8, and the drag force is significantly larger than that for laminar flow.

While laminar boundary-layer theory applies only to a very limited range of the length Reynolds number, the existence of a thin layer in which viscous effects are important has a profound effect on understanding of fluid-mechanical phenomena. Certainly, the most important of these is the d'Alembert paradox, which results from the misconception that as the velocity becomes progressively larger the inertial term (or convective acceleration) in Eq. (27) should become progressively larger than the viscous term. This line of thought is supported by estimates of the type notation (45), which are based largely on dimensional arguments. If the inertial and

$$\rho \mathbf{v} \cdot \nabla \mathbf{v} \sim \rho v_2/\text{length} \qquad \qquad \mu \nabla^2 \mathbf{v} \sim \mu v/(\text{length})^2 \qquad (45)$$

viscous length scales are similar (and they need not be), these estimates lead to Eq. (46), and it

$$\frac{|\rho \mathbf{v} \cdot \nabla \mathbf{v}|}{|\mu \nabla^2 \mathbf{v}|} = \frac{\rho v(\text{length})}{\mu} = \text{Re} \qquad (46)$$

is tempting to neglect the viscous effects in Eq. (27) when the Reynolds number is very large compared to 1. This leads to Euler's equations (47), and under certain circumstances this result,

$$\rho \left(\frac{\partial \mathbf{v}}{\partial t} + \mathbf{v} \cdot \nabla \mathbf{v} \right) = -\nabla p + \rho \mathbf{b} \qquad (47)$$

along with Eq. (28), can provide a reasonable description of the velocity and pressure fields. However, the most interesting aspect of Eq. (47) is that it can be used to prove that the force exerted by a moving fluid on an airfoil such as the one in Fig. 9 consists of nothing but the buoyancy force. This means that if viscous effects were negligible everywhere in the flow field, the lift force on an airfoil would be zero; and this leads to the (correct) conclusion that an airplane could not fly without the aid of viscosity. Clearly, the boundary layer is of great importance in the analysis of aerodynamic flows. SEE BOUNDARY-LAYER FLOW; D'ALEMBERT'S PARADOX.

Inviscid flow. When the Navier-Stokes equations can be approximated by Euler's equations, the flow is often referred to as inviscid, since one can formally progress from Eq. (27) to Eq. (47) by setting the viscosity equal to zero, that is, $\mu = 0$. From a more realistic point of view, there are two conditions that must be met before Eq. (47) can be considered valid: (1) inertial effects are large compared to viscous effects, which is expressed mathematically by inequality (48), and (2) small causes give rise to small effects. A reasonably accurate estimate of the magnitude of the inertial term is given by Eq. (49), in which L_ρ is the inertial length scale, and Δv

$$|\rho \mathbf{v} \cdot \nabla \mathbf{v}| \gg |\mu \nabla^2 \mathbf{v}| \qquad (48) \qquad \qquad \rho \mathbf{v} \cdot \nabla v = \rho v \frac{dv}{ds} \sim \rho v \Delta v / L_\rho \qquad (49)$$

represents the change in the magnitude of the velocity vector that occurs over the distance L_ρ along a streamline. For situations of interest, Δv is often the same order as v, and the estimate takes the form of notation (50). The magnitude of the viscous term is estimated as in notation (51), in which Δv represents the change in the magnitude of the velocity vector that occurs over

$$\rho \mathbf{v} \cdot \nabla v \sim \rho v^2/L_\rho \qquad (50) \qquad \qquad \mu \nabla^2 v \sim \frac{\mu \Delta v}{L_\mu^2} \qquad (51)$$

the viscous length scale L_μ. Once again, the estimate $\Delta v \sim v$ can be used, and the ratio of inertial effects to viscous effects takes the form of notation (52).

$$\frac{\text{Inertial effects}}{\text{Viscous effects}} \sim \frac{\rho v^2/L_\rho}{\mu v/L_\mu^2} = \left(\frac{\rho v L_\mu}{\mu} \right) \left(\frac{L_\mu}{L_\rho} \right) \qquad (52)$$

In a laminar boundary layer, $L_\mu = \delta$ and $L_\rho = x$, so that notation (52) takes the form of notation (53). From the Prandtl boundary-layer equations, it follows that viscous and inertial effects

$$\frac{\text{Inertial effects}}{\text{Viscous effects}} \sim \left(\frac{\rho u_\infty \delta}{\mu} \right) \frac{\delta}{x} \qquad (53)$$

are comparable. Thus notation (53) allows an estimate of the boundary-layer thickness as given by notation (54). This is in good agreement with the exact solution obtained from the Prandtl boundary-layer equations, which is given by Eq. (55). In both these results the symbol ν is used

$$\delta(x) \sim \sqrt{\nu x/u_\infty} \qquad (54) \qquad\qquad \delta(x) = 4.9\sqrt{\nu x/u_\infty} \qquad (55)$$

to represent the kinematic viscosity which is given by $\nu = \mu/\rho$.

In order to use notation (52) to decide when viscous effects are negligible, the Reynolds number is defined by Eq. (56) and notation (52) is used to deduce that inertial effects are large compared to viscous effects when inequality (57) is satisfied. When this inequality is satisfied and

$$\mathrm{Re} = \frac{\rho v L_\mu}{\mu} \qquad (56) \qquad\qquad \mathrm{Re}\left(\frac{L_\mu}{L_\rho}\right) \gg 1 \qquad (57)$$

small causes give rise to small effects, (47) can be used along with the continuity equation to determine the velocity and pressure fields.

Irrotational or potential flow. When viscous effects are negligible and the velocity and pressure fields are described by Eqs. (28) and (47), the flow is often irrotational. This means that a different element of fluid moves without any rotational motion, such as is illustrated in Fig. 9 for the streamline that is outside the boundary-layer region. When the flow is irrotational, the velocity can be expressed in terms of a potential function according to Eq. (58). From continuity

$$\mathbf{v} = -\nabla\Phi \qquad (58)$$

equation (28) it follows that Φ satisfies the laplacian, which can be expressed as in Eq. (59) for

$$\nabla^2\Phi = \frac{\partial^2\Phi}{\partial x^2} + \frac{\partial^2\Phi}{\partial y^2} + \frac{\partial^2\Phi}{\partial z^2} = 0 \qquad (59)$$

rectangular coordinates. Solutions of Eq. (59) are relatively easy to obtain for a variety of situations; thus the inviscid-flow approximation allows for the determination of complex flow fields with little mathematical effort. For the flow past an airflow shown in Fig. 9, viscous forces are important in the boundary layer, and the Prandtl boundary-layer equations can be used to analyze the flow. Outside the boundary layer, viscous effects can generally be neglected, and a solution to Eq. (59) provides the velocity field. By matching the two velocity fields at the outer edge of the boundary layer, the complete solution for the flow past an airfoil can be obtained; however, the process is not simple, and it requires considerable mathematical skill and physical insight. SEE LAPLACE'S IRROTATIONAL MOTION.

Bernoulli's equation. When the flow is steady and incompressible, and viscous effects are negligible, a particularly simple form of the Navier-Stokes equations is available. The above restrictions make it possible to write Eq. (27) as Eq. (60). If the body force **b** represents a conser-

$$\rho\mathbf{v}\cdot\nabla\mathbf{v} = -\nabla p + \rho\mathbf{b} \qquad (60)$$

vative force field, it can be expressed in terms of the gradient of a scalar according to Eq. (61). When **b** represents the gravity vector, ϕ is called the gravitational potential function, and for the coordinate system shown in Fig. 6, ϕ is given to within an arbitrary constant by Eq. (62). Since

$$\mathbf{b} = -\nabla\phi \qquad (61) \qquad\qquad \phi = gz + \text{constant} \qquad (62)$$

the flow is incompressible, Eq. (63) is valid and Eq. (60) takes the form of Eq. (64). The velocity

$$\rho\mathbf{b} = -\rho\nabla\phi = -\nabla(\rho\phi) \qquad (63) \qquad\qquad \rho\mathbf{v}\cdot\nabla\mathbf{v} = -\nabla(p + \rho\phi) \qquad (64)$$

vector can be expressed in terms of the magnitude v and the unit vector $\boldsymbol{\lambda}$ illustrated in **Fig. 10**, so that the left-hand side of Eq. (64) can be expressed as Eq. (65). Here s represents the arc length measured along any one of the streamlines shown in Fig. 10, and the definition of the directional derivative given by Eq. (66) has been used. Substitution of Eq. (65) into Eq. (64) yields a form that does not appear to be suitable for problem solving, Eq. (67). However, the scalar product (or dot

$$\rho\mathbf{v}\cdot\nabla\mathbf{v} = \rho v\boldsymbol{\lambda}\cdot\nabla\mathbf{v} = \rho v\frac{d\mathbf{v}}{ds} \quad (65) \qquad \boldsymbol{\lambda}\cdot\nabla = \frac{d}{ds} \quad (66) \qquad \rho v\frac{d\mathbf{v}}{ds} = -\nabla(p + \rho\phi) \quad (67)$$

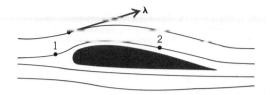

Fig. 10. Streamlines around an airfoil.

product) of this equation with the unit tangent vector is Eq. (68), and a little analysis leads to Eq. (69). This indicates that the quantity ½ $\rho v^2 + p + \rho\phi$ is a constant along a streamline, and Eq.

$$\rho v \lambda \cdot \frac{d\mathbf{v}}{ds} = -\lambda \cdot \nabla(p + \rho\phi) \qquad (68) \qquad \frac{d}{ds}\left(\tfrac{1}{2}\rho v^2 + p + \rho\phi\right) = 0 \qquad (69)$$

(62) can be used to express this idea as Eq. (70). This is known as Bernoulli's equation. For the two points identified on a streamline shown in Fig. 10, Eq. (70) can be used to derive Eq. (71).

$$\tfrac{1}{2}\rho v^2 + p + \rho g z = C \qquad (70) \qquad \tfrac{1}{2}\rho v_2^2 + p_2 + \rho g z_2 = \tfrac{1}{2}\rho v_1^2 + p_1 + \rho g z_1 \qquad (71)$$

For irrotational flows, it can be shown that the constant in Eq. (70) is the same for all streamlines, and most flows for which viscous effects are negligible fall into this category. Under these circumstances, Eqs. (58) and (59) are used to determine the velocity field, and Eq. (70) to determine the pressure field. *See* BERNOULLI'S THEOREM; FLUID FLOW; GAS DYNAMICS; HYDRODYNAMICS.

Bibliography. H. Schlichting, *Boundary Layer Theory*, 7th ed., 1979; C. Truesdell, *Essays in the History of Mechanics*, 1968; W. G. Vincenti and C. H. Kruger, Jr., *Introduction to Physical Gas Dynamics*, 1965, reprint, 1975; S. Whitaker, *Fundamental Principles of Heat Transfer*, 1983; S. Whitaker, *Introduction to Fluid Mechanics*, 1981.

NAVIER-STOKES EQUATIONS
ARTHUR E. BRYSON, JR.

Three scalar partial differential equations that describe conservation of momentum for the motion of a viscous, incompressible fluid. They may be expressed vectorially as one equation, Eq. (1),

$$\rho\frac{\partial \mathbf{v}}{\partial t} + \rho(\mathbf{v}\cdot\nabla)\mathbf{v} = -\nabla p + \rho\mathbf{f} + \mu\nabla^2\mathbf{v} \qquad (1)$$

where ρ is fluid density, $\mathbf{v}$ is fluid velocity vector, p is fluid pressure, $\mathbf{f}$ is body force (such as gravity) per unit mass, μ is fluid viscosity coefficient, and t is time. (These equations are applicable in this form in SI units or any other coherent system of units.) These equations, together with the continuity relation, $\nabla\cdot\mathbf{v}=0$, and suitable boundary conditions determine the flow field; for example, $\mathbf{v}$ and ρ are determined as functions of position in space and of time. One of these boundary conditions is that of no slip at the surface of a body; that is, the fluid immediately at the body surface "sticks" to it and thus has the same velocity as the surface itself. *See* NEWTONIAN FLUID.

Few mathematical solutions to this complicated set of nonlinear partial differential equations are known, except for simple geometries. The importance of viscosity in determining the flow depends on the relative size of the body. Approximations to the Navier-Stokes equations for small Reynolds number Re give good results. For Re < 1, the acceleration forces, those on the left-hand side of the equation, are negligible, leaving only linear terms on the right. Such an approximation is called a Stokes-flow approximation, and one of the most famous applications, made by R. A. Millikan, is to the slow motion of a tiny spherical oil droplet in air. Lubrication theory makes use of the Stokes-flow approximation as well as even further approximations. For Re ≫ 1, the effects of viscosity are confined to a thin layer near the surface of bodies in the fluid. Outside this layer the fluid acts essentially as an inviscid fluid, and this behavior is the reason that inviscid fluid theory is of any use at all. SEE BOUNDARY-LAYER FLOW; D'ALEMBERT'S PARADOX; REYNOLDS NUMBER.

For a compressible, viscous fluid, the viscous term $\mu\nabla^2\mathbf{v}$ must be replaced by the divergence of the viscous stress tensor in which the bulk viscosity coefficient λ occurs. Using rectangular cartesian coordinates x_1, x_2, x_3, the force in the x_i direction ($i = 1, 2, 3$) can be written as notation (2), where Eqs. (3) apply. Here μ and λ are functions of temperature and, in liquids, of pressure. SEE FLUID-FLOW PRINCIPLES.

$$\sum_{j=1,2,3} \frac{\partial \tau_{ij}}{\partial x_j} \tag{2}$$

$$\tau_{ij} = (\lambda - 2/3\mu)\, \delta_{ij}(\nabla \cdot \mathbf{v}) + \mu\left(\frac{\partial u_i}{\partial x_j} + \frac{\partial u_j}{\partial x_i}\right) \qquad \delta_{ij} = \begin{cases} 1 & i = j \\ 0 & i \neq j \end{cases} \tag{3}$$

HYDROKINEMATICS
LOUIS LANDWEBER

Motion of a liquid apart from the cause of the motion. A liquid is treated as a continuum, the elements of which move along continuous paths. At any instant the flow pattern may be delineated by a family of streamlines which are everywhere tangent to the paths, but do not in general coincide with them unless the flow is steady. Complete specification of a flow field requires that the distributions of sources and vorticity be known. SEE FLUID FLOW; HYDRODYNAMICS; STOKES STREAM FUNCTION.

HYDRODYNAMICS
LOUIS LANDWEBER

That branch of continuum mechanics which treats of the laws of motion of an incompressible fluid and of the interactions of the fluid with its boundaries. Partly because of the age-old interest in systems of water works and water-borne vehicles, and partly because of scientific curiosity, the subject has achieved a high state of both practical and analytical development. The present article presents a brief rational sketch of the state of this subject. For more detail on specific parts of the subject SEE FLUID FLOW; FLUID-FLOW PRINCIPLES.

Flow field. Fluids have many physical properties, including, besides thermodynamic and electromagnetic properties, the dynamical ones of density, viscosity, adhesion, and cohesion. Only the dynamical properties will be considered in this article, and furthermore, because capillary effects will not be treated, the last two properties will be of interest only insofar as they affect the inception of cavitation.

A flow field can be studied from either of two points of view, the lagrangian, in which one is concerned with the history of each fluid particle, and the eulerian, in which interest is focused, rather, on the velocity vectors associated with each point. Although the lagrangian method is convenient for demonstrating certain kinematical properties of a flow field, the alternative approach will be used, because it has been found, on the whole, to be simpler and more powerful.

A fruitful concept in hydrodynamics is that of the ideal or inviscid fluid. In such a fluid, across any element of area, the fluid on one side exerts a normal stress or pressure on the fluid on the other side, the magnitude of which is independent of the orientation of the element. Thus there is a scalar pressure field associated with the vector velocity field. A principal problem of hydrodynamics is to determine these fields corresponding to prescribed boundary conditions.

When viscosity is taken into account, the normal stresses across an area element are not, in general, independent of orientation, and in addition, tangential stresses are present. Instead of a scalar pressure field one must consider now a tensor stress field. These stresses, or their integrals, must be determined in order to solve the important problem of obtaining the force and moment acting on a body moving through the fluid. SEE NEWTONIAN FLUID; VISCOSITY.

Hydrokinematics. For the present purpose, the molecular structure of a fluid can be ignored and it can be treated as a continuum. Thus various properties of the fluid will be considered as continuous functions of space and time. If the position of a point is represented in terms of its coordinates x,y,z with respect to a rectangular cartesian coordinate system, the velocity vector field may be expressed in the form $\mathbf{U}(x,y,z,t)$ with components $u(x,y,z,t)$, $v(x,y,z,t)$, $w(x,y,z,t)$ magnitude as in Eq. (1), and the mass density of the fluid (mass per unit volume) as the function $\rho(x,y,z,t)$.

$$V = \sqrt{u^2 + v^2 + w^2} \qquad (1)$$

A basic relation, which expresses the law of conservation of mass, is the equation of continuity which, in regions containing no fluid sources or sinks, may be written in the form of Eq. (2). If ρ is constant, as is assumed hereafter, the equation becomes Eq. (3).

$$\text{div}(\rho\mathbf{U}) = \frac{\partial(\rho u)}{\partial x} + \frac{\partial(\rho v)}{\partial y} + \frac{\partial(\rho w)}{\partial z} = -\frac{\partial \rho}{\partial t} \qquad (2) \qquad \text{div } \mathbf{U} = 0 \qquad (3)$$

From Eq. (3) can be deduced the existence of two families of stream surfaces, as in Eqs. (4), defined by constant values of a and b, the mutual intersections of which define the streamlines

$$\psi(x,y,z) = a \qquad \chi(x,y,z) = b \qquad (4)$$

of the flow pattern, that is, lines tangent at a given instant to the velocity vectors. *See* Streamline flow.

According to this definition, the differential equations of a streamline are as shown in Eq. (5). This should be distinguished from the equations of a path line, Eq. (6), which gives the

$$\frac{dx}{u(x,y,z,t_0)} = \frac{dy}{v(x,y,z,t_0)} = \frac{dz}{w(x,y,z,t_0)} \qquad (5) \qquad \frac{dx}{u(x,y,z,t)} = \frac{dy}{v(x,y,z,t)} = \frac{dz}{w(x,y,z,t)} \qquad (6)$$

path followed by a fluid particle. If the flow is steady (time-independent) the two sets of lines are coincident. In any case they are tangent at the location of a particle.

If the stream functions ψ and χ have been determined for a flow problem, the velocity vector would be given by the vector cross product, as in Eq. (7). These stream functions also have the significant property that the rate of flow Q through a stream channel bounded by the four surfaces ψ_1, χ_1, ψ_2, χ_2 (**Fig. 1**) is as in Eq. (8).

$$\mathbf{U} = \text{grad } \psi \times \text{grad } \chi \qquad (7) \qquad Q = (\psi_2 - \psi_1)(\chi_2 - \chi_1) \qquad (8)$$

In two important special cases the stream functions reduce to a single one. For two-dimensional flow there is the Lagrange stream function $\psi(x,y)$, in terms of which the velocity components are as given by Eqs. (9).

$$u = \frac{\partial \psi}{\partial y} \qquad v = -\frac{\partial \psi}{\partial x} \qquad (9)$$

Constant values of ψ give the streamlines, and the rate of flow q between a pair of streamlines ψ_1 and ψ_2 is given by Eq. (10).

$$q = \psi_2 - \psi_1 \qquad (10)$$

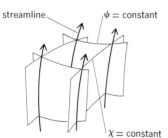

Fig. 1. Stream channel formed by stream surfaces.

The other case is that of axisymmetric flow, for which there is the Stokes stream function $\psi(r,z)$, where the z axis is coincident with the axis of symmetry and r denotes distance perpendicular to it. *See* STOKES STREAM FUNCTION.

The velocity components are now given by Eqs. (11) and the rate of flow between two stream surfaces ψ_1 and ψ_2 is given by Eq. (12).

$$u = -\frac{1}{r}\frac{\partial \psi}{\partial z} \quad w = \frac{1}{r}\frac{\partial \psi}{\partial r} \quad (11) \qquad\qquad Q = 2\pi(\psi_2 - \psi_1) \quad (12)$$

A useful concept in hydrodynamics is that of sources or sinks at various points of a fluid at which fluid is entering or leaving a region. *See* SINK FLOW; SOURCE FLOW.

If $4\pi M$ (where M is called the source strength) denotes the volume rate of entry at a source (or sink, if negative), it is seen, by considering the discharge through a sphere of radius r with its center at the source, that the fluid velocity due to the source is radial and of magnitude $V = M/r^2$. If there are many sources in a region, Gauss' theorem states that the flux of fluid through the surface of this region is the product of the sum of the source strengths by 4π.

If the sources are distributed continuously through a region with a density m per unit volume, so that $m d\tau$, it is necessary to generalize the equation of continuity to become Eq. (13).

$$\text{div } \mathbf{U} = 4\pi m \quad (13)$$

Analysis of the behavior of a fluid element in a stream shows that it is rotating at an angular velocity $\boldsymbol{\omega}/2$, where $\boldsymbol{\omega}$ is defined by Eq. (14) and is called the vorticity vector. This vector

$$\boldsymbol{\omega} = \text{curl } \mathbf{U} \quad (14)$$

is closely related to the circulation, defined as the line integral around a closed curve of the tangential component of the velocity, as in Eq. (15). By application of Stokes' theorem it is readily shown that the circulation is also given by the surface integral of the vorticity over a surface bounded by the curve, as in Eq. (16), where $\mathbf{n}$ is the unit normal vector at a point of the surface,

$$\Gamma = \oint \mathbf{U} \cdot d\mathbf{s} \quad (15) \qquad\qquad \Gamma = \int_S \boldsymbol{\omega} \cdot \mathbf{n} \, dS \quad (16)$$

positive in the sense of advance of a right-hand screw relative to the positive sense of describing the closed curve. *See* VORTEX.

Lines tangent to the vorticity vector at every point are called vortex filaments and a group of vortex lines can form a vortex tube. These have the property that a vortex tube can begin or end only at a boundary, unless it is in the form of a ring. Furthermore, the circulation is constant at all sections of a vortex tube and is called its strength. Two additional laws of vortex motion, for an inviscid fluid on which only conservative forces such as gravitational attraction are acting, are (1) vortex filaments are always composed of the same fluid particles, and (2) the strength of a vortex tube is constant with respect to time.

A straight-line vortex filament of strength Γ induces a velocity $V = \Gamma/r$ tangent to a circle of radius r with center on the line in the plane perpendicular to it.

A flow field can be determined when its distributions of sources and vortices are prescribed. The problem is to determine the velocity $\mathbf{U}$ from Eqs. (17).

$$\text{div } \mathbf{U} = 4\pi m \qquad \text{curl } \mathbf{U} = \boldsymbol{\omega} \quad (17)$$

This is accomplished by putting $\mathbf{U} = \mathbf{U}_1 + \mathbf{U}_2$, as defined by Eqs. (18).

$$\begin{aligned} \text{div } \mathbf{U}_1 &= 4\pi m & \text{curl } \mathbf{U}_1 &= 0 \\ \text{div } \mathbf{U}_2 &= 0 & \text{curl } \mathbf{U}_2 &= \boldsymbol{\omega} \end{aligned} \quad (18)$$

The velocity field $\mathbf{U}_1$ is said to be irrotational, $\mathbf{U}_2$ to be solenoidal. Then Eqs. (19)–(22) hold. The functions ϕ and $\mathbf{A}$ are the scalar and vector potentials.

$$\mathbf{U}_1 = \text{grad } \phi \quad (19) \qquad\qquad \phi = -\int \frac{m(\xi,\eta,\zeta)\, d\xi\, d\eta\, d\zeta}{[(x-\xi)^2 + (y-\eta)^2 + (z-\zeta)^2]^{1/2}} \quad (20)$$

$$\mathbf{U}_2 = \text{curl } \mathbf{A} \quad (21) \qquad\qquad \mathbf{A} = \frac{1}{4\pi}\int \frac{\boldsymbol{\omega}(\xi,\eta,\zeta)\, d\xi\, d\eta\, d\zeta}{[(x-\xi)^2 + (y-\eta)^2 + (z-\zeta)^2]^{1/2}} \quad (22)$$

Hydrokinetics. A fluid element is acted upon by body forces (forces per unit mass) such as gravitational attraction, and stresses on its surface (forces per unit area). Let **B**, with components B_x, B_y, B_z, denote the body force. The stresses can be represented by the symmetric tensor T_{ij}, where the index i indicates that the stress is acting on a plane with positive normal in the direction of increasing x, y, or z according as $i = 1, 2,$ or 3; and $j = 1, 2,$ or 3 indicates the x, y, or z component of the stress (**Fig. 2**). In an inviscid fluid this stress tensor reduces to the scalar pressure field.

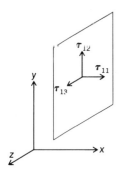

Fig. 2. Normal and tangential stresses.

The force **F** and moment **M** on a body in a fluid then assume the forms given in Eqs. (23) for an inviscid fluid. For a viscous fluid, Eqs. (24) hold.

$$\mathbf{F} = -\int p\mathbf{n}\,dS + \rho\int \mathbf{B}\,d\tau \qquad \mathbf{M} = -\int p\mathbf{r}\times\mathbf{n}\,dS + \rho\int \mathbf{r}\times\mathbf{B}\,d\tau \quad (23)$$

$$X = \sum_{i=1}^{3}\int T_{1i}n_i\,dS + \rho\int B_x\,d\tau \qquad L = \sum_i\int (yT_{3i} - zT_{2i})n_i\,dS + \rho\int (yB_z - zB_y)\,d\tau$$

$$Y = \sum_i\int T_{2i}n_i\,dS + \rho\int B_y\,d\tau \qquad M = \sum_i\int (zT_{1i} - xT_{3i})n_i\,dS + \rho\int (zB_x - xB_z)\,d\tau \quad (24)$$

$$Z = \sum_i\int T_{3i}n_i\,dS + \rho\int B_z\,d\tau \qquad N = \sum_i\int (xT_{2i} - yT_{1i})n_i\,dS + \rho\int (xB_y - yB_x)\,d\tau$$

Application of Newton's laws of motion to an element of an inviscid fluid leads to the Euler equations (25).

$$\frac{\partial u}{\partial t} + u\frac{\partial u}{\partial x} + v\frac{\partial u}{\partial y} + w\frac{\partial u}{\partial z} = B_x - \frac{1}{\rho}\frac{\partial p}{\partial x}$$

$$\frac{\partial v}{\partial t} + u\frac{\partial v}{\partial x} + v\frac{\partial v}{\partial y} + w\frac{\partial v}{\partial z} = B_y - \frac{1}{\rho}\frac{\partial p}{\partial y} \quad (25)$$

$$\frac{\partial w}{\partial t} + u\frac{\partial w}{\partial x} + v\frac{\partial w}{\partial y} + w\frac{\partial w}{\partial z} = B_z - \frac{1}{\rho}\frac{\partial p}{\partial z}$$

Equations (25) may also be written in the vector form as Eq. (26). The form of Eq. (26) is

$$\frac{\partial \mathbf{U}}{\partial t} + \boldsymbol{\omega}\times\mathbf{U} + \tfrac{1}{2}\operatorname{grad}(V^2) = \mathbf{B} - \frac{1}{\rho}\operatorname{grad} p \quad (26)$$

useful when it is desired to transform the equations of motion to other coordinate systems.

It will be supposed that the body force is a conservative one so that it can be expressed in the form $\mathbf{B} = \operatorname{grad}\Omega$ where Ω is a scalar potential function. For example, if z is taken positive upwards at a point on the surface of the Earth, the scalar potential for gravitational attraction is $\Omega = -gz$, where g is the acceleration of gravity.

Probably the most often applied result of hydrodynamics is the Bernoulli equation that can be derived as a first integral of the Euler equations. This assumes different forms for various assumed conditions:

1. If the flow is steady, then for points along a streamline or along a vortex line, Eq. (27) holds.

$$p + \tfrac{1}{2}\rho V^2 + \rho\Omega = \text{constant} \qquad (27)$$

2. In a region in which the flow is irrotational the equation is (28).

$$p + \tfrac{1}{2}\rho V^2 + \rho\Omega + \frac{\partial\phi}{\partial t} = \text{constant} \qquad (28)$$

3. If the flow is steady and the streamlines and vortex lines are parallel, the equation is (29).

$$p + \tfrac{1}{2}\rho V^2 + \rho\Omega = \text{constant} \qquad (29)$$

4. If the coordinate axes are in motion, the origin having the velocity components U, V, W, and the coordinate system the angular velocity Λ, then Eq. (30) holds, in which $\Lambda \cdot \mathbf{r} \times \mathbf{U}$ denotes

$$p + \tfrac{1}{2}\rho[(u-U)^2 + (v-V)^2 + (w-W)^2] + \rho\Omega + \Lambda \cdot \mathbf{r} \times \mathbf{U} + \frac{\partial\phi}{\partial t} = \text{constant} \qquad (30)$$

the triple scalar product of the indicated vectors, and $\mathbf{U}$ and its components u, v, w are velocities of the fluid. For elaboration SEE BERNOULLI'S THEOREM.

If a body is immersed in a steady stream of velocity and pressure V_∞ and p_∞ at infinity, and if V and p denote the velocity and pressure at a point on the body, the Bernoulli equation may be written in the dimensionless form, Eq. (31).

$$\frac{p - p_\infty}{\tfrac{1}{2}\rho V_\infty^2} = 1 - \left(\frac{V}{V_\infty}\right)^2 \qquad (31)$$

At the point of maximum velocity on the body the pressure has a minimum value, denoted by p_m. Because ratio $(p_m - p_\infty)/(\tfrac{1}{2}\rho V^2)$ is a constant for the body, the value of p_m can be reduced either by reducing the ambient pressure p_∞ (as can be done in a variable-pressure water tunnel) or by increasing V_∞. If by either means p_m is reduced to the vapor pressure of a liquid p_v, and the liquid begins to vaporize, the liquid is said to undergo cavitation, and Eq. (32) holds. This is called

$$\sigma_v = \frac{p_\infty - p_v}{\tfrac{1}{2}\rho V_\infty^2} \qquad (32)$$

the vapor-pressure cavitation number. The phenomenon of cavitation is of great technical importance in the design of ship propellers, turbines, and other hydraulic structures because of the erosion caused by the collapsing cavitation bubbles. The inception pressure may be greater than the vapor pressure if a considerable amount of entrained air is present; on the other hand, a specimen of liquid may be able to withstand tensions of thousands of atmospheres if special care has been taken to remove gas nuclei from it. SEE CAVITATION; DIMENSIONLESS GROUPS.

When viscosity is taken into account, the equations of motion become the Navier-Stokes equations (33). SEE NAVIER-STOKES EQUATIONS.

$$\begin{aligned}
\frac{\partial u}{\partial t} + u\frac{\partial u}{\partial x} + v\frac{\partial u}{\partial y} + w\frac{\partial u}{\partial z} &= B_x - \frac{1}{\rho}\frac{\partial p}{\partial x} + \nu\left(\frac{\partial^2 u}{\partial x^2} + \frac{\partial^2 u}{\partial y^2} + \frac{\partial^2 u}{\partial z^2}\right) \\
\frac{\partial v}{\partial t} + u\frac{\partial v}{\partial x} + v\frac{\partial v}{\partial y} + w\frac{\partial v}{\partial z} &= B_y - \frac{1}{\rho}\frac{\partial p}{\partial y} + \nu\left(\frac{\partial^2 v}{\partial x^2} + \frac{\partial^2 v}{\partial y^2} + \frac{\partial^2 v}{\partial z^2}\right) \\
\frac{\partial w}{\partial t} + u\frac{\partial w}{\partial x} + v\frac{\partial w}{\partial y} + w\frac{\partial w}{\partial z} &= B_z - \frac{1}{\rho}\frac{\partial p}{\partial z} + \nu\left(\frac{\partial^2 w}{\partial x^2} + \frac{\partial^2 w}{\partial y^2} + \frac{\partial^2 w}{\partial z^2}\right)
\end{aligned} \qquad (33)$$

Equations (33) may also be written in the vector form as Eq. (34) in which ν is the kinematic viscosity.

$$\frac{\partial \mathbf{U}}{\partial t} + \boldsymbol{\omega} \times \mathbf{U} + \tfrac{1}{2}\, \text{grad}\, (V^2) = \mathbf{B} - \frac{1}{\rho}\, \text{grad}\, p - \nu\, \text{curl}\, \boldsymbol{\omega} \qquad (34)$$

These and the equation of continuity give four equations for solving for u, v, w, and p. If the equations can be solved, the stresses can then be obtained from the fundamental relations between stresses and rates of strain, as shown in Eqs. (35), where μ is the coefficient of dynamic viscosity.

$$\tau_{11} = -p + 2\mu \frac{\partial u}{\partial x} \qquad \tau_{22} = -p + 2\mu \frac{\partial v}{\partial x}$$

$$\tau_{33} = -p + 2\mu \frac{\partial w}{\partial x} \qquad \tau_{23} = \mu \left(\frac{\partial w}{\partial y} + \frac{\partial v}{\partial z} \right) \qquad (35)$$

$$\tau_{31} = \mu \left(\frac{\partial u}{\partial z} + \frac{\partial w}{\partial x} \right) \qquad \tau_{12} = \mu \left(\frac{\partial v}{\partial x} + \frac{\partial u}{\partial y} \right)$$

Two immediate and important consequences of the Navier-Stokes equations are that (1) the circulation in a closed circuit moving with the fluid diminishes at a time rate given by $\nu \oint (\text{curl } \boldsymbol{\omega}) \cdot \mathbf{ds}$, and (2) the energy per unit volume of a fluid diminishes at a time rate given by expression (36). These furnish a measure of the effect of viscosity on the rates of dissipation of both vorticity and energy.

$$\mu \left[2 \left(\frac{\partial u}{\partial x} \right)^2 + 2 \left(\frac{\partial v}{\partial y} \right)^2 + 2 \left(\frac{\partial w}{\partial z} \right)^2 + \left(\frac{\partial w}{\partial y} + \frac{\partial v}{\partial z} \right)^2 + \left(\frac{\partial u}{\partial z} + \frac{\partial w}{\partial x} \right)^2 + \left(\frac{\partial v}{\partial x} + \frac{\partial u}{\partial y} \right)^2 \right] \qquad (36)$$

Irrotational flow. In irrotational flow the vorticity is zero and the velocity is expressible as the gradient of a scalar potential $\mathbf{U} = \text{grad}\, \phi$. The equation of continuity in rectangular coordinates then assumes the form of Laplace's equation, Eq. (37). When the flow is irrotational, the

$$\frac{\partial^2 \phi}{\partial x^2} + \frac{\partial^2 \phi}{\partial y^2} + \frac{\partial^2 \phi}{\partial z^2} = 0 \qquad (37)$$

terms due to viscosity in the Navier-Stokes equations vanish so that they become formally identical with the Euler equations. Thus, in either viscous or inviscid irrotational flow, if Laplace's equation has been solved for the velocity potential, the pressure field can be obtained from the Bernoulli equation.

Elementary solutions of Laplace's equation are: for uniform flow in the direction of a line having direction cosines l, m, n, Eq. (38); for a source of strength M at the origin, Eq. (39); and for a doublet of strength Δ oriented along the x axis, Eq. (40).

$$\phi = V(lx + my + nz) \qquad (38)$$

$$\phi = -\frac{M}{(x^2 + y^2 + z^2)^{1/2}} \qquad (39)$$

$$\phi = -\frac{x\Delta}{(x^2 + y^2 + z^2)^{1/2}} \qquad (40)$$

A doublet may be obtained by letting a source and sink of equal strength approach each other, holding constant the product of the source strength by the distance between them (**Fig. 3**). See DOUBLET FLOW.

By combining these elementary solutions given in Eqs. (38)–(40), many interesting flow patterns can be obtained; for example, Eq. (41), formed by adding the potentials of a uniform

$$\phi = Vx \left[1 + \frac{a^3}{2(x^2 + y^2 + z^2)^{1/2}} \right] \qquad (41)$$

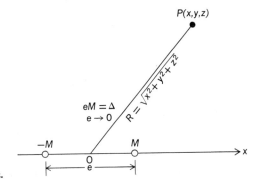

Fig. 3. Definition sketch of doublet.

stream and a doublet, gives the flow about a sphere. Bodies obtained as stream surfaces by trying various combinations of sources, sinks, and doublets in a uniform stream are called Rankine bodies. A necessary condition for such bodies to be closed is that the algebraic sum of the strengths of the sources and sinks be zero.

Methods of solving the direct problem, of finding the flow subject to prescribed boundary conditions, have also been devised. These methods may be classified in five categories.

Separation of variables. This reduces Laplace's equation to several ordinary differential equations, the solutions of which are obtained as sets of orthogonal functions. Combinations of these are then found which satisfy the boundary conditions.

Method of images. This method is suitable when the boundaries are planes, spheres, or circular cylinders. This technique has reached its culmination in the discovery of the so-called sphere and circle theorems which immediately give the modification of the flow when a sphere or circle is introduced into a preexisting flow pattern.

Method of integral equations. Two important classes of problems, the Neumann problem, in which the normal component of the velocity is prescribed on the boundary, and the Dirichlet problem, in which the values of the potential are given on the boundary, can be formulated as Fredholm integral equations. Although it is tedious to solve these equations numerically by hand, programs are available for solving them on moderate- or high-speed automatic computers.

Solution by relaxation. This is a numerical method in which the flow region is divided into a fine network, at the corners of which the values of ϕ are assumed as an initial approximation. The finite-difference form of Laplace's equation and the boundary conditions are then used to adjust the assumed values by trial and error. This method is also tedious.

Conformal mapping. The theory of functions of a complex variable furnishes a remarkably powerful tool for solving two-dimensional irrotational flow problems. Riemann's mapping theorem gives assurance that boundaries of regions can be mapped into circles by means of analytic functions. If the proper function is found, then it also transforms the flow about these boundaries into the much simpler problem of the flow about the circles, which can be solved. The difficulty in conformal mapping lies in finding the appropriate mapping functions. Many flow problems with simple boundaries can be solved by applying the properties of elementary functions of complex variables. For arbitrary shapes, several numerical procedures have been developed.

Viscous flow. Since the viscosity of the most common fluids, air and water, is very small, it is an excellent approximation to treat their flows as inviscid except at very low Reynolds numbers ($Vl/\nu < 10$) and in the neighborhood of the boundaries. Because the velocity of a fluid at a wall is zero relative to that of the boundary (nonslip condition), a so-called boundary layer in which viscous effects are important is present, within which appreciable shear stresses occur. Outside of this boundary layer the flow may be treated as inviscid, and to a good approximation the pressure across the boundary layer may be assumed to be that in the inviscid flow at the outer edge of the boundary layer.

A simplified form of the Navier-Stokes equations, called the boundary-layer equations, yields laminar-flow solutions for the velocity profiles. It is known, however, that boundary-layer flows are often turbulent. A theory of the stability of the laminar boundary layer has been devel-

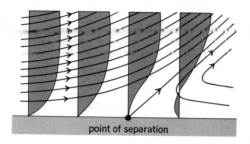

Fig. 4. Boundary-layer separation.

oped, and current research is shedding considerable light on the processes whereby a disturbed laminar flow eventually breaks down into the typical random motions of turbulence, but the mechanisms are not yet clear.

Although turbulent flow is believed to be governed by the Navier-Stokes equations, it has not yet been possible to derive any turbulent-flow solutions from these equations. The difficulty lies in the fact that the Reynolds equations for turbulent flow, which are derived from the Navier-Stokes equations by an averaging process, introduce new unknowns for which additional relations are required. The well-developed theory of homogeneous isotropic turbulence, and the body of accumulated experimental results and hypotheses concerning turbulence in shear flows, have not yet supplied these missing relations.

A phenomenon of real flow that is of great practical concern is that of separation of flow. This occurs in a boundary layer when the fluid in the neighborhood of a wall is brought to rest and caused to reverse its direction by the action of an adverse pressure gradient, that is, increasing pressures in the downstream direction (**Fig. 4**). When the boundary layer is turbulent, separation occurs farther downstream (or not at all) than when it is laminar, because the momentum of the layers near the wall is reinforced by turbulent interchange with layers of higher mean velocity due to the random motions. SEE BOUNDARY-LAYER FLOW; LAMINAR FLOW.

Bibliography. S. Goldstein (ed.), *Modern Developments in Fluid Dynamics*, 1938, reprint 1965; H. Lamb, *Hydrodynamics*, 6th ed., 1945, reprint 1965; B. LaMehante, *An Introduction to Hydrodynamics and Water Waves*, 1975; H. Rouse (ed.), *Advanced Mechanics of Fluids*, 1959, reprint 1976; I. H. Shames, *Mechanics of Fluids*, 2d ed., 1982; V. Streeter and E. B. Wylie, *Fluid Mechanics*, 8th ed., 1985.

HYDROKINETICS
LOUIS LANDWEBER

Forces produced by a liquid as a consequence of its motion. The mass of a liquid gives rise to inertia forces which are manifested as normal stresses or pressures when the liquid flows. The viscosity and cohesion of a liquid give rise to tangential shear forces and also affect the normal stresses. The interaction of these properties on the motion of the fluid is formally expressed in the equations of Navier and Stokes. Because of the difficulty of solving these equations in general, approximate solutions are often obtained by neglecting the less important properties in a particular flow situation. SEE HYDRODYNAMICS.

HYDRAULICS
WILLIAM ALLAN

The behavior of water or other liquids, chiefly when in motion. As a part of fluid mechanics, hydraulics deals with the properties, behavior, and effects of all liquids at rest against, or in motion relative to, boundary surfaces or objects. Its laws also apply to gases when changes in density are small, that is, compressibility effects are negligible.

Applications. Common hydraulic applications encountered in civil engineering include the flow of water in pipes, canals, and rivers, in flood control, in land reclamation, and in hydroelectric power projects. Hydraulics also influences appurtenant structures and devices such as dams, dikes, locks, spillways, weirs, piers, ship hulls, gates, and valves. Mechanical and chemical engineering applications include the flow of gases, oils, or other liquids; lubrication; and fluid machinery such as pumps, turbines, propellers, fans, and fluid power transmission and control devices, including servomechanisms.

The physical laws of hydrostatics govern the effects of fluids at rest. Hydrokinematics covers fluid motion wherein consideration of the forces causing the motion is not required. Hydrodynamics treats of fluid motion, the forces involved, and the accompanying energy changes. Characteristics of particular phenomena in hydraulics include fluid properties; the shapes, sizes, relative roughnesses, and relative motions of the surfaces or objects involved; the relative times and distances (short or long); whether flow is closed as in a pipe or open as in a canal; and whether motion is in a relatively extensive fluid as a submarine deeply submerged or whether motion occurs where there is a free surface as a ship at sea.

Physical laws are expressed by mathematical equations; two are required—an equation of condition (continuity) and one of motion (momentum or energy). When introduced by methods of dimensional analysis, physical quantities which are known, assumed, or sought experimentally may enter the equations in combinations which are dimensionless numbers of ratios. S*ee* D*imensional analysis*; D*imensionless groups*.

Properties. Hydraulics deals with bulk fluids which are regarded as homogeneous; it is not concerned with molecular sizes. As distinct from solids, fluids, such as water, are unable to resist shearing forces while remaining in a state of equilibrium. Among fluids, a liquid mass will have a definite volume, varying only slightly with temperature and pressure, whereas a mass of gas will fill any space available to it, the pressure adjusting itself accordingly.

A common fluid property is density, the mass per unit volume, sometimes expressed dimensionally in slugs per cubic foot, or $[(lb\ force)(s)^2/(ft)^4]$. Specific weight, the weight per unit volume in pounds force per cubic foot, is an alternate way of expressing density.

Viscosity, the ability of a fluid in motion to resist shearing forces, is the seat of all fluid resistance. The coefficient of viscosity (absolute viscosity) is the shearing stress divided by the velocity gradient, that is, the rate at which adjacent fluid layers slip past each other; it has the dimension $[(slug)/(ft)(s)]$ or $[(lb\ force)(s)/(ft)^2]$. The slippage rate is greatest adjacent to a solid boundary. There, in the boundary layer, the velocity rises sharply from zero at the wall toward the value in the mainstream. Where viscosity and density enter a problem, the coefficient of viscosity is divided by density to give the kinematic viscosity, which has the dimension square feet per second. S*ee* V*iscosity*.

Elasticity denotes the relative compressibility of a fluid under pressure. The volume modulus of elasticity is the change in pressure intensity divided by the corresponding unit volumetric change. This is not important in ordinary applications but plays a significant role when gases flow at high velocity or when, in a pipeline carrying a liquid, a valve is closed or opened rapidly, thereby causing a pressure wave. S*ee* W*ater hammer*.

Surface tension is a force per unit of length that binds surface molecules of a liquid to one another or to a solid boundary for which they may have an affinity; it is of minor importance except where thin films with a free surface exist or capillary movements occur. S*ee* H*ydrostatics*.

GAS KINEMATICS
F*rank* H. R*ockett*

The motion of a gas considered by itself, without regard for the cause of the motion. Various flow phenomena arise in widely different applications; however, the phenomena dealt with independently of their causes can be classed into relatively few fundamental patterns. Gas kinematics then deals with these abstracted flows. S*ee* G*as dynamics*; S*ink flow*; S*ource flow*.

GAS DYNAMICS

JOSHUA MENKES, ALI B. CAMDEL, AND LAWRENCE TALBOT

L. Talbot is author of the section Rarefied Gas Dynamics.

The study of the motion of gases which takes into account thermal effects generated by the motion. Gas dynamics combines fluid mechanics and thermodynamics and differs from gas statistics in that there is motion and from gas kinematics in that the forces exerted on or by the gas are considered.

Scope of subject. Several other names are used to define the subject. The most important ones are aerothermodynamics, aerothermochemistry, fluid dynamics, compressible fluid flow, and supersonic aerodynamics. This terminology reflects the fact that in each particular case different aspects of gas dynamics are emphasized. Magnetogasdynamics, which includes effects due to magnetism and electricity, has applications in rocket reentry, propulsion, and astrophysics. SEE AEROTHERMODYNAMICS; COMPRESSIBLE FLOW; FLUID DYNAMICS; FLUID FLOW.

The following discussion deals with the motion of a continuous medium and is not concerned with the behavior of individual atoms or molecules which constitute the gas. At low pressures, however, such as may be encountered at very high altitudes, the particle mean free path is very large and continuum considerations may not be applicable. In that case the methods of rarefied gas dynamics, discussed in the last section of the article, must be applied.

Fundamental relations. The fundamental conservation principles of mechanics and thermodynamics constitute the theoretical basis of gas dynamics. The conservation laws can be derived in lagrangian form with respect to a specific mass of flowing gas or in eulerian form with respect to the rate at which gas enters and leaves a fixed control volume in space. The eulerian conservation laws are expressed by (1) the continuity equation (conservation of mass); (2) the Navier-Stokes equations (conservation of momentum); and (3) the energy equation (conservation of energy).

A list follows of the principal notations used in the field.

A = cross-sectional area
a = acoustic velocity
c = speed of light
c_p = specific heat at constant pressure
c_v = specific heat at constant volume
f = friction factor, as in Eq. (53)
G = mass velocity, $G = \rho V$
G' = mass flow per unit time
h = enthalpy
k = thermal conductivity
L = characteristic length
m = molecular mass
n_i = number of particles of component i per unit volume
p = pressure
p_0 = total or stagnation pressure
Q_m = heat
R, R' = gas constant
r_h = hydraulic radius
T = temperature
T_0 = total or stagnation temperature
t = time
u = internal energy
V = average gas velocity
$\mathbf{V}$ = gas vector velocity
$\mathbf{v}_i$ = particle vector velocity
w_i = rate of production of species i
x = fractional dissociation
α = Mach angle
γ = specific heat ratio, c_p/c_v
ν = kinematic viscosity
ρ = density
∇ = gradient operator
$(\ )^*$ = critical values
$(\ )_0$ = reservoir condition

Continuity equation. Two types of continuity equation can be written: global and species conservation. The global continuity equation is expressed by Eq. (1), where the first term defines the rate of change of the mass flow with respect to the space coordinates, whereas the second term, usually called the source term, indicates changes with respect to time within the control volume. If the flow is steady and there are no sources present, the continuity equation reduces to Eq. (2). For incompressible flow, it is given simply by Eq. (3). It is often convenient in steady irrotational flow to introduce a function φ satisfying Eq. (4).

$$\nabla \cdot (\rho \mathbf{V}) + \frac{\partial \rho}{\partial t} = 0 \quad (1) \qquad \nabla \cdot (\rho \mathbf{V}) = 0 \quad (2)$$

$$\nabla \cdot \mathbf{V} = 0 \quad (3) \qquad \nabla \varphi = \mathbf{V} \quad (4)$$

The incompressible continuity equation (3) becomes Eq. (5). The identity $\nabla \times (\nabla \varphi) = 0$

$$\nabla^2 \varphi = 0 \quad (5)$$

shows that the flow is irrotational. Equation (5) is known as Laplace's equation. Solutions to Laplace's equation which satisfy the appropriate boundary conditions are solutions of gas kinematics since they describe a flow without regard to the forces maintaining it.

If the gas undergoes chemical changes, the concentration of species is altered, which affects the spatial distribution of energy since each species has its particular velocity $\mathbf{v}_i$. Thus, in order to properly keep track of the distribution of energy, one must account for the rate of change of each species or component by a continuity equation of the form shown in Eq. (6). The term on

$$\frac{\partial n_i}{\partial t} + \nabla \cdot (n_i \mathbf{V}_i) = w_i \quad (6)$$

the right-hand side represents the rate of species production. When Eq. (6) is summed for all species, one arrives at the continuity equation, (1).

Frequently a one-dimensional approach is followed in gas dynamics, in which case the properties are assumed to vary mainly in the flow direction. The integral of the steady global continuity equation is then given by Eq. (7).

$$\rho A V = \text{constant} = G' \quad (7)$$

Momentum equation. The momentum equation expresses the conservation of momentum. It must take into consideration the effects of friction and of external body forces such as gravity, magnetism, and possibly others. Written in vector form, it is given by Eq. (8). The term D/Dt represents the mobile operator which is defined by Eq. (9). The first term on the right-hand

$$\frac{D\mathbf{V}}{Dt} = -\frac{1}{\rho} \nabla p + \nu \nabla^2 \mathbf{V} + 1/3 \nu \nabla (\nabla \cdot \mathbf{V}) + \sum_{l=1}^{N} \mathbf{F}_l \quad (8) \qquad \frac{D}{Dt} = (\partial/\partial t) + \mathbf{V} \cdot \nabla \quad (9)$$

side of Eq. (8) represents the forces exerted on the fluid at the control volume boundaries; the second and third terms on the right-hand side are the viscous stresses; and F_l stands for any body force such as gravity, electric, and magnetic forces. Equation (8) in its vectorial form is quite general and applies to classical gas dynamics and aerothermochemistry as well as to magnetohydrodynamics. The specialization to particular coordinate systems, for example, cartesian coordinates, is obtained by standard vector manipulation methods.

It may not be always necessary, however, to consider all terms. For example, if electromagnetic effects are not present but viscous effects are included, the Navier-Stokes equation is obtained. SEE NAVIER-STOKES EQUATIONS.

A general solution for the Navier-Stokes equation has not been found, and only a few particular solutions exist. When the momentum equation is simplified to exclude viscous effects as well as forces on the body, Euler's equation (10) is obtained. Its one-dimensional form is given by Eq. (11).

$$\frac{D\mathbf{V}}{Dt} = -\frac{1}{\rho} \Delta p \quad (10) \qquad \frac{dp}{\rho} + V\, dV = 0 \quad (11)$$

Equation (11) can be integrated for incompressible flow (that is, $\rho = $ constant) to yield Bernoulli's equation (12). SEE BERNOULLI'S THEOREM.

$$\frac{p}{\rho} + \frac{V^2}{2} = \text{constant} \quad (12)$$

For compressible flow, the thermodynamics of the flow must be considered, which is done through the energy equation.

Energy equation. The energy equation expresses the principle of conservation of energy (first law of thermodynamics) as applied to a flowing gas. There are a large number of equivalent forms in which this equation can be written. The apparent differences arise from the use of such subsidiary thermodynamic relations as, for example, the equation of state $p = \rho RT$ for a perfect gas, or p/ρ^γ = constant for an isentropic process, to express one state variable in terms of another. A fairly common form of the energy equation is given in Eq. (13). The left-hand side of Eq. (13)

$$\underbrace{\frac{\rho D(u + 1/2\, V^2)}{Dt}}_{} = \underbrace{\nabla \cdot p\mathbf{V}}_{(I)} + \underbrace{\rho \sum_{l=1}^{N} \mathbf{F}_l \cdot \mathbf{V}}_{(II)} + \underbrace{\nabla \cdot k\nabla T}_{(III)} + \underbrace{\sum_{m=1}^{M} Q_m}_{(IV)} \qquad (13)$$

gives the change in internal and kinetic energy, which is balanced on the right-hand side by the rate at which work is done by (I) the pressure forces and (II) the body forces and (III) by the heat conducted across the boundary. Term (IV) accounts for any other energy-transfer mechanism, such as the transfer of heat generated by the dissipative action of viscosity and electrical conductivity or radiative heat transfer or transfer of electromagnetic energy.

If the flow is steady, Eq. (14) is obtained. Furthermore, if the flow is adiabatic, then the

$$\rho \mathbf{V} \cdot \nabla \left(u + 1/2\, V^2 + \frac{p}{\rho} \right) = \nabla \cdot k\nabla T + \rho \sum \mathbf{F} \cdot \mathbf{V} + \sum Q \qquad (14)$$

work done by the shear forces Φ does not leave the system but simply raises the internal energy of the gas and is, therefore, accounted for by u. In the absence of body forces, all terms on the right-hand side of Eq. (14) are zero, and the energy equation can be integrated to yield Eq. (15). Moreover, if the gas is perfect, Eq. (16) is obtained.

$$u + 1/2\, V^2 + \frac{p}{\rho} = \text{constant} \qquad (15) \qquad \frac{\gamma}{\gamma - 1}\frac{p}{\rho} + \frac{V^2}{2} = c_p T + \frac{V^2}{2} = \text{constant} \qquad (16)$$

The idealizations introduced by the concept of a perfect gas are approximately satisfied at moderate temperatures and pressures. At very high pressures or temperatures or both, the real nature of the gas molecules must be taken into consideration. Research in gas dynamics is not ordinarily concerned with very high pressure; but high temperatures, and the attendant effects on the gas, are regularly encountered in missile flight, combustion, nuclear reactors, and many other technological applications. A very simple correction to the perfect equation of state can be made if the gas molecules dissociate. The equation of state then can be written as Eq. (17), where

$$p = R'Tn(1 + x) \qquad (17)$$

$R'/m = R$ and x is the fraction of the gas which is dissociated—the bookkeeping is done by the conservation of species equations.

At sufficiently high temperatures, the electrons may become so energetic as to leave their orbits and become free electrons. When this happens, the gas is said to be ionized and is called a plasma.

In a fairly complex situation where the gas can consist of molecules, atoms, ions, and electrons, the equation of stage may have to be modified to account for more than just one dissociating species, as in Eq. (18). Another consequence of the nonideal nature of the gas is the

$$p = R'Tn(1 + \sum x_i) \qquad (18)$$

fact that it takes time for a molecule to dissociate, to ionize, and to combine with another species. The time that it takes to effect such a change is called relaxation time. When the relaxation time is short compared to a characteristic flow time, which might be the time it takes a fluid element to pass through a shock, a state of thermodynamic equilibrium is said to exist; this state is characterized by Damköhler's first ratio, given by Eq. (19).

$$\text{Da}_\text{I} = \frac{t_\text{transit}}{t_\text{relaxation}} \gg 1 \qquad (19)$$

At equilibrium, the degree of ionization X is given by the Saha equation (20), where E_0 is a characteristic parameter of the gas.

$$\frac{x^2}{1-x^2} = \text{constant} \, \frac{T^{5/2}}{p} \exp\left[-(E_0/RT)\right] \tag{20}$$

In the case $Da_I \ll 1$, a state of frozen flow is said to exist. Here the relaxation time is so large that the gas behaves nearly like a perfect gas.

When relaxation and transit times are comparable, the necessary accounting of all the possible species depends on a detailed knowledge of all possible chemical processes and the controlling rates; these rates have been established for a large number of reactions. However, the details of even a relatively simple phenomenon such as the burning of a Bunsen burner are still not completely understood.

Real gases deviate in other important aspects from a perfect gas. In the latter, the specific heat at constant pressure c_p and at constant volume c_v are constants. In a real gas, the manner in which the molecules "store" the heat must be considered. A gas that obeys the perfect gas thermal equation of state $p = \rho RT$ has the caloric equation of state given by Eq. (21), where c_{p0}

$$\frac{c_p}{c_{p0}} = 1 + \frac{\gamma - 1}{\gamma}\left[\left(\frac{\theta}{T}\right)^2 \frac{e^{\theta/T}}{(e^{\theta/T} - 1)^2}\right] \tag{21}$$

and γ are constant reference values. For $\gamma = 1.4$, the correction at 3000 K (5000°F) is approximately 26%. θ refers to molecular vibrational-energy constant. For air between 300 and 2500 K (80 and 4000°F), the value of $\theta \approx 2800$ K (4600°F).

Dimensionless parameters. Of the large number of dimensionless parameters that can be formed, there are a few that are particularly useful in effecting simplifications in the equations of motion. The simplifications are generally the consequence of one or more parameters being very large or very small. SEE DIMENSIONLESS GROUPS.

The Knudsen number and Damköhler's ratio have been mentioned already. The Reynolds number $\text{Re} = VL/\nu$ is the dominating parameter when the effects of viscosity and the inertia of the fluid both contribute to the gas motion. The phenomena which are peculiar to gas dynamics, however, can best be categorized in terms of the Mach number. The Mach number is the ratio of the flow speed to the speed of sound. The speed of sound is the propagation velocity not only of audible sound but of any weak pressure disturbance. SEE DIMENSIONAL ANALYSIS; DYNAMIC SIMILARITY; MACH NUMBER.

Gas-dynamic flow regimes can then be classified as follows.

$M < 1$ subsonic flow	$M > 1$ supersonic flow
$M = 1$ sonic flow	$M > 5$ hypersonic flow
$0.9 < M < 1.1$ transonic flow	

Speed of sound. Consider a tube with insulated walls filled with a compressible nonviscous gas. Neglecting all external forces, Euler's equations in one dimension, Eq. (22), and in continuity, Eq. (23), describe the motion.

$$\frac{\partial V}{\partial t} + \frac{V \partial V}{\partial x} = -\frac{1}{\rho}\frac{\partial p}{\partial x} \tag{22} \qquad \frac{\partial \rho}{\partial t} + \frac{\partial(\rho V)}{\partial x} = 0 \tag{23}$$

For the case of small perturbations, the equation of state is given by Eq. (24), where ρ_0 is a reference density.

$$\rho = \rho_0(1 + \epsilon) \tag{24}$$

Substituting Eq. (24) into Eqs. (22) and (23) yields Eqs. (25) and (26).

$$\frac{\partial V}{\partial t} = -\frac{\rho}{\rho_0}\frac{dp}{d\rho}\frac{\partial \epsilon}{\partial x} \tag{25} \qquad \frac{\partial \epsilon}{\partial t} = -\frac{\partial V}{\partial x} \tag{26}$$

Eliminating ϵ between Eqs. (25) and (26) yields the one-dimensional wave equation (27), where Eq. (28) defines the velocity of propagation of weak disturbances in general and sound in particular. If the transformation is assumed to be isentropic, which is quite reasonable, Eq. (29) is obtained.

$$\frac{\partial^2 V}{\partial t^2} = a^2 \frac{\partial^2 V}{\partial x^2} \quad (27) \qquad a^2 = \left.\frac{dp}{d\rho}\right|_{\rho=\rho_0} \quad (28) \qquad a^2 = \gamma RT \quad (29)$$

The derivation of the speed of sound is illustrative of a whole class of problems that can be treated by the simple wave equation. The term waves is used in gas dynamics not only in the classical sense but also to denote wavefronts.

Consider a pulsating pressure source moving with velocity V. This source starts pulsating at time $t = 0$. In a time interval t_1, the source travels a distance Vt_1, while the signal reaches the surface of a sphere of radius at_1 (the surface of this sphere constitutes the wavefront at time t_1). At a later time t_2, the source point is at Vt_2, while the signal front is at at_2. Thus, as long as $V/a = M < 1$, the source point is always inside the outermost wavefront (**Fig. 1**).

Still another use of the term wave refers to a wave envelope. If the point source moves so fast that $V > a$, the wavefront spheres will no longer contain the source. The condition is shown in **Fig. 2**. The envelope to this family of spheres is a cone, known as the Mach cone, and the Mach angle is such that $\sin \alpha = 1/M$.

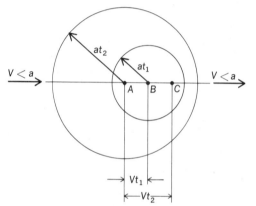

Fig. 1. Wavefronts produced by a point source moving at subsonic velocity. (*After A. B. Cambel and B. H. Jennings, Gas Dynamics, McGraw-Hill, 1958*)

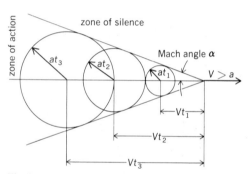

Fig. 2. Rule of forbidden signals from a point source moving at supersonic velocity. (*After A. B. Cambel and B. H. Jennings, Gas Dynamics, McGraw-Hill, 1968*)

The Mach line constitutes a demarcation. In Fig. 2 the fluid outside the Mach line will receive no signal from the source. T. von Kármán has appropriately called this phenomenon the rule of forbidden signals and designated the region ahead of the Mach line the zone of silence and the region inside the Mach line the zone of action.

Shock waves. In the same manner in which a Mach wave is the envelope of infinitesimal disturbances, a shock wave is the envelope of finite disturbances. The steady conditions on either side of a standing shock wave can be obtained by applying the conservation laws (**Fig. 3**) expressed by Eqs. (30), (31), and (32), where $h = u + p/\rho$ is termed enthalpy, which for a perfect

$$\rho_1 V_1 = \rho_2 V_2 \quad \text{(continuity)} \quad (30)$$

$$p_1 + \rho_1 V_1^2 = p_2 + \rho_2 V_2^2 \quad \text{(momentum)} \quad (31)$$

$$h_1 + \frac{V_1^2}{2} = h_2 + \frac{V_2^2}{2} \quad \text{(energy)} \quad (32)$$

gas is given by $c_p T$. By a simple rearrangement of these equations, one obtains for a perfect gas the approximate expressions for the shock Mach number, Eq. (33) for weak shocks and Eq. (34) for strong shocks.

$$M_s \approx 1 + \frac{\gamma + 1}{4\gamma} \frac{p_2 - p_1}{p_1} \quad (33) \qquad M_s \approx \left(\frac{\gamma + 1}{2\gamma} \frac{p_2}{p_1}\right)^{1/2} \quad (34)$$

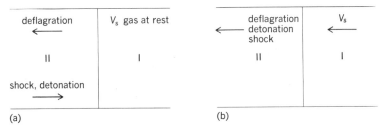

Fig. 3. Gas dynamic discontinuity. (a) Moving discontinuity, (b) Stationary discontinuity. (*After A. B. Cambel and B. H. Jennings, Gas Dynamics, McGraw-Hill, 1958*)

In a sound wave, $p_2 - p_1 \cong 0$ and, therefore, $M_s = 1$; for $p_2/p_1 = 4$, the shock speed is roughly twice the speed of sound, as shown in Eq. (34). SEE SHOCK WAVE.

Detonation and deflagration waves. Other interesting gas dynamic waves are characterized by the same continuity and momentum equations. The energy equation, however, is modified to include a term which accounts for chemical heat release; such waves are either detonations or deflagrations.

Eliminating the kinetic energy from the energy equation by the use of the momentum equation yields Eq. (35).

$$h_1 - h_2 + Q = 1/2(p_1 - p_2)\left(\frac{1}{\rho_2} + \frac{1}{\rho_1}\right) \qquad (35)$$

For a given Q, zero or nonzero, and given p_1 and ρ_1 (which through the equation of state gives h_1) and one additional variable behind the wave, for example, V_2 or h_2, the locus of all possible combinations of ρ_2 and p_2 can be plotted on a so-called Hugoniot diagram (**Fig. 4**). The lines OJ and OK are tangent to the Hugoniot curve; O is the point $p_1, 1/\rho_1$. Points J and K separate strong and weak waves. Flows corresponding to J and K are characterized by the fact that all the

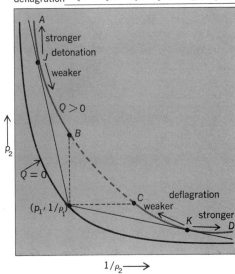

	M_1	M_2	p_2/p_1	V_2/V_1	ρ_2/ρ_1	T_2/T_1
detonation	>1	≤1	>1	<1	>1	>1
deflagration	<1	<1	<1	>1	<1	>1

Fig. 4. Hugoniot diagram and Chapman-Jouget conditions. (*After A. B. Cambel and B. H. Jennings, Gas Dynamics, McGraw-Hill, 1958*)

thermodynamic and fluid-mechanic variables have an extremum. Transitions from B to C or vice versa involve a decrease in entropy and are therefore forbidden. The slope of the tangent is connected through the momentum equation to the flow velocity by Eq. (36). The derivative at constant entropy (since the entropy is stationary near J) is given by Eq. (37).

$$\frac{p_2 - p_1}{\frac{1}{\rho_1} - \frac{1}{\rho_2}} = (\rho_2 V_2)^2 = \frac{\Delta p}{\Delta\left(\frac{1}{\rho}\right)}\bigg|_2 \quad (36) \qquad \rho^2 \frac{dp}{d\rho}\bigg|_2 = \rho_2^2 a_2^2 \quad (37)$$

In other words, the wave has a Mach number of unity with respect to the gas behind it. The points are appropriately named Chapman-Jouget points after the men who discovered their unique properties. The significance of these points lies in the fact that a stable detonation will eventually reach the point J and a deflagration point K.

Hydromagnetic (Alfvén) waves. Illustrative of the interaction of an electromagnetic field with a flowing plasma is the hydromagnetic, or Alfvén, wave. It is assumed that the velocity has only one component, for example, V_y, and the applied magnetic field B_x is perpendicular to it; the electric field E_z is perpendicular to B_x, and the current density j_z is parallel to the electric field. In order to simplify matters, it is assumed that the medium is incompressible so that the energy equation need not be considered. The fluid, moreover, is assumed to be inviscid and possesses infinite electrical conductivity. The equations of motion are then given by Eqs. (38) and (39).

$$\rho \frac{\partial V_y}{\partial t} = j_z B_x \quad (38) \qquad\qquad E_z - V_y B_x = 0 \quad (39)$$

Substituting these equations into Maxwell's equations yields a wave, Eq. (40), with the propagation velocity given by expression (41), called the Alfvén speed, where c is the speed of light.

$$\frac{\partial^2 E_z}{\partial x^2} = \left(1 + \frac{4\pi\rho c^2}{B_x^2}\right) \frac{1}{c^2} \frac{\partial^2 E_z}{\partial t^2} \quad (40) \qquad \frac{c}{(1 + 4\pi\rho c^2/B_x^2)^{1/2}} \quad (41)$$

Mach number functions. In many applications it is reasonable to assume that the gas is perfect—both thermally and calorically—and that the flow is adiabatic. It then becomes very useful to express all the dependent variables in terms of the Mach number.

One usually starts with the energy equation and defines a stagnation temperature T_0 by expression (42). Physically, T_0 represents the temperature the gas would have if all its kinetic

$$\frac{V^2}{2} + c_p T = c_p T_0 \quad (42)$$

energy were transformed into thermodynamic enthalpy. The stagnation enthalpy can be measured by a thermometer immersed into a gas stream.

From Eq. (42), one simply obtains Eq. (43), where T is now called the static temperature.

$$\frac{T_0}{T} = 1 + \frac{\gamma - 1}{2} M^2 \quad (43)$$

This temperature is measured with a thermometer at rest with respect to the gas. The stagnation temperature is constant in any adiabatic flow, even through a shock, and thus provides an excellent reference parameter. By using the isentropic relation $p = \rho^\gamma$ and the perfect gas law, a reference stagnation pressure and density can be defined by means of Eqs. (44) and (45).

$$\frac{p_0}{p} = \left(1 + \frac{\gamma - 1}{2} M^2\right)^{\gamma/\gamma - 1} \quad (44) \qquad \frac{\rho_0}{\rho} = \left(1 + \frac{\gamma - 1}{2} M^2\right)^{1/\gamma - 1} \quad (45)$$

When $M^2 \ll 1$, one may expand Eq. (44) by the binomial theorem to obtain Eq. (46).

$$p_0 = p + 1/2\, \rho V^2 \left(1 + \frac{M^2}{4} + \cdots\right) \quad (46)$$

The deviation from Bernoulli's equation (12) due to compressibility at $M = 0.5$ is only 6%.

Flows can be classified as internal flow and external flow. Internal flow refers to the cases where the gas is constrained by a duct of some sort. Characteristically external flow is flow over an airplane or missile.

Internal one-dimensional flow. Internal flows are conveniently characterized by (1) the shape of the duct and its variation, (2) the heat transfer through the walls of the duct and internal heat sources, and (3) frictional effects. By varying one of these characteristics at a time, the essential features of internal flow can be discussed most simply.

Variable area flow. A device to accelerate the flow of a gas or liquid is termed a nozzle. In most engineering applications the contour of the nozzle is first converging and then diverging; it thus has a minimum cross section, called a throat. *See Nozzle.*

For isentropic flow in a convergent-divergent nozzle in which the flow is supersonic in the divergent section, the velocity at the throat is sonic, that is, $M^* = 1$. The throat pressure is then said to be critical p^* and is given by Eq. (47). Velocity and pressure are related in this case by Eq. (48).

$$p^* = p_0 \left(\frac{2}{\gamma + 1}\right)^{\gamma/(\gamma - 1)} \quad (47) \qquad V = \frac{2\gamma R}{\gamma - 1} T_0 [1 - (p/p_0)^{(\gamma - 1)/\gamma}] \quad (48)$$

Consider a convergent-divergent deLaval nozzle inserted between two reservoirs as in **Fig. 5**. There will be no flow if the ratio of exit pressure p_e to reservoir pressure p_0 is $p_e/p_0 = 1$ (case a in **Fig. 6**). If p_e is reduced so that it is slightly less than the entrance pressure, the nozzle will act like a conventional venturi, as represented by curve b in Fig. 6. For this case, the flow is always subsonic and resembles incompressible flow. When the exit pressure is reduced further, the critical pressure can be reached at the throat, as curve g shows. In this case, the velocity is sonic at the throat but is never supersonic within the nozzle, even though the pressure at the throat corresponds to the critical. The minimum pressure which can exist in the nozzle outlet is depicted by point d, for which the pressure at the throat will be the critical. Here the velocity in the converging section is subsonic, in the diverging section it is supersonic, and at the throat it is sonic. For the range of exit pressures from p_d to p_g, the rate-of-flow curve is the same, for a given reservoir pressure p_0, and is plotted in **Fig. 7**. The flow rate reaches a maximum value and remains there over this wide range of exit pressures.

Even in the absence of friction, isentropic flow can exist only for the range of exhaust pressures from p_a to p_g and at the pressure reached along curve d, but not at intermediate pressures. The pressures p_g and p_d are the significant design pressures for a given nozzle. For exhaust pressures in the range between p_d and p_g, shocks will occur in the nozzle, raising the pressure from f to h (or f' to h'), followed by a pressure rise after the shock points to an exit pressure such

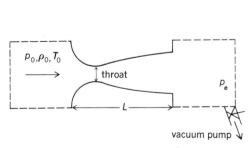

Fig. 5. Convergent-divergent nozzle. (*After A. B. Cambel and B. H. Jennings, Gas Dynamics, McGraw-Hill, 1958*)

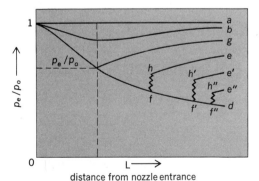

Fig. 6. Pressure distribution in convergent-divergent nozzle between two reservoirs. (*After A. B. Cambel and B. H. Jennings, Gas Dynamics, McGraw-Hill, 1958*)

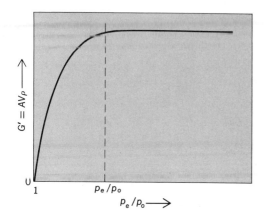

Fig. 7. Flow rate for given p_0, through convergent-divergent nozzle between two reservoirs. (After A. B. Cambel and B. H. Jennings, Gas Dynamics, McGraw-Hill, 1958)

as p_e. If p_e is less than p_d, the jet leaving the nozzle is said to be underexpanded and will drop in pressure after leaving the mouth of the nozzle. The velocity V_1 in front of a shock is supersonic, and the velocity V_2 behind a normal in contrast to an oblique shock is always subsonic. For a normal shock $V_1 V_2/a^{*2} = 1$, where a^* is the critical speed of sound corresponding to $M = 1$. Thus for $V_1 > V_2$, $V_1/a^* > 1$ and therefore $V_2/a^* < 1$. See Isentropic flow.

Diabatic flow. Heat exchangers and combustion chambers are devices in which heat transfer occurs. The equations describing nonadiabatic or diabatic processes are complicated; consequently, certain limiting assumptions are usually required to make possible analytical solutions of the equations.

These assumptions are that (1) the flow takes place in a constant-area section, (2) there is no friction, (3) the gas is perfect and has constant specific heats, (4) the composition of the gas does not change, (5) there are no devices in the system which deliver or receive mechanical work, and (6) the flow is steady.

Equations which conform to these requirements are called Rayleigh equations, and the associated flow is designated as Rayleigh flow. Designating by $Q_{1\to 2}$ the quantity of heat introduced between stations 1 and 2, Eq. (49) is obtained for the energy equation.

$$Q_{1\to 2} = c_p(T_2 - T_1) + \frac{(V_2^2 - V_1^2)}{2} = h_2 - h_1 + \frac{(V_2^2 - V_1^2)}{2} \tag{49}$$

If the stagnation enthalpy is introduced, then $Q_{1\to 2}$ is given by Eq. (50), which can be expressed in terms of stagnation temperatures by Eq. (51).

$$Q_{1\to 2} = h_{02} - h_{01} \tag{50} \qquad Q_{1\to 2} = c_p(T_{02} - T_{01}) \tag{51}$$

Because $Q_{1\to 2} \neq 0$ for diabatic flow and because $c_p > 0$ always, it follows that $T_{02} \neq T_{01}$. This inequality states that in diabatic flow the stagnation temperature is not solely determined by the reservoir conditions, as is the case with adiabatic flow. Heating raises the stagnation temperature; cooling lowers it.

The locus of points of properties during a constant-area, frictionless flow with heat exchange is called the Rayleigh line. By definition, along the Rayleigh line the continuity equation and the momentum equation must apply. Equation (30) applies to steady flow in a constant-area duct. Mass velocity by definition is $G = \rho V$ and, from Eq. (31), the momentum relation is $p + \rho V^2 = C$, where C is a constant. Consequently, Eq. (52) is obtained, which is one of the many Rayleigh-line equations.

$$p + \frac{G^2}{\rho} = C \tag{52}$$

The variations of pressure, temperature, and density with Mach number for Rayleigh flow are plotted in **Fig. 8**. The fact that the curve for T_0/T_0^* in Fig. 8 reaches a maximum at a Mach

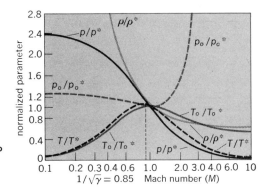

Fig. 8. Diabatic flow parameters for specific heat ratio of 1.4. Asterisk is where $M = 1$. (After A. B. Cambel and B. H. Jennings, Gas Dynamics, McGraw-Hill, 1958)

number of unity indicates that it is impossible to pass from one flow domain into the other by the same heat-transfer process. Thus, if heat is added to a subsonic flow, the flow can be accelerated only until its Mach number becomes unity. Further addition of heat will not further accelerate the gas but will result in choking of the flow. As a consequence, the flow must readjust itself, which it will do by lowering its initial Mach number. **Table 1** summarizes some of the Rayleigh flow phenomena.

Table 1. Variation of flow properties for Rayleigh flow

Property	Heating		Cooling	
	$M > 1$	$M < 1$	$M > 1$	$M < 1$
T_0	Increases	Increases	Decreases	Decreases
p	Increases	Decreases	Decreases	Increases
p_0	Decreases	Decreases	Increases	Increases
V	Decreases	Increases	Increases	Decreases
T	Increases	Increases when $M < 1/\sqrt{\gamma}$	Decreases	Decreases when $M < 1/\sqrt{\gamma}$
		Decreases when $M > 1/\sqrt{\gamma}$		Decreases when $M > 1/\sqrt{\gamma}$

Flow with friction. In long pipes the effects of friction may result in a significant pressure drop. Over a length dx, this pressure drop dp is given by the Fanning equation (53), where f is a friction factor that must be determined experimentally. To solve friction-flow problems analytically, certain simplifying assumptions are made and the resulting hypothetical flow is called Fanno flow. The Fanno flow assumptions are the same as those for Rayleigh flow except that the assumption that there is no friction is replaced by the requirement that the flow be adiabatic. Numerous Fanno flow equations may be written by combining the energy and the continuity equations in accordance with these assumptions. In **Fig. 9** may be seen the variation of properties during Fanno flow. **Table 2** summarizes the trends of the most important properties during subsonic and supersonic flow.

$$dp = -f \frac{\rho V^2}{2 r_h} dx \qquad (53)$$

When the Fanno and Rayleigh lines are plotted for the same constant mass velocity $G = \rho V$, the curves appear as in **Fig. 10**. The Rayleigh and Fanno lines have two points of intersection, denoted by a and b; a normal shock connects these two points. The flow through a shock wave is irreversible; thus, associated with it is an increase in entropy. Point b, thus, always lies to the right of point a.

Table 2. Fanno flow phenomena

Property	Initial flow is subsonic	Initial flow is supersonic
M	Increases	Decreases
V	Increases	Decreases
p	Decreases	Increases
T	Decreases	Increases
ρ	Decreases	Increases

There is another interesting point about the Fanno curve. If frictional flow continues along the subsonic portion of the Fanno line, the Mach number tends to increase toward unity, whereas if it continues along the supersonic portion, the Mach number decreases toward unity. As in the case of Rayleigh flow, it is impossible, by virtue of the second law of thermodynamics, to pass from one flow regime to the other (subsonic into supersonic or conversely) unless the mass velocity is readjusted.

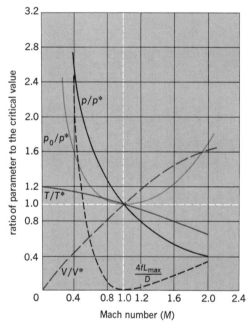

Fig. 9. Functions for constant-area flow with friction ($k = 1.4$). Asterisk is where $M = 1$. (*After A. B. Cambel and B. H. Jennings, Gas Dynamics, McGraw-Hill, 1958*)

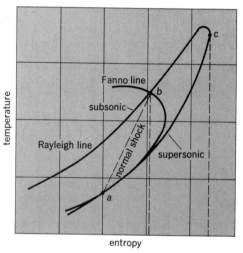

Fig. 10. Rayleigh and Fanno lines. (*After A. B. Cambel and B. H. Jennings, Gas Dynamics, McGraw-Hill, 1958*)

External flow. Boundary layers and wakes are the centers of interest in external flows. Here the effects of compressibility are substantially more difficult to analyze than in internal flows, if for no other reason than the inapplicability of a one-dimensional approach.

Boundary layers. Ballistic missiles and space vehicles enter the Earth's atmosphere with velocities typically of 4 mi/s (6–7 km/s). The corresponding Mach number, depending on the altitude, is of the order of 20. The energy that maintains the bow shock and the work done to overcome the viscous shear stresses, Φ, reduce the kinetic energy of the vehicle, and it decelerates. This loss of kinetic energy, which is the work of the drag forces, reappears in part as the increased enthalpy and temperature of the fluid near the vehicle surface. The rate of heat transfer driven by the large enthalpy gradient $(h_g - h_s)/\Delta$, where h_g is the gas enthalpy at some suitably defined distance Δ from the surface and h_s is the enthalpy right at the surface, is so large that special protection must be afforded the vehicle. To this purpose, the vehicle can be covered with a material designed to char, melt, or gasify and in so doing absorb much of the heat that would otherwise penetrate into the structure. This "ablation" process, as it is called, introduces large amounts of material (some of it chemically active, some of it ionized, and some of it radiating) into what ordinarily would be called a boundary layer. Unfortunately, many of the assumptions that make it possible to introduce the boundary-layer simplifications are violated here. Very complex computer programs have been developed, however, that yield reasonably accurate estimates of these effects. SEE BOUNDARY-LAYER FLOW.

Wakes. Hypersonic wakes were observed long before the space age. The tails of "shooting stars" are luminous wakes of meteors as they enter the atmosphere and burn up. Meteor velocities range between 12 and 45 mi/s (20 and 70 km/s), and temperatures as high as 5800°F (3500 K) have been estimated to be necessary for vaporization. Since meteor trails, as well as reentry-vehicle wakes, contain electrons which scatter electromagnetic energy, the trails can be observed with radar. The estimation of the decay rate of these electrons provides a simple example of wake chemistry. SEE WAKE FLOW.

The probability of capture per second of an electron by an ion is $\beta_i n_e$, where β_i is the recombination coefficient of the particular ion and n_e is the electron density. The capture rate dn_e/dt between an electron and an ion is then given by Eq. (54) for a single ionized species. In a neutral plasma $n_e = n_i$, and Eq. (54) can be integrated to yield Eq. (55), where n_0 is the ion density at time zero.

$$\frac{dn_e}{dt} = n_e \beta_i n_i \quad (54) \qquad n_e = \left(\frac{1}{n_0} + \beta_i t\right)^{-1} \quad (55)$$

Rarefied gas dynamics. Rarefied gas dynamics is that branch of gas dynamics dealing with the flow of gases under conditions where the molecular mean free path is not negligibly small compared to some characteristic dimension of the flow field. Rarefied flows occur when the gas density is extremely low, as in the cases of vacuum systems and high-altitude flight, but also when gases are at normal densities if the characteristic dimension is sufficiently small, as in the case of very small particles suspended in the atmosphere.

The dimensionless parameter which describes the degree of rarefaction existing in a flow is the Knudsen number, $\mathrm{Kn} = \lambda/L$, defined as the ratio of the mean free path λ to some characteristic dimension L of the flow field. Depending on the situation, L might be chosen, for example, as the diameter of a duct in a vacuum system, the wavelength of a high-frequency sound wave, the diameter of a suspended submicrometer-size particle, the length of a high-altitude rocket, or the thickness of a boundary layer or a shock wave. The mean free path λ, which is the average distance traveled by a gas molecule between successive collisions with other molecules, is equal to the molecular mean speed, given by Eq. (56), divided by the collision frequency ν_c: thus, Eq. (57) is satisfied. However, it is often more convenient in evaluating the Knudsen number to use the viscosity-based mean free path given by Eq. (58), where ν is the kinematic viscosity. SEE VISCOSITY.

$$\bar{C} = \sqrt{\frac{8}{\pi} RT} \quad (56) \qquad \lambda = \bar{C}/\nu_c \quad (57) \qquad \lambda \cong \frac{2\nu}{\bar{C}} \quad (58)$$

Flow regimes. It is convenient to divide rarefied flows into three flow regimes, according to the range of values of the appropriate Knudsen numbers. The regime of highly rarefied flow, which obtains for Kn much greater than 1 (typically greater than 10), is called collisionless or free-molecule flow, while the regime of slight rarefaction, where Kn is much less than 1 (typically less than 0.1), is called slip flow. Flows at Knudsen numbers intermediate to these limiting values are termed transition flows. The phenomena and methods of analysis associated with the three regimes are in general quite dissimilar, so the classifications are helpful.

Collisionless flow. As the name implies, collisionless flows are ones for which intermolecular encounters are very rare. Thus, in the case of a vehicle in high-altitude flight, such as a satellite, molecules impinging on a surface will travel, after reflection, very far before colliding with other molecules, with the consequence that the incoming flux of molecules is unperturbed by the reflected flux. For many applications, such high-altitude aerodynamics, it can be assumed that the molecular velocity distribution f_i of the incident molecules is a drifting maxwellian given by Eq. (59), where m is the molecular mass, n the number density, k Boltzmann's constant, T the

$$f_i = n\left(\frac{m}{2\pi kT}\right)^{3/2} e^{-mC^2/2kT} \tag{59}$$

free stream temperature, and the peculiar molecular velocity **C** is the difference between the absolute molecular velocity **ξ** and the mean gas velocity **V**, which in the case of an aerodynamic body in steady flight would be its flight velocity. With this assumption, the incident fluxes of mass, momentum, and energy on a unit area of a body can be evaluated readily in terms of appropriate moments of f_i.

The net fluxes to the surface element are determined by the nature of the velocity distribution f_r of the reflected molecules. Unfortunately, f_r is generally not known, and recourse must be had to empirical models of the molecular reflection process. One limiting case is specular reflection, in which the molecular velocity component normal to the surface is reversed, and the tangential velocity component remains unchanged. However, this is not characteristic of real gas-solid surface interactions. Another limiting case is diffuse reflection, for which the velocities of the molecules after reflection are independent of their incident velocities, and are in a half-range maxwellian distribution at the wall temperature T_W. The degree to which an actual surface behaves in a diffusive manner is customarily measured by three surface interaction parameters, or accommodation coefficients.

Calculation of the incident and reflected fluxes of energy and momentum leads to evaluation of the local pressure, shear stress, and energy transport to a surface element of a body. Then by integration over the entire body surface, overall lift, drag, and heat-transfer characteristics are obtained. The calculations are straightforward for simple convex shapes such as flat plates, spheres, and cylinders.

The analysis of collisionless flows becomes more complex if molecules can undergo multiple surface interactions, as in flows in ducts and over concave surfaces.

Slip flow. This is the regime of slight rarefaction, and is manifested initially as an alteration of the boundary conditions associated with the basic flow equations, the Navier-Stokes equations. The typical length scale L defining the Knudsen number is the boundary-layer thickness δ, although it could be the diameter of a pipe in the case of an internal flow. Boundary layers in the slip flow regime are usually laminar, but may exhibit strong compressibility and heat-transfer effects in high-speed flows. Slip relations are applicable only under conditions of small rarefaction, typically $\lambda/L \leq 0.1$.

The phenomena of slip and of temperature jump arise from the fact that molecules arriving at a stationary wall from a mean-free-path distant region of moving gas will carry with them the mean energy (temperature) and velocity characteristics of that region. When these are averaged with the properties of the molecules reflected from the wall, a finite-bulk gas slip velocity at the wall results, in contrast with the zero-slip velocity boundary condition employed with the Navier-Stokes equations in the continuum regime. Likewise, the temperature of the gas immediately adjacent to the wall is found to be different from that of the wall itself; thus the terminology temperature jump. The wall layer where these effects occur, the Knudsen layer, has been the subject of many detailed kinetic theory analyses based on the Boltzmann equation or models of it. These analyses have yielded Eqs. (60) and (61) for the fluid-solid surface boundary conditions at a

$$u_0 = C_m \lambda_0 \left(\frac{\partial u}{\partial y}\right)_0 + \frac{C_s v_0}{T_0}\left(\frac{\partial T}{\partial x}\right)_0 \quad \text{(velocity slip)} \tag{60}$$

$$T_0 - T_W = C_t \lambda_0 \left(\frac{\partial T}{\partial y}\right)_0 \quad \text{(temperature jump)} \tag{61}$$

plane wall ($y = 0$), with x and velocity u in the direction of the flow, where the subscript 0 indicates that the quantities are evaluated in the gas at $y = 0$. Currently accepted kinetic theory values for the coefficients are $C_m = 1.14$, $C_s = 1.1.7$, and $C_t = 2.18$, for diffuse reflection in a monatomic gas. The first term in Eq. (60) is the velocity slip due to finite wall velocity gradient, while the second is the thermal creep term arising from a temperature gradient in the gas in the direction of flow. Thermal creep is responsible for the phenomenon of thermophoresis, where small suspended particles experience a force in the direction opposite to a temperature gradient (although the phenomenon persists throughout the entire range of rarefaction, to the collisionless limit). Slip and temperature jump have the effects of reducing skin friction and heat transfer, but these effects are often obscured by other low Reynolds number and compressibility effects, and it has been found in general that slip flow corrections to the Navier-Stokes equation have a very limited range of applicability.

Transition flow. Numerous attempts have been made to develop higher-order continuum equations to replace the Navier-Stokes equations for regimes of moderate rarefaction. The results, which for the most part have been based on expansions of the Boltzmann equation, such as the Chapman-Enskog method, have proved not to be measurably superior to the Navier-Stokes equations. More success has been achieved through the use of simple interpolation between the slip flow and near-collisionless flow limits, or by solving the Boltzmann equation or models of it, or by direct simulation. This last method appears to be the most powerful and versatile of those mentioned, and will be described briefly.

The direct-simulation method models a real gas flow by computing the trajectories and positions of thousands of simulated molecules, and storing their positions and velocities as a function of time. With each time step, the molecules are advanced along their current trajectories, representative collisions with other molecules or with solid surfaces are calculated, the new velocities and positions are stored, and the process is repeated for the next time step. The flow is always unsteady, since at time $t = 0$ the molecules are introduced into the flow field in an arbitrary fashion, but a steady flow field is obtained as the long-time average limit of the unsteady flow. The method has proved successful in modeling shock-wave structure, the free jet expansion of a gas into a vacuum, multidimensional flows about bodies, and polyatomic and chemically reacting gas flows.

Bibliography. R. P. Benedict, *Fundamentals of Gas Dynamics*, 1983; G. A. Bird, *Molecular Gas Dynamics*, 1976; S. Chapman and T. G. Cowling, *The Mathematical Theory of Non-Uniform Gases*, 3d ed., 1970; J. E. John, *Gas Dynamics*, 2d ed., 1985; M. N. Kogan, *Rarefied Gas Dynamics*, 1969; H. Oguchi, *Rarefied Gas Dynamics*, 1985; M. Shalan and A. Nasser, *Gas Dynamics*, 1985; C. M. Van Atta, *Vacuum Science and Engineering*, 1965; W. G. Vincenti and C. H. Kruger, *Introduction to Physical Gas Dynamics*, 1965, reprint 1975; M. J. Zucrow and J. D. Hoffman, *Gas Dynamics*, 2 vols., 1976, 1977.

LOW-PRESSURE GAS FLOW
Zbigniew C. Dobrowolski

Flow of gases below atmospheric pressures, particularly gases following the perfect gas laws, in pipes, fittings, and other common configurations. *See* Gas.

Flow regimes. The flow regimes commonly encountered at subatmospheric pressures are shown in **Fig. 1**. The laws of gas flow at subatmospheric pressures in the turbulent, transition, and laminar flow range are the same as the laws of fluid flow at any pressures.

The laws of flow in the molecular range are based on the assumption that the molecules have a Maxwell-Boltzmann distribution of velocities. *See* Gas dynamics; Laminar flow; Turbulent flow.

Measurement of pressure. Pressure below atmospheric is usually expressed in reference to perfect vacuum or absolute zero of pressure (**Fig. 2**). Like absolute zero of temperature, it cannot be achieved, but it provides a convenient reference datum. Standard atmospheric pressure is 14.695 psia (pounds per square inch absolute) or 101.325 kilopascals, 29.92 in. of mercury absolute, or 760 mm of mercury of density 13.595×10^3 kg/m^3 where acceleration due to gravity $g = 9.80665$ m/s^2. One millimeter of mercury, which equals 1 torr (133 pascals), is the most commonly used unit of absolute pressure P. Derived units, micrometer of mercury and millitorr,

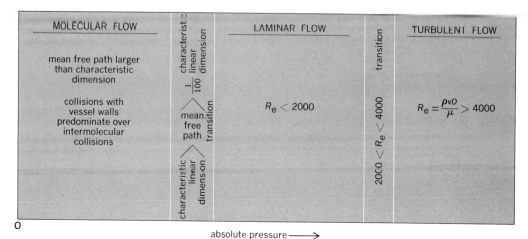

Fig. 1. Flow regimes that are commonly encountered in vacuum systems.

representing 1/1000 of 1 mmHg and 1/1000 of 1 torr, are also used. Vacuum, usually expressed in inches of mercury, is the depression of pressure below atmospheric level. Absolute zero of pressure corresponds to vacuum of 29.92 in. or 760 mm of mercury. SEE PRESSURE MEASUREMENT.

Pumping speeds. Vacuum-producing devices maintain a system at a desired pressure or remove gas from a vacuum system at a rate S_p, which is defined as a volume flow of gas per unit of time, $S_p = dV/dt$. Common units used for pumping speed are liters per second, cubic feet per minute (cfm), or cubic meters per second, and volumetric flow is specified at the existing pressure rather than at a standard pressure.

Throughput. A volume of gas at a known pressure and temperature which crosses a plane or enters a pumping device in a unit of time is characterized by the throughput, as in Eq. (1). The common units are torr liters/s, torr cfm, or Pa·m³/s and express the volumetric gas flow

$$Q = S_p P_{in} = SP = \frac{dV}{dt} P \tag{1}$$

at an absolute pressure P of 1 torr or 1 Pa, respectively.

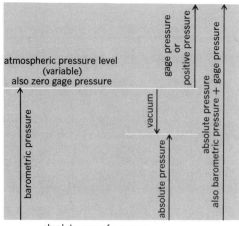

Fig. 2. Relationship of gage, absolute, and subatmospheric pressures (vacuum).

The throughput is proportional to the mass flow of gas since under steady flow conditions Eq. (2) holds, where w, M, R, and T are given in the list below. Thus, the mass flow is given by Eq. (3). Maximum throughput in a pipe is limited by sonic velocity of flow propagation.

$$Q = P\frac{dV}{dt} = \frac{d}{dt}\left(\frac{w}{M}RT\right) \quad (2) \qquad \frac{dw}{dt} = \frac{M}{RT}Q \quad (3)$$

Conductance. The flow of gas entering the pump generally passes through pipes or conduits, which present a resistance to flow, and thus a pressure difference along the flow path is established. By analogy with an electric circuit, Eqs. (4) are obtained. Hence, Eqs. (5) and (6) hold.

$$Q = (P_1 - P_2)C \quad (4) \qquad Q = S_1 P_1 = S_2 P_2 \quad (4)$$

$$Q = \left(\frac{Q}{S_1} - \frac{Q}{S_2}\right)C \quad (5) \qquad \frac{1}{S_1} = \frac{1}{S_2} + \frac{1}{C} \quad (6)$$

The pumping speed at some other point in the system can be obtained from the known pumping speed and the conductance of the portion of the system (pipes, valves, and apertures) in between.

The combined or effective pumping speed at any one place is given by Eq. (7), where S_p

$$S = \frac{S_p C}{S_p + C} \quad (7)$$

is the pumping speed at the inlet to the pump. Similarly, for an arrangement of pipes and conductances in series, Eq. (8) holds; for parallel arrangement, Eq. (9) is valid.

$$\frac{1}{C_{series}} = \frac{1}{C_1} + \frac{1}{C_2} + \cdots \quad (8) \qquad C_{parallel} = C_1 + C_2 + \cdots \quad (9)$$

Design criteria. In order to establish flow characteristics of a vacuum system, the parameters involved in the equation $Q = C(P_1 - P_2) = SP$ must be known. The total gas flow into the system, if not given, must be determined by calculations or assumption. The pumping speed and conductance are then selected in such a fashion as to maintain the desired operating pressure.

The total gas flow is often established from design pressure and known performance characteristics of a pump. This flow can be calculated from orifice flow (vacuum chucking, house vacuum), temperature equilibrium, vapor-pressure considerations (distillation and drying), gas-solubility considerations (deaeration, metallurgy), and outgassing characteristics.

In the turbulent and laminar flow range, for economy reasons (cost of pumping equipment), pressure drop in the pressure range above 0.1 torr (13 Pa) is rarely assumed to be more than 10–20% of the operating pressure, the permitted pressure drop being higher at lower pressures.

If the pumping system is used not only for maintaining a predetermined operating pressure but also for pumpdown of the system from atmospheric pressure, the design consideration should be based on the lowest operating pressure. If the conduit is designed for the lowest anticipated pressure, losses due to gas flow will be negligible during pumpdown.

It is often more convenient to determine the pressure drop by the "equivalent" pipe-length method than by the conductance method. A convenient first approximation for pipe-size calculations is the assumption that the opening into the pumping system should have a conductance 5 to 10 times higher than the pumping speed of the pump.

Nomenclature and units. The following symbols are used in low-pressure gas-flow technology.

A = cross-sectional area in square meters
a = cross-sectional area in square inches (1 in.^{2} = 0.645 × 10^{-3} m^2)
C = conductance in m^3/s
c = conductance in cfm (1 cfm = 0.472 × 10^{-3} m^3/s)
C_F = nozzle coefficient

D = internal diameter in meters
D' = internal diameter in feet (1 ft = 0.3048 m)
d = internal diameter in inches (1 in. = 0.0254 m)
g = acceleration due to gravity, 32.2 ft/s^2 (9.81 m/s^2)

k = gas constant per molecule (Boltzmann constant), 1.38×10^{-23} joule/kelvin
L = length of pipe or conduit in meters
l = length of pipe or conduit in feet (1 ft = 0.3048 m)
m = molecular mass in kilograms
M = molecular weight
n = molecular gas density per cubic meter
P = pressure in pascals (1 Pa = 1N/m^2)
p = pressure in torr (1 torr = 133.3 Pa)
$\bar{P}$ = average pressure in pascals
$\bar{p}$ = average pressure in torrs
Q = throughput in Pa·m^3/s
q = throughput in torrs cfm (1 torr cfm = 0.06289 Pa·m^3/s = 0.06289 W)
R = gas constant, 8.314 joules/(mole·kelvin)
S = pumping speed or volumetric flow in m^3/s at flowing conditions
s = pumping speed in cfm (1 cfm = 0.472 $\times 10^{-3}$ m^3/s)
t = time interval in seconds
T = absolute temperature in kelvins
V = mean velocity of flow in m/s
v = mean velocity of flow in ft/s (1 ft/s = 0.3048 m/s)
w = mass of gas in kilograms
α = molecular diameter in meters (α = 0.372×10^{-9} m for air)
$\gamma = \dfrac{C_p}{C_v} = \dfrac{\text{specific heat at constant pressure}}{\text{specific heat at constant volume}}$
μ = absolute viscosity in Pa·s (1 Pa·s = 10 poise)
μ_e = absolute viscosity in lb/ft s (1 lb/ft s = 1.488 kg/m·s = 1.488 Pa·s)
Λ = mean free path of molecules in meters
λ = mean free path of molecules in inches (1 in. = 0.0254 m)
ρ = density of fluid in lb/ft^3 at flowing conditions (1 lb/ft^3 = 16.02 kg/m^3)
Σ = sum

Turbulent flow in circular conduits. This occurs for Reynolds number Re, as defined in Eq. (10), greater than 4000. Gas throughput can be expressed in terms of Reynolds number, as in Eq. (11), and for air at 68°F (20°C), the Reynolds number can be represented in terms of throughput, as in Eqs. (12), in SI or English units.

$$\text{Re} = \frac{\rho v D'}{\mu_e} \quad (10) \qquad Q = SP = \frac{\pi}{4} \frac{R_0 T \mu}{M} \text{Re} D \quad (11)$$

$$\text{Re} = 0.828 \frac{Q}{D} \quad (12a) \qquad \text{Re} = 2.04 \frac{q}{d} \quad (12b)$$

Since pressure losses are higher in the turbulent flow range, the transition range can be treated together with the turbulent flow range for the purpose of calculation, so that the change from the turbulent flow to laminar flow is given by the critical throughput, $Q_c = 2407 D$ or $q_c = 975 d$. SEE REYNOLDS NUMBER.

Pipes. In a circular pipe the pressure drop due to friction expressed in feet of fluid is given by the Darcy-Fanning equation (13) or by Eq. (14), where the friction factor f is obtained from Moody diagram and s is in cfm. SEE PIPE FLOW.

$$h_f = 12f \frac{l}{d} \frac{v^2}{2g} \quad (13) \qquad P_1 - P_2 = 0.624 f \frac{l\rho s^2}{d^5} \quad (14)$$

Valves. These can be divided into two main groups on the basis of resistance to flow: globe valves, which represent high resistance to flow, and gate valves, which represent a very modest resistance to flow. Most other valves fall between these two groups.

Fittings. These can be classified as branching (tee), deflecting (elbows), and reducing or expanding (bushings).

Pressure losses in valves and fittings are usually represented as pressure losses in an equivalent length of straight pipe which will cause the same pressure drop as the valve, under the same flow conditions. The numerical values of these losses are obtained empirically and vary from source to source. Typical values are obtained from the nomograph in **Fig. 3**. For convenience of tabulation, equivalent losses are sometimes expressed in terms of resistance coefficient as shown below, where the ratio L/D is the equivalent length in pipe diameters of straight pipe.

$$K = f \frac{L}{D}$$

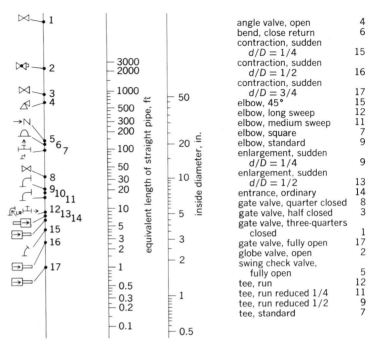

Fig. 3. Nomograph for equivalent length (l equivalent) of valves, fittings, enlargements, and contractions. The numerical values are obtained empirically and vary from source to source. (*After G. G. Brown, Unit Operations, John Wiley and Sons, 1950*)

Flow characteristics of control valves are expressed in terms of a flow coefficient C_v, defined as the flow of water at 60°F (15.6°C) in gallons per minute (1 gallon/min = 0.0631 liter/s) at a pressure drop of 1 psi (6.895 kPa) across the valve.

The relationship between these coefficients is given in Eq. (15). The pipe length in Eq. (13)

$$C_v = \frac{29.9d^2}{\sqrt{f\frac{L}{D}}} = \frac{29.9d^2}{\sqrt{K}} \tag{15}$$

represents the sum of the length of straight pipe sections, as well as the equivalent length of valves and fittings, entrance losses, exit losses, and losses due to sudden contractions, as in Eq. (16).

$$l = \sum l \text{ pipe} + \sum l \text{ equivalent} \tag{16}$$

Because of the characteristics of pumps used in this flow range (Roots pumps, water-ring pumps, water-ejector pumps, oil-sealed mechanical pumps, vane pumps, and piston compressors), the base pressure obtainable by these pumps need not be much lower than the operating pressure. The inleakage into a system is usually not very important, but should be less than 10% of available pumping speed. Leaks have to be generally greater than 10^{-5} atm cm^3/s (10^{-6} watt or Pa·m^3/s) to exhibit turbulent flow characteristics. Piping length can be quite long.

Laminar flow. Poiseuille's law of flow for fluids can be expressed as shown in Eq. (17), where $\bar{P} = P_1 + P_2/2$ is average pressure in the circular conduit.

$$Q = \frac{\pi D^4}{128\,\mu L} \bar{P}(P_1 - P_2) \tag{17}$$

Conductance for any gas is given by Eq. (18). For air at 68°F (20°C) conductance of a pipe

$$C = \frac{Q}{P_1 - P_2} = \frac{\pi}{128} \frac{D^4}{\mu L} P \qquad (18)$$

is given by Eqs. (19). Unlike the derivation for the turbulent flow range, the derivation for the

$$C = 1343 \frac{D^4}{L} \bar{P} \qquad (19a) \qquad\qquad c = 517 \frac{d^4}{l} \bar{p} \qquad (19b)$$

laminar flow is rational, meaning that it can be derived from first principles.
The expression for pressure drop in terms of volumetric flow given by Eq. (20a) is indepen-

$$P_1 - P_2 = 40.7 \frac{\mu L S}{D^4} \qquad (20a) \qquad\qquad p_2 - p_1 = 1.9 \times 10^{-3} \frac{sl}{d^4} \qquad (20b)$$

dent of pressure in the range for Reynolds number less than 2000 and for mean free path smaller than 1/100 of critical linear dimension. For air at 68°F (20°C), Eq. (20b) holds. Pressure losses in valves and fittings become relatively negligible, but can be extrapolated from the turbulent range by Eq. (21). Subscript T refers to equivalent length of pipe determined from tests in the turbulent

$$(l \text{ equivalent})_L = \frac{\text{Re}}{1000} (l \text{ equivalent})_T \qquad (21)$$

flow range. The minimum equivalent length is the length along the center line of the actual flow path.

Pumping equipment used in the laminar flow range consists of oil-sealed vacuum pumps, waterjet pumps, water-ring pumps, Roots pumps in series with the above pumps, and steam ejectors. Leaks into the vacuum system and pressure losses in piping become important, particularly at pressures approaching the transition range between laminar and molecular flow.

Laminar flow characteristics are exhibited by small leaks in the range of 10^{-1} to 10^{-6} atm cm³/s (10^{-2} to 10^{-7} watt or Pa·m³/s). Inleakage into the system should not represent more than 10% of the available pumping speed at the operating pressure. The base pressure of the pumps when isolated from the vacuum system should be 5–10 times lower than the operating pressure. SEE STREAMLINE FLOW.

Flow through nozzles and orifices. The rate of flow through nozzles and thin apertures in the subcritical range for compressible fluids as a function of pressure difference across the orifice is given by Eq. (22), where W = rate of flow in lb/h (1 lb/h = 1.26×10^{-4} kg/s),

$$W_{\text{subcritical}} = 1891 \, C_F d^2 Y \sqrt{(p_1 - p_2) \rho_1} \qquad (22)$$

C_F = flow coefficient, d = diameter in inches (1 in. = 2.54 cm), Y = expansion factor, p' = pressure in psia (1 psi = 6.895 kPa), and ρ_1 = weight density of fluid in lb/ft³ at upstream conditions (1 lb/ft³ = 16.02 kg/m³). Y in Eq. (22) is defined by Eq. (23), where $R = P_2/P_1 \leq 1$ and $\beta = d_0/d_2$,

$$Y = \sqrt{\frac{\gamma}{\gamma - 1} R^{2/\gamma} \frac{1 - R\gamma^{-1/\gamma}}{1 - R}} \sqrt{\frac{1}{1 - \beta^4 R^{2/\gamma}}} \qquad (23)$$

d_0 and d_2 being the diameters of orifice and downstream pipe, respectively. The value of the nozzle coefficient C_F varies with Reynolds number and with the geometry of aperture. For Re > 4000 the coefficients are approximately constant. Representative values are 0.7 for square and sharp-edge orifices (for example, entrance to a pipe, $\beta = 1$), 0.86 for a rounded orifice, 0.9 for a 22° bevel orifice, and 0.98 for flow nozzles.

For gases in the critical flow range given in Eq. (24), the flow of gas is given by Eq. (25),

$$\left(\frac{P_2}{P_1}\right)_{\text{critical}} = \left(\frac{2}{\gamma + 1}\right)^{\gamma/\gamma - 1} \qquad (24) \qquad\qquad W_{\text{critical}} = 1891 \, C_F d^2 Z \sqrt{p'_1 \rho_1} \qquad (25)$$

with units as shown for subcritical flow. Z in Eq. (25) is defined in Eq. (26).

$$Z = \frac{\left(\frac{2}{\gamma + 1}\right)^{1/\gamma - 1} \sqrt{\frac{\gamma}{\gamma + 1}}}{\sqrt{1 + \beta^4 \left(\frac{2}{\gamma + 1}\right)^{2/\gamma - 1}}} \quad (26)$$

For $d_0/d_2 \ll 1$, $\beta = 0$ and $Z = 0.484$; for an air flow into a pipe $\beta = 1$ and $Z = 0.626$. The flow coefficient is that of a square-edge orifice unless specified otherwise. The volumetric flow of air at 68°F (20°C) into an orifice in the critical flow range can be represented by Eqs. (27), where C_F is the flow coefficient.

$$S_{0(\text{critical})} = 270 \, C_F a \quad (27a) \qquad S_{0(\text{critical})} = 197 \, C_F A \quad (27b)$$

Transition range. The transition between the laminar and molecular flow range is given by expression (28) where K is the Knudsen number. The conductance of a pipe in cubic meters

$$10 \geq K \geq \frac{1}{100} \quad (28)$$

per second in the transition range is given by Eq. (29). Calculations for the transition range are

$$C = \left(\frac{\pi D^4 \overline{P}}{128 \, \mu \, L}\right) + \left(\frac{1}{6} \frac{D^3}{L} \sqrt{\frac{2\pi \, kT}{m}}\right) \left(\frac{1 + \sqrt{\frac{m}{kT}} \frac{D\overline{P}}{\mu}}{1 + 1.24 \sqrt{\frac{m}{kT}} \frac{D\overline{P}}{\mu}}\right) \quad (29)$$

rarely justified and are of consequence only if the operating process takes place in the transition range.

Molecular flow range. This is characterized by a mean free path of molecules longer than the smallest linear dimension in the conduit. Collisions with vessel walls predominate over intermolecular collisions. The mean free path for air is given by Eqs. (30). For other homogeneous gases Eq. (31) holds.

$$\Lambda = \frac{6.6}{P} \times 10^{-3} \quad (30a) \qquad \lambda = \frac{\alpha}{p} \times 10^{-3} \quad (30b) \qquad \Lambda = \frac{kT}{\sqrt{2}\pi P a^2} \quad (31)$$

The pressure losses in pipes and ducts become exceedingly high. Calculations are based not on assumed pressure losses, but rather on the most economic combination of ducts and pumps necessary for the teaching and maintaining of operating pressure. The effective pumping speed at the entrance to the process chamber may often be only 20% of the designed speed of the pumps at inlet.

Gas loads due to outgassing characteristics of the pumping manifolding and process chamber must be included in gas flow calculations. **Figures 4–8** give conductances of commonly used shapes and connections. The value of conductance for any shape is given as $C = KC_0$, where C_0 is the conductance of an aperture equal in area to the opening into the shown configuration and is calculated by using Eqs. (35). Figures 4–8 were calculated by using Monte Carlo methods and assuming that the flow is steady state and diffuse so that molecules are reflected from the walls of the vessel according to the cosine law.

The conductance of a long circular pipe in the laminar and molecular flow range, based on V.O. Knudsen's work, as given by Eq. (29) can be rearranged as in Eq. (32). Conductance calcu-

$$C = \frac{D^3}{L} \left[0.0245 \, \frac{\overline{P}D}{\mu} + 38.1 \sqrt{\frac{T}{M}} \left(\frac{1 + 1.1 \sqrt{\frac{M}{T}} \frac{\overline{P}D}{\mu}}{1 + 1.24 \sqrt{\frac{M}{T}} \frac{\overline{P}D}{\mu}} \right) \right] \quad (32)$$

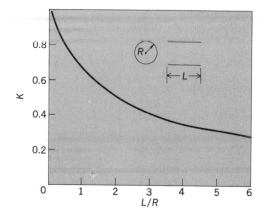

Fig. 4. K, where $C_e = KC_o$ and C_o is calculated by Eq (35). (*After L. L. Levenson, N. Milleron, and D. H. Davis, Optimization of Molecular Flow Conductance, University of California Radiation Laboratory, Publ. UCRL-6014, 1960*)

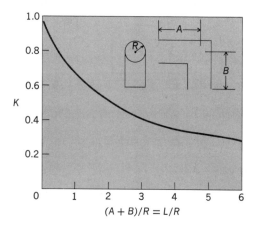

Fig. 5. K for 90° elbow, where K is Knudsen number. (*After L. L. Levenson, N. Milleron, and D. H. Davis, Optimization of Molecular Flow Conductance, University of California Radiation Laboratory, Publ. UCRL-6014, 1960*)

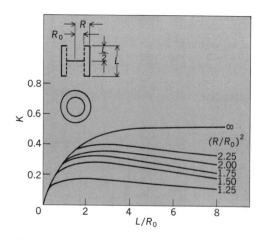

Fig. 6. K for cold trap, baffle, and valve-baffle. (*After L. L. Levenson, N. Milleron, and D. H. Davis, Optimization of Molecular Flow Conductance, University of California Radiation Laboratory, Publ. UCRL-6014, 1960*)

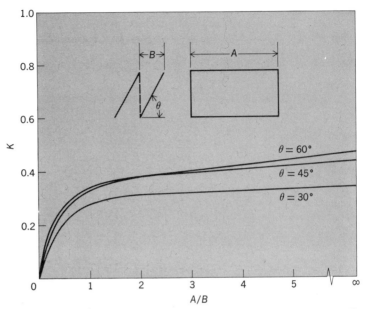

Fig. 7. *K* for louver geometries (cold traps and baffles). (*After L. L. Levenson, N. Milleron, and D. H. Davis, Optimization of Molecular Flow Conductance, University of California Radiation Laboratory, Publ. UCRL-6014, 1960*)

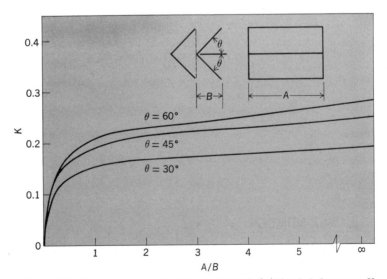

Fig. 8. *K* for chevron geometry (cold traps and baffles). (*After L. L. Levenson, N. Milleron, and D. H. Davis, Optimization of Molecular Flow Conductance, University of California Radiation Laboratory, Publ. UCRL-6014, 1960*)

lations based on this equation are in good agreement with calculations based on Monte Carlo methods. The equation is valid provided the flow is not turbulent and the pressure drop in the pipe is small, so the flow regime does not change along the pipe. In the molecular flow range where pressures are low, the terms including average pressure $\bar{P}$ become negligible, and the conductance is independent of pressure. Conductances in m³/s are given by Eqs. (33a)—

$$C = 38.1 \sqrt{\frac{T}{M} \frac{D^3}{L}} \quad (33a) \qquad c = 13.8 \frac{d^3}{l} \quad (33b)$$

$$C = 38.1 \sqrt{\frac{T}{M} \frac{D^3}{L} + \frac{4D}{3}} \quad (34a) \qquad c = 13.8 \frac{d^3}{l + 0.11d} \quad (34b)$$

$$C = 36.4 \sqrt{\frac{T}{M}} A \quad (35a) \qquad c = 159a \quad (35b)$$

$$C = 97.1 \sqrt{\frac{T}{M} \frac{a^2 b^2}{(a+b)L + 8/3\, ab}} \quad (36a) \qquad c = 35.2 \frac{a^2 b^2}{(a+b)l + 0.22ab} \quad (36b)$$

$$C = 38.1 \sqrt{\frac{T}{M} \frac{(D_2^2 - D_1^2)(D_2 - D_1)}{L + 4/3(D_2 - D_1)}} \quad (37a) \qquad c = 13.8 \frac{(d_2^2 - d_1^2)(d_2 - d_1)}{l + 0.11(d_2 - d_1)} \quad (37b)$$

(37a), or for air at 68°F (20°C) in ft³/min by Eqs. (33b)–(37b).

For a long pipe Eqs. (33) hold. For a short pipe including an end correction Eqs. (34) hold. For an aperture Eqs. (35) hold. For a rectangular duct with cross section = $a \times b$ Eqs. (36) hold, where the dimensions of the sides of the duct are expressed in meters in Eq. (36a) and in inches in Eq. (36b). For an annulus between two concentric tubes (where $D_2 > D_1$) Eqs. (37) hold. Combination of the above formulas allows the approximate calculation of the effective conductance even for complex shapes.

Pumping equipment in the molecular flow range consists of diffusion pumps backed by mechanical pump combinations, getter-ion pumps, molecular pumps, sorbtion pumps, and cryopumps.

For the high vacuum range (10^{-3} to 10^{-7} torr or 10^{-1} to 10^{-5} Pa), leaks should be kept below 10^{-8} atm cm³/s (10^{-9} Pa·m³/s), and in the ultrahigh vacuum range (below 10^{-7} torr or 10^{-5} Pa), they should be kept below 10^{-8} to 10^{-13} atm cm³/s (10^{-9} to 10^{-14} Pa·m³/s). Leaks should not contribute more than 10% of the gas load at the operating pressure, and high-vacuum connections must be as short as physically possible. Leaks into the system smaller than 10^{-5} atm cm³/s (10^{-6} watt or Pa·m³/s) exhibit molecular flow characteristics. Pumping equipment should be inherently capable of producing a base pressure an order of magnitude lower than the desired processing pressure. *See* F*luid flow*.

Bibliography. ASME, *Compressors and Exhausters*, PTC 10, and *Flow Measurement*, PTC 19.5, 1974; J. F. O'Hanlon, *A User's Guide to Vacuum Technology*, 1980; J. L. Vossen and D. W. Kern, *Thin Film Processes*, 1978; G. L. Weissler and R. W. Carlson (eds.), *Methods of Experimental Physics*, vol. 14: *Vacuum Physics and Technology*, 1979.

KNUDSEN NUMBER
Frank H. Rockett

In fluid mechanics, the ratio l/L of the mean-free-path length l of the molecules of the fluid to a characteristic length L of the structure in the fluid stream. When the mean free path of the fluid particles is short relative to the size of the object being considered, the fluid can be treated as a continuum. If the path length between molecular encounters is comparable to or larger than a significant dimension of the flow region, the gas must be treated as consisting of discrete particles. The usual classifications of flow according to Knudsen number are as follows: For $l/L \leqslant 0.01$ the flow can be dealt with by the methods of gas dynamics; for $l/L \approx 1$ the behavior is termed slip flow; for $l/L \geqslant 10$ the behavior is termed free-molecular flow or rarefied gas dynamics. SEE GAS DYNAMICS.

GAS KINETICS
FRANK H. ROCKETT

The effects of gases due to motion. The simplest gas phenomena are those associated with gas at rest. Additional phenomena contribute to the effects when the gas is in motion. The method of science is to isolate one phenomenon at a time for study, holding all else constant. However, in actual gas flows, motion is accompanied by other phenomena, so that for practical purposes a more comprehensive view is necessary. SEE BERNOULLI'S THEOREM; BUOYANCY; COMPRESSIBLE FLOW; EULER'S MOMENTUM THEOREM; FLUID STATICS; GAS DYNAMICS; VISCOSITY.

AERODYNAMICS
WILLIAM C. WALTER

The branch of aeromechanics dealing with the properties and characteristics of, and the forces exerted by, air and other gases in motion. The field of aerodynamics includes the science of a gas itself in motion and the science of bodies immersed in a gas between which there exists a relative motion. SEE GAS DYNAMICS.

Aerodynamics is a broad field with numerous specializations and applications, some of which extend into apparently unrelated fields of science and engineering. Perhaps the most frequently practiced function of the aerodynamicist is the analysis of the forces and moments exerted on a solid body in motion through the air.

Of fundamental significance in the term aerodynamics is the prefix aero-, which refers to the air of the Earth's atmosphere. Until flight can be achieved within the atmospheres of other planets, the limits of aerodynamic flight and the majority of practical considerations will be confined to the limits of the Earth's atmosphere as defined in aerodynamic terms.

Figure 1 presents the practical limits of aerodynamic flight within the Earth's atmosphere based on a quantitative analysis of the governing factors. To a large extent the significant aerodynamic reactions of missiles in passing through the atmosphere also occur below the upper boundary shown in Fig. 1.

Two overlapping flight regimes are shown. The upper regime, defined as the aerospace-vehicle flight regime, indicates the operating region of reentry vehicles and space vehicles which use aerodynamic lift during their descent through the atmosphere. Its upper and lower boundaries are defined by a wing loading of 20 lb/ft^2 (1.0 kilopascal) and 40 lb/ft^2 (1.9 kilopascals), respectively. The lower flight regime in Fig. 1 is defined as the cruising-vehicle flight regime (sometimes referred to as the corridor of continuous flight). The upper and lower boundaries here are 10 lb/ft^2 (0.5 kPa) and 200 lb/ft^2 (10 kPa), respectively. The portion of this flight regime penetrated in the first 50 years of powered flight is also indicated.

The aerodynamic heating limit cuts off access to a large portion of the cruising flight regime between 12,000 and 24,000 ft/s (3.7 and 7.3 km/s). Actually, this temperature represents the approximate upper extreme of human engineering ability to penetrate what is sometimes called the thermal thicket; it cannot be accurately described as a barrier.

Strictly ballistic reentry vehicles could not reach the surface of the Earth wholly within the boundaries of the flight regimes just defined. The ballistic path of reentry in Fig. 1 would pass through both regimes almost vertically at high speed, the greatest reduction in velocity occurring in the denser atmosphere below the cruising-vehicle flight regime. As a result, ballistic reentry vehicles suffer far greater extremes of aerodynamic heating during reentry as a result of air friction than do aerodynamic reentry vehicles. SEE AEROTHERMODYNAMICS.

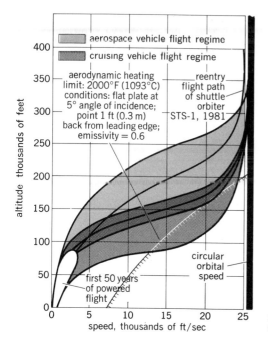

Fig. 1. Practical limits of aerodynamic flight within the atmosphere. *(Aero/Space Eng.)*

Figure 2 is a plot similar to Fig. 1 except that the flight envelopes previously described have been subdivided into three speed regions: subsonic, supersonic, and hypersonic. The aerodynamic heating region or thermal thicket is shaded, the shades deepening with rising temperature.

At velocities below the speed of sound (Mach 1) air may be considered to be incompressible. The laws governing flow phenomena in this speed range comprise subsonic aerodynamics.

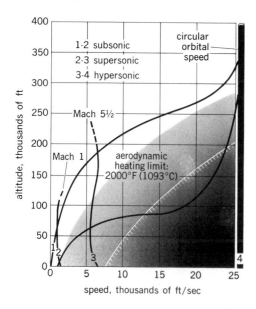

Fig. 2. The subsonic, supersonic, and hypersonic flight regimes within the atmosphere. 1 ft = 0.3 m.

Between Mach 1 and approximately Mach 5.5 is the region of supersonic aerodynamics. Supersonics differs markedly from subsonics because the air becomes compressible. The narrow transition region between subsonic and supersonic flight is characterized by a rapid drag rise often referred to as the sound barrier. *See* MACH NUMBER.

Above a speed of about Mach 5.5 another branch of aerodynamics (hypersonics) has been defined to categorize phenomena which differ markedly from supersonic flow. At hypersonic speeds, properties of the gaseous medium in which a vehicle is immersed begin to differ significantly from those of air at lower speeds. High-temperature gas phenomena must now be included in the aerodynamic analysis.

Bibliography. J. D. Anderson, Jr., *Introduction to Flight*, 2d ed., 1985; A. M. Kuethe and C.-Y. Chow, *Foundations of Aerodynamics*, 3d ed., 1976.

AEROTHERMODYNAMICS
SHIH-YUAN CHEN

Flow of gases in which heat exchanges produce a significant effect on the flow. Traditionally, aerodynamics treats the flow of gases, usually air, in which the thermodynamic state is not far different from standard atmospheric conditions at sea level. In such a case the pressure, temperature, and density are related by the simple equation of state for a perfect gas; and the rest of the gas's properties, such as specific heat, viscosity, and thermal conductivity, are assumed constant. Because fluid properties of a gas depend upon its temperature and composition, analysis of flow systems in which temperatures are high or in which the composition of the gas varies (as it does at high velocities) requires simultaneous examination of thermal and dynamic phenomena. For instance, at hypersonic flight speed the characteristic temperature in the shock layer of a blunted body or in the boundary layer of a slender body is proportional to the square of the Mach number. These are aerothermodynamic phenomena.

Two problems of particular importance require aerothermodynamic considerations: combustion and high-speed flight. Chemical reactions sustained by combustion flow systems produce high temperatures and variable gas composition. Because of oxidation (combustion) and in some cases dissociation and ionization processes, these systems are sometimes described as aerothermochemical. In high-speed flight the kinetic energy used by a vehicle to overcome drag forces is converted into compression work on the surrounding gas and thereby raises the gas temperature. Temperature of the gas may become high enough to cause dissociation (at Mach number ≥ 7) and ionization (at Mach number ≥ 12); thus the gas becomes chemically active and electrically conducting.

In order to describe aerothermodynamic problems more fully, three specific phenomena are discussed in the following sections: internal flow, external flow, and aerodynamic heating.

Internal flow. Movement of a fluid, whether confined or not, is characterized by two types of flow, laminar and turbulent. In laminar flow each streamline or particle of fluid moves generally parallel to the flow of the main body of the fluid. In turbulent flow each streamline of the fluid undergoes irregular fluctuations of motion and velocity that have components perpendicular, as well as parallel, to the average flow of the fluid. The dominant characteristic of turbulent flow is the presence of many vortices or eddies. *See* LAMINAR FLOW; TURBULENT FLOW.

When fluid moves along a surface, the streamlines or particles of fluid flow adjacent to the wall are retarded by viscous forces (friction). This layer of retarded flow is known as the boundary layer. The velocity gradient and the temperature profile across the boundary layer are pronounced.

In internal flow the gas is confined by the walls of a duct. Aerothermodynamic effects in this case are caused either by gases, such as air at high temperatures, or by combustion. The internal flow of gases at high temperature is a phenomenon largely confined to laboratory equipment, such as the hypersonic wind tunnel, the shock tube, the hot-shot tunnel, or the plasma jet used to simulate flight conditions for testing models. The rocket, ramjet, and the turbojet engine also involve combustion processes in which aerothermodynamic effects are important.

The shock tube in its various forms is a laboratory device in which aerothermodynamic effects play an important role. The shock tube consists basically of a long pipe, divided by a diaphragm into two compartments and closed at both ends. Gases at different pressures are placed in the two sections of pipe, the diaphragm is ruptured, and a shock wave propagates into the quiescent low-pressure gas. The gas behind the shock wave is accelerated, compressed, and heated to a high temperature. This region of high-velocity and high-temperature air can be used to simulate high-speed flight conditions. The radiation from the hot gas, chemical kinetics, heat transfer to simple shapes, forces on simple shapes, and boundary-layer transition can all be studied by this experimental device. *See Shock wave.*

For the propulsion units normally used in various vehicles that rely on ambient air as a source of oxygen and as a working fluid, the free-stream tube of air entering an inlet must be decelerated to a low subsonic velocity before entering the combustion chamber. In practical applications the deceleration of a supersonic stream (flow traveling faster than the local speed of sound) is not possible without the formation of discontinuities, or shock waves. The formation of shock waves in a stream always results in an increase in entropy; that is, the available energy in the stream is diminished as the flow proceeds across the shock wave.

Deceleration of the supersonic airstream to subsonic speed before it enters the combustion chamber is accomplished most efficiently by an oblique shock diffuser. The oblique shock diffuser, which consists of a cone or a wedge, produces a shock wave system of one or more oblique shock waves followed by one normal shock wave. Because the Mach number of the air following a normal shock wave is always subsonic, the downstream air is at the required low velocity. As the upstream Mach number increases, the subsonic Mach number following the normal shock decreases. *See Mach number.*

In combustion engines aerothermodynamic effects are important in several respects. First there is the diffusion and mixing process of the fuel and oxidizer. In some cases fuel is introduced in liquid form, as in liquid-propellant rocket engines, the combustors of ramjet and turbojet or fanjet engines, and afterburners. In such cases the process involves the breakup of the fuel spray, evaporation of the liquid drops, and mixing of fuel and oxidizer.

In the combustion process itself chemical reactions are complex and turbulence is high. One important problem is to stabilize the flame. To produce useful work, the high-temperature products of combustion are expanded, either in a turbine or in a nozzle.

As the flight speed increases into the high supersonic and hypersonic regimes, the diffusion process in the inlet becomes very inefficient. The flow that is slowed down by the normal shock causes an increase in both static temperature and static pressure. As a result of the high static temperature, air begins to dissociate appreciably, causing a substantial energy loss. In addition, both high static temperatures and static pressures may, in turn, require a heavier structure. These problems can be considerably minimized if the combustion process is allowed to occur in the supersonic stream. However, the energy losses due to heat addition processes in a supersonic stream can be appreciable.

In a turbine the design problems are twofold: the analysis of flow over airfoils in cascade, and heat transfer to the turbine blades. Flow is usually treated by ideal-gas techniques, with little consideration given to real-gas effects. Because the turbine blades are constantly exposed to hot combustion gases, it is necessary to develop materials able to withstand higher temperatures and to devise ways of cooling the turbine blades.

In the nozzle, high-temperature and high-pressure products of combustion are expanded to a high velocity as the temperature and pressure decrease. The flow is complicated by the complex chemical composition of the gas. The efficiency of the nozzle is affected by its contour, the amount of heat lost through its wall, the degree of incomplete combustion, and the presence of nonequilibrium thermal and chemical states, flow separation, shock waves, and turbulence. Heat transfer to the nozzle walls, especially at the throat, is important because of its effect on the structural integrity of the walls. Other energy losses can be reduced by proper nozzle design. *See Nozzle.*

External flow. Aerothermodynamic effects in the external flow about bodies occur during high supersonic or hypersonic flight. Any real body has a finite radius of curvature at the leading edge or nose. This radius results in a bow shock, which is detached and nearly normal to the flow at the stagnation region. The air between the normal shock and the stagnation region of the body is compressed to a high temperature. This compressed region is the primary source of high-temperature gas in the external flow about a body. For the purpose of discussion and analysis, the external flow field about a body (between the shock wave and the body surface) is divided into three regions, namely, the inviscid external flow, the viscous boundary layer, and the wake or jet in the base region.

Both the compression and the temperature rise that a gas experiences as it passes through the bow shock increase with Mach number and shock wave angle. The high-temperature gas transfers heat from the boundary layer to the body. A large amount of heat is transferred in the stagnation region of a nose and on the leading edge of a wing, not only because the gas pressure is maximum, but also because the boundary layer has a minimum thickness in such regions. Heat energy is also transferred to the bottom surface of the vehicle, as well as to the side, top, and rear surfaces.

The wake is a separation region that occurs near the aft portion of a body. The jet is a region comprising hot combustion products from an engine.

In designing a vehicle, it is necessary to apply information obtained from the solution of the equations of the external flow field to such problems as structural design (in the form of applied forces and heat input), guidance and control (in the form of applied forces), and communication and detection (in the form of electromagnetic disturbances). The applied forces are usually determined by solving the aerodynamic equations for an ideal inviscid gas and correcting for real-gas effects. Heat transfer, on the other hand, is estimated by a consideration of the thin viscous boundary layer.

At high temperatures the gas in the shock layer emits electromagnetic radiation. While this phenomenon increases heat transfer, it is more important as a source of electromagnetic noise. Although this electromagnetic noise can provide the basis for a detection system, it can also have an adverse effect on communication with the vehicle. At still higher temperatures the gas ionizes, producing electromagnetic forces in addition to the radiation.

Aerodynamic heating. Aerodynamic heating is a severe problem at high flight speeds and at all altitudes at which air forms a continuum, that is, up to approximately the altitude at which the mean free path is of the same order of magnitude as the diameter of the vehicle's nose. Such an altitude might be about 350,000 ft (107 km) for a nose having a diameter of about 1 ft (0.3 m).

To investigate aerodynamic heating on a missile, aircraft, spacecraft, or other flying object, one needs detailed information about flight profiles, such as the altitude, velocity, angle of attack, and bank-angle history of the vehicle as a function of time. For most supersonic aircraft, aerodynamic heating can be estimated by using the cruise-design conditions. Under cruise conditions skin temperature generally reaches a steady state, and the thermal design can then be based on the fixed-equilibrium skin temperature (the temperature at which heat input equals heat output). For a missile, equilibrium is not reached; because of its short flight time and its rapid acceleration and deceleration, the missile's skin temperature will generally always be changing. This transient nature of missile flight trajectories complicates the temperature analysis. For an atmospheric-entry spacecraft, skin temperature will in some cases reach equilibrium because of its longer flight time, but the equilibrium skin temperature may change with time because of changes in altitude, speed, angle of attack, and bank angle.

The atmospheric-entry characteristics of spacecraft and of ballistic missiles differ in both the total heat load experienced and in maximum heating rates. Total heat load is greater for the spacecraft, whereas the maximum heating rate is greater for the ballistic missile because of its steeper entry path. Therefore methods of heating analysis and of thermal protection may be different for each kind of vehicle. For spacecraft with a high lift-to-drag ratio ($L/D \approx 3$), the severity of heating can be reduced by trajectory selection, configuration design, and radiation cooling. For spacecraft with medium and low lift-to-drag ratios ($L/D \approx 2$ or less) and for the ballistic entry vehicle, ablation cooling techniques have been used extensively to protect interior structure. Changing the nose configuration alone cannot completely solve the structural heating problem for a ballistic missile.

In general, heating analysis is conducted by dividing the vehicle into components; appropriate heat-transfer formulas are then applied to each component to estimate its heating rate. These components are nose cap or nose tip, conical section, leading edge, lower surface, side surface, upper surface, leading edge of the fin, side surface of the fin, and base. Various factors

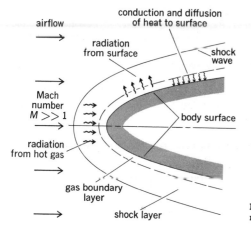

Heat transfer in stagnation region of blunted nose cap.

in the analysis of a nose cap component are shown in the **illustration**. Analysis can be further complicated by boundary-layer transition, boundary-layer interaction, separation flow, shock impingement, mass transfer, nonequilibrium phenomena, gap effects, and rough or wavy surfaces.

Accurate estimates of aerodynamic heating also require knowledge of fluid properties of the gas, such as composition, thermal and transport properties, reaction rates, relaxation times, and radiation characteristics within the fluid around the hypersonic vehicle. Under standard conditions air is composed chiefly of diatomic molecules. These molecules can rotate about only two axes perpendicular to the line joining the centers of the two atoms. Motions of translation and rotation are excited at room temperature. As the temperature of air increases, vibration begins, increasing in amplitude as temperature increases, until eventually the intramolecular bond is broken. The molecule is then said to be dissociated. Still higher energy levels are excited as the temperature increases further; eventually, electrons are separated from their parent atoms and the gas becomes ionized.

Bibliography. A. H. Shapiro, *The Dynamics and Thermodynamics of Compressible Fluid Flow*, vols. 1 and 2, 1953–1954, vol. 2 reprint, 1983; Y. S. Touloukian and C. Y. Ho (eds.), *Thermophysical Properties of Selected Aerospace Structual Materials*, pt. 1: *Thermal Radiative Properties*, pt. 2: *Thermophysical Properties*, 1977; W. G. Vincenti and C. H. Kruger, Jr., *Introduction to Physical Gas Dynamics*, 1965, reprint 1975; A. Walz, *Boundary Layers of Flow and Temperature*, 1969.

4

NONVISCOUS FLOW

Euler's momentum theorem	106
Bernoulli's theorem	106
D'Alembert's paradox	107
Kelvin's circulation theorem	108
Laplace's irrotational motion	108
Potential flow	109
Kelvin's minimum-energy theorem	109
Electrolytic tank	110
Hydraulic analog table	110
Crocco's equation	112

EULER'S MOMENTUM THEOREM
Arthur E. Bryson, Jr.

A principle of fluid mechanics which states that momentum of the particles in a moving frictionless, or inviscid, fluid is conserved. This theorem is expressed by Euler's hydrodynamical equations, a set of nonlinear partial differential equations. When viscous shear forces are included, the momentum equations are called the Navier-Stokes equations. The eulerian method of viewing fluid motion is to consider it as a velocity-pressure field; that is, the velocity of the fluid particles is considered as a vector function of position in space and time, $\mathbf{v} = \mathbf{v}(\mathbf{r},t)$, and the pressure as a scalar function of position and time, $p = p(\mathbf{r},t)$, where $\mathbf{r}$ is the position vector locating a point in space and t is time. Euler's equations are essentially a restatement of Newton's single-particle law, force equals mass times acceleration, adapted to the many-particle continuum concept of fluid motion. In such a flowing continuum the acceleration of a particle of fluid at a given point in the fluid is given by Eq. (1), where $(\partial \mathbf{V}/\partial t)\,dt$ represents the velocity change in the time interval

$$\mathbf{a} = d\mathbf{v}/dt = \partial \mathbf{v}/\partial t + (\mathbf{v} \cdot \nabla)\mathbf{v} \tag{1}$$

dt at a fixed point and $(\mathbf{v}\,dt \cdot \nabla)\mathbf{v}$ represents the velocity change of the particle in traveling the distance $\mathbf{v}\,dt$ from one point in the fluid to another point during time dt. Consider a volume V in the fluid with a surface S and let $\mathbf{n}$ be a unit vector normal to the surface, positive outward. One of the forces acting on this volume is the resultant of the pressure forces given by expression (2) which, by the analog to Gauss' divergence theorem, is the equivalent of expression (3). Thus, the

$$-\iint_S p\mathbf{n}\,dS \tag{2} \qquad -\iiint_V \nabla p\,dV \tag{3}$$

negative of pressure gradient $-\Delta p$ may be regarded as a force per unit volume acting on a fluid particle. There may also be body forces such as gravity and electromagnetic forces (if the fluid is electrically conducting). Let the amount of body force per unit mass be $\mathbf{F}$; then the force per unit volume is $\rho \mathbf{F}$, where ρ is the density. Euler's equations (as one vector equation) then become Eq. (4), where the left-hand side is (mass per unit volume) times acceleration and the right-hand side

$$\rho[(\partial \mathbf{v}/\partial t) + (\mathbf{v} \cdot \nabla)\mathbf{v}] = -\nabla p + \rho \mathbf{F} \tag{4}$$

represents the forces per unit volume. In rectangular cartesian coordinates (x,y,z) with velocity components (u,v,w) and body force components (X,Y,Z), the relations are Eqs. (5). [Equations (4)

$$\rho\left[\frac{\partial u}{\partial t} + u\frac{\partial u}{\partial x} + v\frac{\partial u}{\partial y} + w\frac{\partial u}{\partial z}\right] = -\frac{\partial p}{\partial x} + \rho X \tag{5a}$$

$$\rho\left[\frac{\partial v}{\partial t} + u\frac{\partial v}{\partial x} + v\frac{\partial v}{\partial y} + w\frac{\partial v}{\partial z}\right] = -\frac{\partial p}{\partial y} + \rho Y \tag{5b}$$

$$\rho\left[\frac{\partial w}{\partial t} + u\frac{\partial w}{\partial x} + v\frac{\partial w}{\partial y} + w\frac{\partial w}{\partial z}\right] = -\frac{\partial p}{\partial z} + \rho Z \tag{5c}$$

and (5) are applicable in SI units or any coherent system of units.] An integral of these equations can be found for flows starting from rest or uniform motion. In this way the problem of determining the velocity field is reduced to purely kinematic considerations. SEE BERNOULLI'S THEOREM; GAS DYNAMICS; KELVIN'S CIRCULATION THEOREM; LAPLACE'S IRROTATIONAL MOTION; NAVIER-STOKES EQUATIONS; UNITS OF MEASUREMENT.

BERNOULLI'S THEOREM
Frank M. White

An idealized algebraic relation between pressure, velocity, and elevation for flow of an inviscid fluid. Its most commonly used form is for steady flow of an incompressible fluid, and is given by the equation shown, where p is pressure, ρ is fluid density (assumed constant), V is flow velocity,

$$\frac{p}{\rho} + \frac{V^2}{2} + gz = \text{constant}$$

g is the acceleration of gravity, and z is the elevation of the fluid particle. The relation applies along any particular streamline of the flow. The constant may vary across streamlines unless it can be further shown that the fluid has zero local angular velocity. SEE KELVIN'S CIRCULATION THEOREM; STREAMLINE FLOW.

The above equation may be extended to steady compressible flow (where changes in ρ are important) by adding the internal energy per unit mass, e, to the left-hand side. SEE COMPRESSIBLE FLOW.

The equation is limited to inviscid flows with no heat transfer, shaft work, or shear work. Although no real fluid truly meets these conditions, the relation is quite accurate in free-flow or "core" regions away from solid boundaries or wavy interfaces, especially for gases and light liquids. Thus Bernoulli's theorem is commonly used to analyze flow outside the boundary layer, flow in supersonic nozzles, flow over airfoils, and many other practical problems. SEE AERODYNAMICS; BOUNDARY-LAYER FLOW.

According to the above equation, if velocity rises at a given elevation, pressure must fall. This principle is used in Bernoulli-type flow meters, such as orifice plates, venturi throats, and pitot tubes. A flow obstruction is deliberately created to cause a local pressure drop which can be calibrated with the flow rate. SEE FLOW MEASUREMENT.

By invoking additional restrictive assumptions, the Bernoulli theorem can be extended to unsteady flow with zero or constant angular velocity, flow in rotating machinery, and piping systems with frictional losses. SEE PIPE FLOW.

The theorem is named after Daniel Bernoulli, who hinted at a "proportionality" between pressure and velocity in a 1738 hydrodynamics textbook. However, the above equation was actually derived by Leonhard Euler in a series of papers in 1755. SEE FLUID FLOW.

Bibliography. L. M. Milne-Thomson, *Theoretical Hydrodynamics*, 5th ed., 1968; V. L. Streeter and E. B. Wylie, *Fluid Mechanics*, 8th ed., 1985; F. M. White, *Fluid Mechanics*, 2d ed., 1986; C. S. Yih, *Fluid Mechanics*, rev. ed., 1979.

D'ALEMBERT'S PARADOX
ARTHUR E. BRYSON, JR.

A theorem in fluid mechanics which states that no forces act on a body moving at constant velocity in a straight line through a large mass of incompressibile, inviscid fluid which was initially at rest (or in uniform motion). This seemingly paradoxical theorem can be understood by first realizing that inviscid fluids do not exist. If such fluids did exist, there would be no internal physical mechanism for dissipating energy into heat; hence there could be no force acting on the body, because work would then be done on the fluid with no net increase of energy in the fluid.

The viscosity of many fluids is very small, but it is essential in explaining the forces that act on bodies moving in them. The action of viscosity creates a rotation of the fluid particles that come near the surface of a moving body. This vorticity, as it is called, is convected downstream from the body so that the assumption or irrotationality of the fluid motion, made in the proof of D'Alembert's theorem, does not correspond to reality for any known fluid. For a winglike body this viscous action sets up a circulation around the body which creates a lifting force. Viscosity also gives rise to tangential stresses at the body surface, called skin friction, which result in a drag force on the body. The work done on the fluid by moving a body through it shows up first as kinetic energy in the wake behind the body, which is gradually dissipated into heat by further action of viscosity.

D'Alembert's theorem does not preclude the possibility of a couple acting on the body, and in fact the irrotational, inviscid fluid theory does predict such a couple. This couple is almost always such as to cause the body to present its greatest projected area in the direction of motion. SEE FLUID-FLOW PRINCIPLES.

KELVIN'S CIRCULATION THEOREM
Arthur E. Bryson, Jr.

A theorem in fluid dynamics that pertains to an incompressible, inviscid fluid. A direct consequence of this theorem is a great simplification in understanding and analyzing a large class of fluid flows called irrotational flows. The circulation Γ about a closed curve in a fluid is defined as the line integral of the component of velocity along the contour, as in Eq. (1), where $\mathbf{v}$ is the fluid

$$\Gamma = \oint \mathbf{v} \cdot d\mathbf{s} \qquad (1)$$

velocity and $d\mathbf{s}$ is a length element along the curve. (Circulation bears a strong analogy to the work done in moving a particle around a closed curve in a force field.) Kelvin's theorem states that in an incompressible, inviscid fluid the circulation along a closed curve, always consisting of the same fluid particles, does not change with time. An important consequence of this theorem relates to fluid motions starting from rest or uniform motion; in such flows the circulation is initially zero for every possible closed curve and hence remains equal to zero thereafter according to Kelvin's theorem. This implies that the line integral of the velocity taken from fixed point A to fixed point B is independent of the path from A to B. Consider, for example, two different paths, C_1 and C_2, from A and B; then $C_1 - C_2$ forms a closed contour, and the line integral around it vanishes, showing that the line integrals along C_1 and C_2 are equal. If the line integral is independent of path, the integrand must be a perfect differential; that is, the velocity must equal the gradient of some scalar function $\mathbf{v} = \nabla \phi$ so that Eq. (2) holds, where ϕ is called the velocity

$$\int_A^B \mathbf{v} \cdot d\mathbf{s} = \int_A^B \nabla \phi \cdot d\mathbf{s} = \phi_B - \phi_A \qquad (2)$$

potential and is a function of position in the fluid and time, $\phi = \phi(\mathbf{r},t)$. Because Eq. (3) is valid,

$$\text{curl } \mathbf{v} = \text{curl (grad } \phi) = 0 \qquad (3)$$

the fluid motion is irrotational. The fluid motion is also divergence-free (solenoidal) from consideration of continuity. Thus it follows that div $\mathbf{v}$ = div (grad ϕ) = 0, which is Laplace's equation. SEE BERNOULLI'S THEOREM; FLUID-FLOW PRINCIPLES; LAPLACE'S IRROTATIONAL MOTION.

LAPLACE'S IRROTATIONAL MOTION
Arthur E. Bryson, Jr.

Laplace's equation for irrotational motion of an inviscid, incompressible fluid is partial differential equation (1), where x_1, x_2, x_3 are rectangular cartesian coordinates in an inertial reference frame, and Eq. (2) gives the velocity potential. The fluid velocity components u_1, u_2, u_3 in the three

$$\frac{\partial^2 \phi}{\partial x_1^2} + \frac{\partial \phi}{\partial x_2} + \frac{\partial^2 \phi}{\partial x_3^2} = 0 \qquad (1) \qquad \phi = \phi(x_1,x_2,x_3,t) \qquad (2)$$

respective rectangular coordinate directions are given by $u_i = \partial \phi / \partial x_i$, $i = 1, 2, 3$. More generally, in any inertial coordinate system, the equation is div (grad ϕ) = 0 and the velocity vector is $\mathbf{v}$ = grad ϕ.

Irrotational motion implies that the fluid particles translate without rotation (like the cars on a ferris wheel) and is stated mathematically by saying curl $\mathbf{v} = 0$ where $\mathbf{v} = \mathbf{v}(\mathbf{r},t)$ is the velocity vector, $\mathbf{r}$ is the position vector of a particular point in the fluid flow, and t is the time. If the fluid motion is at any time irrotational it will stay irrotational. Thus any motion starting from rest will be irrotational. If curl $\mathbf{v} = 0$ then $\mathbf{v}$ may be written as grad ϕ because curl (grad ϕ) is identically zero. For an incompressible fluid, the continuity equation is div $\mathbf{v} = 0$; hence, combining this relation with irrotationality gives Laplace's equation, div (grad ϕ) = 0. SEE FLUID-FLOW PRINCIPLES; KELVIN'S CIRCULATION THEOREM.

The velocity field $\mathbf{v}(\mathbf{r},t)$ in a certain region is determined by Laplace's equation with a boundary condition given on the entire surface surrounding the region. The two most common boundary conditions are those a solid surface and at a free surface. At a solid surface the fluid velocity normal to the surface must match the velocity of the surface normal to itself, $\mathbf{v} \cdot \mathbf{n} = \mathbf{v}_s \cdot \mathbf{n}$; that is, $\partial \phi/\partial n = \mathbf{v} \cdot \mathbf{n}$ is given on the boundary. At a free surface, such as one occurring between two fluids of different density, the pressure must be continuous; this boundary condition, in general, involves the use of the nonstationary Bernoulli equation and usually leads to wave motion. SEE BERNOULLI'S THEOREM; WAVE MOTION IN FLUIDS.

POTENTIAL FLOW
SHIH I. PAI

Fluid flow which can be specified by a velocity potential. In contrast to creeping flow, it represents a condition where inertia is controlling and viscous forces are negligible. When the effect of viscosity of a fluid is negligible, in most cases the fluid flow will be irrotational at all times if it starts from rest. For irrotational flow, there exists a velocity potential such that the velocity vector is the gradient of the velocity potential. For an incompressible fluid, the velocity potential satisfies the Laplace equation, which has been investigated very thoroughly from a mathematical point of view. For compressible fluid flow, potential flow exists for subsonic flow. In transonic flow over a thin body, potential flow may be considered as a good approximation. However, in supersonic flow, behind a strong shock wave, potential flow does not exist. SEE CREEPING FLOW; FLUID FLOW; FLUID-FLOW PRINCIPLES; HYDRODYNAMICS; LAPLACE'S IRROTATIONAL MOTION.

The potential flow satisfies the Navier-Stokes equations of a viscous and incompressible fluid, but not the boundary conditions at a solid wall for viscous flow. Hence the potential flow applies to most flow problems of a fluid of small viscosity, such as air and water, far away from a solid wall but not in the boundary-layer flow near the wall. SEE BOUNDARY-LAYER FLOW; NAVIER-STOKES EQUATIONS.

Bibliography. H. Lamb, *Hydrodynamics*, 6th ed., 1932; S. I. Pai, *Introduction to the Theory of Compressible Flow*, 1959; H. Schlichting, *Boundary Layer Theory*, 7th ed., 1979; S. Schreier, *Compressible Flow*, 1982.

KELVIN'S MINIMUM-ENERGY THEOREM
ARTHUR E. BRYSON, JR.

A principle of fluid mechanics which states that the irrotational motion of an incompressible, inviscid fluid occupying a simply connected region has less kinetic energy than any other fluid motion consistent with the boundary condition of zero relative velocity normal to the boundaries of the region. This remarkable theorem is easily proved as follows. Let T be the kinetic energy of the irrotational motion with velocity potential ϕ. Let T_1 be the kinetic energy of another motion with velocity field $\mathbf{v} = \nabla \phi + \mathbf{v}_0$. From continuity it follows that $\nabla \cdot \mathbf{v}_0 = 0$, and from the boundary condition, $\mathbf{n} \cdot \mathbf{v}_0 = 0$ on the boundary, where $\mathbf{n}$ is unit vector normal to the boundary surface. It follows that Eq. (1) is valid (in SI units or any other coherent system of units). However $\mathbf{v}_0 \cdot \nabla \phi =$

$$T_1 = T + \frac{\rho}{2} \iiint \mathbf{v}_0 \cdot \mathbf{v}_0 \, d\tau + \rho \iiint \mathbf{v}_0 \cdot \nabla \phi \, d\tau \qquad (1)$$

$\nabla \cdot \phi \mathbf{v}_0$ because $\phi \nabla \cdot \mathbf{v}_0 = 0$, and applying Gauss' divergence theorem to the last integral above, it becomes integral (2), which vanishes, since $\mathbf{v}_0 \cdot \mathbf{n} = 0$ on the boundary. The second integral is

$$\rho \iint \phi \mathbf{v}_0 \cdot \mathbf{n} \, dA \qquad (2)$$

a positive quantity, since its integrand is everywhere positive, thus proving that $T_1 > T$. SEE D'ALEMBERT'S PARADOX; FLUID-FLOW PRINCIPLES; LAPLACE'S IRROTATIONAL MOTION.

ELECTROLYTIC TANK
William B. Brower, Jr.

A special type of computing machine based on the fact that in ideal fluid flow the velocity potential φ, and in planar flow the stream function ψ, satisfy Laplace's equations. *See* Fluid flow; Hydrodynamics; Laplace's irrotational motion.

Laplace's equations in rectangular cartesian coordinates are written as Eqs. (1), with velocity components as in Eqs. (2).

$$\frac{\partial^2 \varphi}{\partial x^2} + \frac{\partial^2 \varphi}{\partial y^2} + \frac{\partial^2 \varphi}{\partial z^2} = 0 \qquad \frac{\partial^2 \psi}{\partial x^2} + \frac{\partial^2 \psi}{\partial y^2} = 0 \qquad (1)$$

$$u = \frac{\partial \varphi}{\partial x} = \frac{\partial \psi}{\partial y} \qquad v = \frac{\partial \varphi}{\partial y} = -\frac{\partial \psi}{\partial x} \qquad w = \frac{\partial \varphi}{\partial z} \qquad (2)$$

Because the voltage V for an electrical flow through a homogeneous isotropic conducting medium is likewise governed by an equation of the form of Eq. (1), either the velocity potential or stream function or both can be related to the voltage through constants m and n called the scale factor as in Eqs. (3).

$$\text{Analogy A: } \varphi = mV \qquad \text{Analogy B: } \psi = nV \qquad (3)$$

By use of the previous relations the fluid velocity components are proportional to the respective voltage gradients so that $u = m(\partial V/\partial x) = n(\partial V/\partial y)$. Every ideal fluid-flow problem, therefore, has an electrical counterpart; for this reason the electrolytic tank is also known as an electrical tank analogy or potential flow analyzer. Its solution includes construction of a scaled electrical-flow model, its installation in an electrical tank with proper simulation of the physical boundary conditions both on the model contour and on the field boundaries, measurement of the electrical variables as required, and finally the translation of the measurements by numerical computation into meaningful fluid-flow terms.

Apparatus. For the most accurate measurements, a tank constructed of a nonconducting material such as slate is used. A typical two-dimensional tank geometry is 80 in. long, 60 in. wide, and 5 in. deep (200 cm × 150 cm × 13 cm). Brass electrodes are clamped to opposite sides of the tank, depending on which analogy is employed, and the tank is filled with water until it is flush with the top surface of the model, which must be a cylinder of the desired cross section.

The electrical circuit requires a signal generator (usually 1000 Hz), a power amplifier (20 W), and a precision voltage divider, all connected to a null indicator (oscilloscope). A platinum-wire probe and follower allows either streamlines (analogy B) or potential lines (analogy A) to be traced. By plotting φ or ψ, calculated from Eq. (2), as a function of the distance along a specified boundary (for example an airfoil surface), the velocity distribution can be obtained by finite differencing, or graphical differentiation. When the velocity is known, the corresponding pressure variation can be calculated from Bernoulli's equation. *See* Bernoulli's theorem.

Application. It is possible to construct special tanks which allow analysis of two-dimensional subsonic flows, axisymmetric flows, three-dimensional flows, conformal mappings, and flows in the hodograph plane. The electrolytic tank allows investigations of highly complex geometries with only slightly more complexity than elementary cases. However, since 1960 the extraordinary development of high-speed computational techniques has greatly reduced the relative importance of electrolytic-tank computation in fluid mechanics.

HYDRAULIC ANALOG TABLE
Donald R. F. Harleman

An experimental facility which makes use of the analogy between water flow with a free surface and two-dimensional compressible gas flow. The water flows over a smooth horizontal surface and is bounded by vertical walls geometrically similar to the boundaries of the corresponding compressible gas flow. The change in water depth between a reference station and any other point in the flow is related to the change in density, pressure, and temperature in the analogous gas flow. The analog table is an effective low-cost research tool; flow patterns are easily observed and boundary changes may be made rapidly and comparatively inexpensively during exploratory studies.

Theoretical basis. The basic assumptions are that both liquid and gas flows are inviscid (no energy dissipation) and that the vertical pressure distribution in the free surface flow is hydrostatic. Under these conditions the analogy may be obtained by writing the energy and continuity equations for the water and gas. SEE FLUID FLOW; FLUID-FLOW PRINCIPLES.

From the energy equation for water, the velocity is given by Eqs. (1), where h is the water depth and the subscript 0 refers to the value at a stagnation point ($V = 0$).

$$V = \sqrt{2g(h_0 - h)} \qquad V_{max} = \sqrt{2gh_0} \qquad (1)$$

The corresponding equations for a gas are Eqs. (2), where c_p is the specific heat at constant pressure and T is the temperature.

$$V = \sqrt{2gc_p(T_0 - T)} \qquad V_{max} = \sqrt{2gc_p T_0} \qquad (2)$$

If the ratios V/V_{max} are equated for water and gas, Eqs. (3) are obtained. If u and v are the

$$\frac{h_0 - h}{h_0} = \frac{T_0 - T}{T_0} \quad \text{or} \quad \frac{h}{h_0} = \frac{T}{T_0} \qquad (3)$$

x and y components for velocity V, the continuity equations for water and gas are given by Eqs. (4). Hence, it follows that depth h and density ρ are also analogous quantities; thus Eq. (5) holds.

$$\frac{\partial(uh)}{\partial x} + \frac{\partial(vh)}{\partial y} = 0 \text{ (water)} \qquad \frac{\partial(u\rho)}{\partial x} + \frac{\partial(v\rho)}{\partial y} = 0 \text{ (gas)} \qquad (4)$$

$$\frac{h}{h_0} = \frac{\rho}{\rho_0} = \frac{T}{T_0} \qquad (5)$$

However, in an isentropic gas flow Eq. (6) is valid. Therefore, the analogy requires that the adiabatic exponent $k = 2$. SEE ISENTROPIC FLOW.

$$\frac{\rho}{\rho_0} = \left[\frac{T}{T_0}\right]^{1/(k-1)} = \left[\frac{p}{p_0}\right]^{1/k} \qquad (6)$$

The relations of the classical analogy may therefore be summarized in Eq. (7). In water the

$$\frac{h}{h_0} = \frac{\rho}{\rho_0} = \frac{T}{T_0} = \sqrt{\frac{p}{p_0}} \qquad (7)$$

speed of propagation of a small disturbance is $c = \sqrt{gh}$, and in a gas the corresponding quantity is the acoustic velocity $c = \sqrt{kp/\rho}$.

The ratio $V/\sqrt{gh}$ is known as the Froude number in water, and in a gas the analogous ratio $V/\sqrt{kp/\rho}$ is the Mach number. SEE FROUDE NUMBER; MACH NUMBER.

Applications. An analogy has been shown to exist for a gas with $k = 2$; however the applications are primarily for air where $k = 1.4$. Various proposed modifications to the classical analogy have shown the analog table to be a quantitatively useful tool for application to many aerodynamic problems.

In the realm of transonic flow, experimental results from the water table can be interpreted for a hypothetical gas with $k = 2$ and transferred to air by means of the transonic similarity laws.

In supersonic flow, the classical analogy is no longer strictly valid because the presence of shock waves invalidates the condition of negligible energy dissipation. For the indicated classical relationships between the water depth and the gas density, temperature, and pressure, only the depth and density relation remains useful for supersonic airflow. A shadowgraph of an airfoil taken in an analog table is shown in the **illustration**. SEE SHOCK WAVE.

Shadowgraph of airfoil in hydraulic analog table.

The importance of the hydraulic analog table as a quantitative tool is decreasing with the introduction of computer-aided numerical calculations.

Bibliography. H. M. Morris and J. M. Wiggert, *Applied Hydraulics in Engineering*, 2d ed., 1972.

CROCCO'S EQUATION
ARTHUR E. BRYSON, JR.

A relationship between vorticity and entropy gradient for the steady flow of an inviscid compressible fluid. Crocco's equation, given below, pertains to isoenergetic flow, which is a common type

$$\mathbf{v} \times \boldsymbol{\omega} = -T \text{ grad } s$$

of flow where the total energy, or stagnation enthalpy, of the fluid per unit mass is constant throughout the fluid. In the equation $\mathbf{v}$ is the fluid velocity vector, $\boldsymbol{\omega} = \text{curl } \mathbf{v}$ is the vorticity vector, T is the fluid temperature, and s is the entropy per unit mass of the fluid. (The equation in this form is applicable in SI units or any other coherent system of units.) Thus entropy gradients can occur only at right angles to $\mathbf{v}$ and $\boldsymbol{\omega}$. An irrotational flow is one where $\boldsymbol{\omega} = 0$ throughout the flow, and it follows from Crocco's equation that it must also be an isentropic flow, that is, one with constant entropy throughout. An example of a steady, inviscid, isoenergetic flow that is rotational, and hence nonisentropic, is that behind the curved shock wave at the nose of a blunt body traveling at supersonic speeds; the flow is uniform and isentropic ahead of the shock wave, but because the entropy rise through the shock depends on the inclination of the shock front to the oncoming stream, the flow is nonisentropic behind the curved shock. SEE FLUID-FLOW PRINCIPLES; ISENTROPIC FLOW; LAPLACE'S IRROTATIONAL MOTION.

5

VISCOUS FLOW

Viscosity	114
Newtonian fluid	118
Non-Newtonian fluid	118
Non-Newtonian fluid flow	123
Creeping flow	126
Rheology	126

VISCOSITY
Norman H. Nachtrieb

The resistance that a gaseous or liquid system offers to flow when it is subjected to a shear stress. Viscosity is a measure of the internal friction that arises when there are velocity gradients within the system. For fluids (gases and liquids) its meaning is conceptually and operationally well defined. In the regime of laminar or streamline flow the force required to maintain a velocity gradient, (dv/dx), between planes of fluid of area A is described by Newton's equation (1) and **Fig. 1**. The

$$f = \eta A \left(\frac{dv}{dx}\right) \qquad (1)$$

proportionality constant η is called the viscosity coefficient. Its dimensions are (mass)(length)$^{-1}$ (time)$^{-1}$, and in the cgs system the unit of viscosity is the poise (1 g · cm^{-1} s^{-1}). It is the force per unit area (dynes cm^{-2}) required to sustain a unit velocity gradient (cm s^{-1} cm^{-1}) normal to the flow direction. In the International System (SI) the unit of viscosity is kg · m^{-1} s^{-1}, and is hence larger than the poise by a factor of 10; conversely 1 kg · m^{-1} s^{-1} = 10 poise. In the British

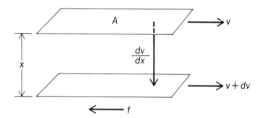

Fig. 1. Viscous shear in fluids.

absolute system of units, the unit of viscosity is 1 lbf · s · ft^{-2} = 1 slug · ft^{-1} · s^{-1} = 47.88 kg · m^{-1} s^{-1} = 478.8 poise. Conversely, 1 kg · m^{-1} s^{-1} = 2.088 × 10^{-2} lbf · s · ft^{-2} and 1 poise = 2.088 × 10^{-3} lbf · s · ft^{-2}.

Simple gases typically have viscosities in the range of 100 to 200 micropoise at standard temperature and pressure (273 K, 1 atm or 101,325 pascals), whereas simple liquids under the same conditions have coefficients of viscosity about two orders of magnitude larger. The **table**

Coefficients of viscosity of selected gases and liquids

Substance	Temperature, °F (°C)	η, poise*
Hydrogen	32 (0)	84.2 × 10^{-6}
Helium	32 (0)	186 × 10^{-6}
Nitrogen	32 (0)	167 × 10^{-6}
Oxygen	32 (0)	181 × 10^{-6}
Water (liquid)	68 (20)	10.1 × 10^{-3}
Ethyl alcohol	68 (20)	12.0 × 10^{-3}
Diethyl ether	68 (20)	2.5 × 10^{-3}
Carbon tetrachloride	68 (20)	9.8 × 10^{-3}
Mercury	68 (20)	15.5 × 10^{-3}
Glycerin	68 (20)	10.69
Glass	752 (400)	10^{13}
	1472 (800)	10^{7}

*1 poise = 0.1 kg · m^{-1} s^{-1}.

lists values for the coefficients of viscosity of selected gases and liquids. The flow characteristics of gases and simple liquids such as water, carbon tetrachloride, and ethyl alcohol are accurately described by Eq. (1), and such fluids are called newtonian fluids. Aqueous suspensions, such as clays, gelatin, and agar, are termed non-newtonian fluids because their viscosities may depend upon the rate of shear and prior treatment. Hydrophilic sols often form extended networks involving water, and their nonnewtonian behavior is believed to be due to the breakdown of their structure under shear. SEE FLUID FLOW; NEWTONIAN FLUID.

Molecular basis of viscosity in gases. The origin of internal friction (viscosity) at the molecular level is the net transfer of momentum between layers of fluid moving with different velocities in parallel flow by the mechanism of molecular collisions. In this process the directed energy of fluid flow is degraded to random thermal energy (heat).

It was one of the early triumphs of the kinetic theory of gases that established the relationship between the viscosity of a hard-sphere gas and the mean speed of its molecules. If x in Fig. 1 represents the mean free path λ of molecules in hypothetical planes that move with velocities equal to v and v', respectively, an exchange of molecules between the planes will result in the net transfer of momentum per unit time equal to $\frac{1}{3} An\bar{c}(mv - mv')$, where n is the number of molecules per unit volume, m is the mass of a molecule, mv and mv' are the additional momenta of molecules in the planes in consequence of their shear velocities, and $\bar{c}$ is the mean speed of molecules $(8kT/\pi m)^{1/2}$, where k is Boltzmann's constant and T is the absolute temperature. Since the gas density if $\rho = nm$, the retarding force that resists the shear is given by Eq. (2). Combining Eqs. (1) and (2) gives Eq. (3).

$$f = \tfrac{1}{3} A \rho \bar{c} \left(\frac{dv}{dx}\right) \lambda \qquad (2) \qquad\qquad \eta = \tfrac{1}{3} \bar{c} \rho \lambda \qquad (3)$$

Substitution for the mean free path $[\lambda = (n\sqrt{2}\pi d^2)^{-1}$, where d is the average diameter of a molecule] permits restatement of Eq. (3) as Eq. (4).

$$\eta = \frac{m\bar{c}}{3\sqrt{2}\pi d^2} \qquad (4)$$

More refined calculations for hard-sphere gases replace the factor $\frac{1}{3}$ in Eq. (4) by 0.499, but the functional form is correct. It predicts that the viscosity of gases should be independent of pressure because the mean free path and gas density are affected in opposite ways by pressure, and this prediction is in accord with experiment up to moderately high pressures. Equation (4) also predicts a $T^{1/2}$ dependence of the gas viscosity, which is in fair agreement with experiment. Real gases show a somewhat stronger temperature dependence because their molecules are not ideal hard spheres. Equation (4) has had widespread application to the determination of the diameters of molecules, and the agreement is good but not perfect. Such minor discrepancies as do exist with molecular diameters determined by other methods (such as molar refraction, equation of state, and electron diffraction) are due to the fact that each method probes a somewhat different region of the potential surface of real molecules.

Viscosity of liquids. Momentum transfer between shearing layers also underlies the viscous behavior of simple liquids, but since the mean free path has little meaning for liquids, no simple relation such as Eq. (4) exists for them. In contrast to the behavior of gases, temperature decreases the viscosities of simple liquids and its effect is much larger. The temperature dependence of the viscosity of simple liquids bears no simple relationship to gas kinetic theory, but instead generally follows an exponential law of the form of Eq. (5), where A and B are parameters

$$\eta = A \exp(B/RT) \qquad (5)$$

characteristic of the liquid and are reasonably constant over finite ranges of temperature; R is the gas constant, 8.314 J · mol^{-1} · K^{-1}. The form of Eq. (5) is the same as that typically found for transport properties in the defect crystalline state, where the concept of a simple thermally activated process is generally accepted. This similarity in temperature dependence has in the past led to various "hole" theories of the liquid state. These have been based upon analogy with the vacancy model of defect crystals, and the parameter E has been identified with the energy required to create a void of molecular dimensions in the liquid and to move a nearby molecule into it. Such hole theories are now generally thought to be oversimplified, and viscous flow, like diffusion in liquids, is thought to be a highly complex process in which many molecules participate.

Measurements of viscosity as a function of pressure likewise show completely different behavior for gases and liquids. Whereas the former show little dependence of viscosity on pressure in the low-density region, very high hydrostatic pressure generally increases the viscosity of liquids, sometimes quite markedly. In the region of laminar flow, nevertheless, Newton's equation accurately describes the viscous behavior of simple liquids, and the presumption is that the transfer of momentum between shearing layers involves a high degree of correlated molecular motions.

When the shear velocity exceeds a critical value in vessels of a given radius, streamline flow is replaced by turbulent flow. The criterion for the onset of turbulence in a tube of radius r is that the Reynolds number $2r\rho v/\eta$ exceed a certain value (approximately 2000 for normal liquids). *See* REYNOLDS NUMBER; STREAMLINE FLOW; TURBULENT FLOW.

Measurement of viscosity. The laminar flow of both gases and liquids in long narrow tubes is described by Poiseuille's equation (6), where η is the fluid viscosity (lbf · s · ft^{-2}, or poise

$$\eta = \frac{\pi(p_1 - p_2)r^4 t}{8Vl} \tag{6}$$

or kg · m^{-1} s^{-1}), where r is the radius of the tube (ft, or cm or m), l is its length (ft, or cm or m), $(p_1 - p_2)$ is the pressure drop (lbf · ft^{-2}, or dynes cm^{-2} or Pa) across the tube, and V is the volume of fluid (ft^3, or cm^3 or m^3) that flows through the tube in time t (s). Poiseuille's equation is based on the assumption that the layer of fluid in contact with the tube wall is stationary. It provides a basis for the absolute measurement of the viscosity of both gases and liquids. *See* BOUNDARY-LAYER FLOW.

More commonly for liquids, relative measurements of viscosity are made by use of either the Ostwald viscometer or the falling-sphere viscometer. The former (**Fig. 2**) consists of two glass bulbs separated by a length of capillary tubing. Liquid is drawn up into the upper bulb, and the time required for its meniscus to fall between calibration marks above and below the upper bulb

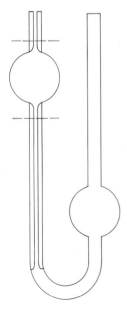

Fig. 2. Ostwald viscometer.

is accurately measured. A similar measurement is made with a liquid of known viscosity. From Eq. (6), Eq. (7) follows. Here η_1 and η_2 are the viscosities of the two liquids, ρ_1 and ρ_2 are their

$$\frac{\eta_1}{\eta_2} = \frac{\rho_1 t_1}{\rho_2 t_2} \tag{7}$$

densities, and t_1 and t_2 are the corresponding flow times. Equation (7) takes an even simpler form if the viscosity is divided by the density of the liquid. This quantity, called the kinematic viscosity, is measured in units termed stokes (cm^2 s^{-1}) in cgs units, in units of m^2 s^{-1} in SI, and in units of ft^2 s^{-1} in the British absolute system.

The falling-sphere viscometer is based upon Stokes' law for the frictional force on a spherical body of radius r falling with constant velocity in a fluid of viscosity η in an unbounded space, Eq. (8). This force is equal and opposite to the net force of gravity acting on the sphere, as in Eq. (9), where ρ and ρ' are the densities of a metal sphere and the fluid, and g is the acceleration of

$$f = 6\pi\eta vr \tag{8} \qquad f = \tfrac{4}{3}\pi r^3(\rho - \rho')g \tag{9}$$

gravity. Equations (8) and (9) lead to the absolute viscosity of the fluid, Eq. (10). As with the Ostwald viscometer, it is simpler to compare the times of fall of the sphere in fluids of known and unknown viscosity, and to use Eq. (11).

$$\eta = \frac{2gr^2(\rho - \rho')}{9v} \tag{10} \qquad \frac{\eta_1}{\eta_2} = \frac{t_1(\rho - \rho'_1)}{t_2(\rho - \rho'_2)} \tag{11}$$

Other methods for the absolute or relative measurement of fluid viscosities are based upon the determination of the torque exerted upon a cylinder immersed in a fluid when a coaxial cylinder is rotated with constant velocity, or the damping of the amplitude of an oscillating disk

Flow behavior of complex fluids. Many fluids display flow behavior that deviates profoundly from that of simple gases and liquids. This subject is normally treated by the field of rheology. A few examples will serve to indicate the complexity of flow behavior in some fluids. Monoclinic sulfur, whose molecules consist of puckered rings of eight sulfur atoms, melts at 203.9°F (95.5°C) to form a simple liquid (S_λ) of the same molecularity. Its viscosity is low enough to classify it as a normal liquid, and its viscosity decreases with temperature in the normal manner. Between 320 and 356°F (160 and 180°C), however, the viscosity increases dramatically by many orders of magnitude, and it appears that ring opening occurs followed by the formation of long-chain polymers. Above this temperature interval, the viscosity again decreases as thermal energy breaks up the long chains into smaller units. The process is highly irreversible, and crystalline sulfur may be recovered only by condensation from sulfur vapor.

Various colloidal dispersions of solids in oil or aqueous media decrease their viscosity when stirred at constant temperature, and revert to their former state of higher viscosity when the shear stresses are reduced. This phenomenon of thixotropy is an essential property of paints that contain solid pigments.

The flow of blood in mammalian vascular systems is non-newtonian, and Poiseuille's law is not obeyed. In part this behavior is attributable to the presence of red corpuscles and other suspended bodies, but the phenomenon is very complex.

Glasses are amorphous solids, structurally much closer to liquids than to crystals. Even at ordinary temperatures they deform under stress over long periods of time, and their viscosity varies over tens of orders of magnitude as the temperature is raised to the softening point. Profound structural changes in the random three-dimensional network and of the dynamical modes of local structural elements take place as the temperature of a glass is increased.

Some adhesives exhibit flow along directions that are not parallel to the direction of stress. Such fluids are anisotropic, and their flow properties are tensors. SEE LIQUID; NON-NEWTONIAN FLUID; NON-NEWTONIAN FLUID FLOW; RHEOLOGY.

Bibliography. P. W. Atkins, *Physical Chemistry*, 3d ed., 1985; J. O. Hirschfelder, C. F. Curtiss, and R. B. Bird, *The Molecular Theory of Gases and Liquids*, 1964; E. A. Moelwyn-Hughes, *Physical Chemistry*, 2d ed., 1964; W. J. Moore, *Basic Physical Chemistry*, 1983.

NEWTONIAN FLUID
Arthur E. Bryson, Jr.

A fluid in which the state of stress at any point is a linear function of the time rate of strain at that point. The fluid thus bears a direct analogy to a hookean solid, for which the state of stress is a linear function of the strain. Many gases and liquids are closely newtonian over a wide range of pressures and temperatures.

The simplest example of newtonian fluid flow is Couette flow, the low-speed steady motion of a viscous fluid between two infinite plates moving parallel to each other with relative velocity U, as in the **illustration**. The shear stress τ_{21} in the fluid is constant and equals $\mu(\partial u_1/\partial x_2) = \mu(U/\delta)$. The time rate of strain at a point is a tensor quantity given by Eq. (1), where u_i are velocity

$$\epsilon_{ij} = \frac{1}{2}\left(\frac{\partial u_i}{\partial x_j} + \frac{\partial u_j}{\partial x_i}\right) \qquad (1)$$

components of fluid and x_i are rectangular cartesian coordinates with $i = 1, 2, 3$. Fluids are inherently isotropic so that the most general linear relationship between stress τ_{ij} and ϵ_{ij} is given by Eq. (2). Here μ is the ordinary viscosity coefficient, λ is the bulk or volume viscosity coefficient,

$$\tau_{ij} = -P\delta_{ij} + (\lambda - \tfrac{2}{3}\mu)\delta_{ij}\epsilon_{mm} + 2\mu\epsilon_{ij} \qquad (2)$$

and P is the pressure. In gases, $\lambda = 0$ if the molecules have no internal degrees of freedom or if the internal motions are not excited. For low Mach numbers, $\epsilon_{mm} = 0$, so λ does not appear in

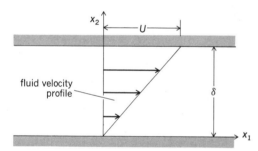

Newtonian fluid. Top plate moves relative to bottom one to produce Couette flow of intervening viscous fluid.

the stress relationship. For most liquids, μ decreases with temperature and increases with pressure; for gases it increases with temperature and is almost independent of pressure. See Fluid-flow principles; Fluids.

NON-NEWTONIAN FLUID
J. Harris and W. L. Wilkinson

A fluid whose flow behavior departs from that of an ideal newtonian fluid. In a newtonian fluid the rate of shear in the fluid under isothermal conditions is proportional to the corresponding stress at the point under consideration. Consider, for example, two flat plates of area A containing a layer of fluid of thickness y and caused to move parallel to each other at a relative velocity u as in **Fig. 1**, where F is shearing force. The shear stress is then F/A, and the rate of shear is u/y. For a newtonian fluid, Eq. (1) can be written, where μ, the constant of proportionality, is called

$$\frac{F}{A} \propto \frac{u}{y} \quad \text{or} \quad \frac{F}{A} = \mu\frac{u}{y} \qquad (1)$$

the newtonian viscosity and completely characterizes the fluid; that is, the relationship between the shear stress and the rate of shear is linear, as shown in **Fig. 2**. SEE NEWTONIAN FLUID; VISCOSITY.

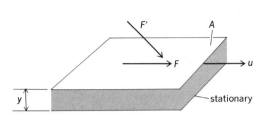

Fig. 1. Illustration of shear stress.

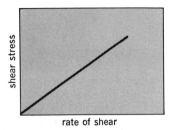

Fig. 2. Newtonian shear stress–shear rate curve.

A non-newtonian fluid is one for which this rather strict requirement is not met. There are several ways in which real fluids can depart from the relatively simple newtonian condition, but it is better to think in terms of non-newtonian regions rather than non-newtonian fluids because many fluids exhibit both newtonian and non-newtonian behavior, depending on the conditions of flow, and cannot be strictly classified as being exclusively one type or the other.

Non-newtonian fluids can be broadly classified into three types as follows:

1. Time-independent fluids for which the rate of shear at any point in the fluid is some function of the shear stress at that point and depends on nothing else. (A newtonian fluid is a special case of a fluid in this category.)

2. More complex time-dependent fluids for which the relationship between shear stress and shear rate depends on the time the fluid has been sheared, that is, on its previous history.

3. Fluids which have the characteristics of both viscous liquids and elastic solids and exhibit partial elastic recovery after deformation, the so-called viscoelastic fluids.

Time-independent fluids. The rheological equation for these fluids is Eq. (2), where $\dot{\gamma}$, the rate of shear, is a function only of τ, the corresponding shear stress. These fluids may be

$$\dot{\gamma} = f(\tau) \qquad (2)$$

conveniently divided into three distinct types, depending on the nature of the function in Eq. (2). These types, the flow curves for which are shown in **Fig. 3**, are as follows.

Bingham plastics. These exhibit a yield stress τ_y which is the stress that must be exceeded before flow starts. Thereafter the rate-of-shear curve is linear. This behavior is found in such materials as toothpaste, oil paints, and drilling muds. There are other materials which also exhibit a yield stress, but the flow curve is thereafter not linear. These are usually called generalized Bingham plastics.

Pseudoplastic fluids. These show no yield value, but the ratio of shear stress to the rate of shear, which may be termed the "apparent viscosity," falls progressively with shear rate. This phenomenon of shear thinning is characteristic of suspensions of asymmetric particles or solutions of polymers such as cellulose derivatives.

Dilatant fluids. These are similar to pseudoplastics in that they show no yield stress, but the apparent viscosity for dilatant materials increases with increasing rates of shear. Such "shear thickening" is observed with suspensions of solids at high solids content (approaching point of tightest packing). Corn-flour pastes can be dilatant.

Time-dependent fluids. The apparent viscosity of some fluids depends not only on the rate of shear but also on the time the shear has been applied. These fluids may be subdivided into two classes—thixotropic fluids and rheopectic fluids—according to whether the shear stress decreases or increases with time when the fluid is sheared at a constant rate.

Thixotropic materials. The structure breakdown of these materials is a function of the duration of shear as well as of the rate of shear. Consider a thixotropic material which is sheared

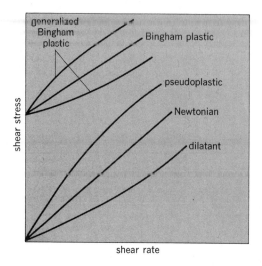

Fig. 3. Non-newtonian flow curves.

at a constant rate after a period of rest. The structure will be progressively broken down, and the apparent viscosity will decrease with time. The rate of breakdown of structure during shearing at a given rate will depend on the number of linkages available for breaking and therefore must decrease with time. (This could be compared with the rate of a first-order chemical reaction.) Also, the simultaneous rate of reformation of structure will increase with time as the number of possible new structural linkages increases. Eventually a state of dynamic equilibrium is reached when the rate of buildup of structure equals the rate of breakdown. This equilibrium position depends on the rate of shear and moves toward greater breakdown at increasing rates of shear.

As an example, consider the material confined in the annular gap of a cylindrical viscometer. After the material has been resting for a long time, one of the cylinders is rotated at a constant speed. The torque on the other cylinder then decreases with time, as shown in **Fig. 4**. The rate of decrease and final torque both depend on the speed, that is, on the rate of shear. Thixotropy is a reversible process, and after resting, the structure of the material builds up again gradually. This type of behavior leads to a kind of hysteresis loop on the shear stress–shear rate curve if the curve is plotted for the rate of shear increasing at a constant rate followed by the curve for the rate of shear decreasing at a constant rate.

Rheopectic fluids. These are fluids for which the structure builds up on shearing, and this phenomenon can be regarded as the reverse of thixotropy—in fact, rheopectic behavior is often referred to as antithixotropy. Such materials are comparatively rare, and few if any are of industrial significance.

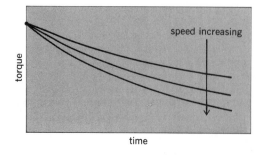

Fig. 4. Cylinder torque decrease due to thixotropy.

Pseudoplastic fluids. These are analogous to thixotropic fluids but with the difference that in the case of pseudoplastic fluids the time required for structure breakdown is negligible; that is, the time effect is not observable in the apparatus being used for the testing of the so-called pseudoplastic fluids. The difference, then, is only a matter of degree.

In the same way rheopectic fluids are superficially similar to their time-independent counterparts, dilatant fluids, in which the time for structure buildup is insignificantly small. Here, however, the analogy is not so close because rheopexy is a case where buildup is often brought about by small shearing rates only. In some cases there is an upper limit to the shear rate beyond which the analogy breaks down.

Viscoelastic fluids. A viscoelastic material is one which possesses both viscous and elastic properties; that is, although the material might be viscous, it exhibits a certain elasticity of shape and is capable of storing energy of deformation. This is easily demonstrated in the laboratory by making the shearing force F in Fig. 1 a function of time, as in Eq. (3), where F_0 is the force amplitude, ω is the circular frequency of oscillation, and t is the elapsed time. The corresponding time-dependent velocity is then given by Eq. (4). The relation between F_0 and U_0 is illustrated on the phase diagram in **Fig. 5**.

$$F = F_0 \cos \omega t \qquad (3) \qquad\qquad U = U_0 \cos(\omega t + \phi) \qquad (4)$$

If the substance is purely viscous, no energy of deformation can be stored and the directions of F_0 and U_0 coincide. If the response of the material is purely elastic, no energy can be dissipated and F_0 will lag U_0 by 90°. In the case of viscoelastic materials, some energy is stored and $0 < \phi < 90°$; that is, the material possesses both viscous and elastic properties.

The constitutive equation for a viscoelastic material is a relationship between stress and strain and their time derivatives. One simple equation which describes linear viscoelasticity and approximates the behavior of some real fluids is that proposed by Maxwell, Eq. (5), where τ is the

$$\tau + \lambda \frac{d\tau}{dt} = \mu \dot{\gamma} \qquad (5)$$

stress and $\dot{\gamma}$ is the rate of shear. The constant λ, which has dimensions of time, is referred to as the relaxation time of the fluid. It is the time constant of the exponential decay of stress at constant strain; that is, if the motion is suddenly stopped, the strain will decay as $\exp(-t/\lambda)$. In general, after a suddenly imposed strain, which is subsequently held constant, the stress in viscoelastic materials decays in time according to an equation of the form of Eq. (6), where $G(t - $

$$d\tau(t) = G(t - t') \frac{d\alpha}{dt'}(t')dt' \qquad (6)$$

t') is a time-dependent rigidity modulus, $\tau(t)$ is the current stress, and $d\gamma(t')$ is the strain increment at some previous time t'.

For an arbitrary strain history, integration over the entire strain history is required and may extend to $-\infty$, as in Eq. (7).

$$\tau(t) = \int_{-\infty}^{t} G(t - t') \frac{d\gamma(t')}{dt'} dt' \qquad (7)$$

This type of behavior, which is typical of solutions of macromolecules and molten polymers, is analogous to mechanical springs and dashpots in series.

Real materials behave as though they comprised a large number of such spring-dashpot combinations, each with its own relaxation time defined as the ratio of viscosity to the corresponding rigidity modulus in the combination. Much experimental work is devoted to finding the viscosity contribution of each flow unit with a given relaxation time to the total bulk viscosity of the material from steady-flow measurements.

Nonlinear viscoelasticity. In the steady flow of solutions of macromolecules and polymer melts, nonlinear elastic effects manifest themselves quite readily. The first of them to occur is that of the forces F in Fig. 1 moving out of the planes of shear and assuming directions inclined to these planes, as F', even though the planes themselves continue to move parallel to each other.

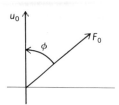 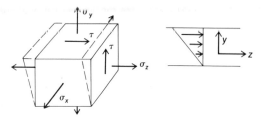

Fig. 5. Phase relationship between the applied force and motion for viscoelastic materials.

Fig. 6. State of stress on a fluid element.

This phenomenon arises from the state of anisotropic stress which exists at each point within the flow. The stress state on an element of fluid is illustrated in **Fig. 6**. If the fluid is newtonian, then Eq. (8) holds, where $\sigma_x = \sigma'_x - p$, $\sigma_y = \sigma'_y - p$, and $\sigma_z = \sigma'_z - p$. Generally, σ'_x, σ'_y, σ'_z, the so-

$$\sigma'_x = \sigma'_y = \sigma'_z = 0 \tag{8}$$

called extra stresses, are generated by the shearing motion while p is an isotropic stress whose value is determined by the boundary conditions in specific problems. Note that on one pair of faces of the cube in Fig. 6 no shear stresses are present.

A further nonlinear effect appears if the shear rate is increased, namely, that of variable viscosity. A simple proportionality between shear stress and shear rate no longer applies; instead, the viscosity varies as shown in **Fig. 7**. In the region of the low-shear-rate newtonian viscosity, the extra stresses σ'_x, σ'_y, σ'_z are a function of the square of the shear rate and vanish at a faster rate than the shear stress which is an odd function in shear rate. However, in the high-shear-rate newtonian viscosity zone, normal stress would be expected to be present always.

There are many theories available in the literature today which accommodate the above nonlinear effects; none have achieved more than limited acceptance.

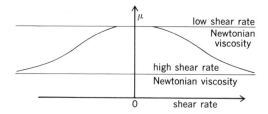

Fig. 7. Variation of viscosity with shear rate for solutions of macromolecules.

Experimental characterization. There are two main viscometric methods for the determination of the rheological properties of time-independent fluids, as follows:

1. Determination of the shear stress–shear rate relationship directly by subjecting the entire sample to a uniform rate of shear in a suitably designed instrument and measuring the corresponding shear stress. These viscometers are usually rotational instruments of the coaxial cylinders or cone and plate type.

2. To infer the shear stress–shear rate relationship indirectly from observations on the pressure gradient and volumetric flow rate in a straight pipe or capillary tube viscometer. In these instruments the rate of shear is not constant but rises from zero at the center of the pipe to a maximum at the wall, and consequently the interpretation of the results is not easy.

Viscoelastic fluids are difficult to characterize quantitatively. Much useful information can be obtained by subjecting the material to sinusoidal shearing motions and following the amplitude and phase shift of the induced stress, that is, a frequency-response analysis. Normal stresses can also be measured, for example, in a cone and plate viscometer.

The characterization of time-dependent fluids of the thixotropic type is also difficult, and only qualitative results can be obtained by testing in steady shear in capillary or rotational viscometers. Oscillatory testing offers more promise, but this is a newer technique attempted on this type of fluid. SEE FLUIDS; NON-NEWTONIAN FLUID FLOW; RHEOLOGY.

Bibliography. E. W. Billington and A. Tate, *The Physics of Deformation and Flow*, 1981; F. R. Eirich, *Rheology*, vols. 1–5, 1956–1970; J. Harris, *Rheology and Non-Newtonian Flow*, 1978; W. R. Schowalter, *Mechanics of Non-Newtonian Fluids*, 1978.

NON-NEWTONIAN FLUID FLOW
J. HARRIS AND W. L. WILKINSON

The flow behavior of non-newtonian fluids. Many important flow problems of practical significance cannot be treated within the framework of classical fluid mechanics, which rests on the application of the Navier-Stokes equations, comprising the continuum equations of motion together with Newton's law of viscosity. Non-newtonian flow, by definition, is concerned with fluids which do not have a constant viscosity in accordance with Newton's law. Some other rheological equation is required which describes the behavior of the fluid, and this is invariably more complicated. SEE FLUID FLOW; FLUID-FLOW PRINCIPLES; NAVIER-STOKES EQUATIONS; NON-NEWTONIAN FLUID; VISCOSITY.

The flow behavior of non-newtonian fluids cannot be predicted quantitatively at present except in relatively simple situations. In the case of time-independent non-newtonian fluids the behavior under laminar flow in tubes can be treated rigorously, but turbulent flow is not well understood. Although time-dependent fluids, particularly of the thixotropic type, are of considerable commercial importance, there is little progress of engineering significance on problems where the changes of fluid properties with time are important. The steady flow of viscoelastic fluids can be reasonably well predicted, but when unsteady conditions exist, such as the flow through valves and fittings and in the whole field of turbulent flow, the elastic properties of the fluid are of major importance and quantitative treatments are not yet well developed. At a free surface, viscoelastic effects can manifest themselves in the form of the Weissenberg effect, in which an elastic liquid tends to climb up a rotating rod, as shown in **Fig. 1**a, or as the swelling of a jet on extrusion from a die, as in Fig. 1b. These phenomena are the result of the normal stresses set up in viscoelastic fluids under shear.

Laminar flow, time-independent. If the fluid properties are independent of time, the rheological equation relating the rate of shear and the shear stress can be written in the form of Eq. (1), where $\dot{\gamma}$ is the rate of shear (that is, du/dr in pipe flow, where u is the velocity at radius

$$\dot{\gamma} = f(\tau) \tag{1}$$

r) and τ is the shear stress at the corresponding point. This shear stress is given by $r\Delta p/2L$ in the case of pipe flow, where r is the radius under consideration and Δp is the pressure drop over the

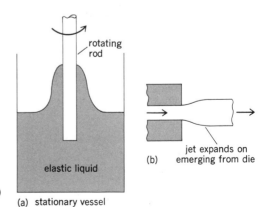

Fig. 1. Viscoelastic effects at a free surface. (*a*) Weissenberg effect. (*b*) Die-swell effect.

Fig. 2. Illustration of shear stress.

length L as illustrated in **Fig. 2**. The shear stress at the pipe wall is similarly given by $D\Delta p/4L$, where D is the pipe diameter.

Under these circumstances it can be shown that for all time-independent fluids the flow rate Q is given by Eq. (2), no matter how complicated the functional relationship $f(\tau)$ may be,

$$Q = \frac{\pi D^3}{8\tau_w^3} \int_0^{\tau_w} \tau^2 f(\tau) d\tau \qquad (2)$$

providing there is no slip at the walls. The integral is a function only of its limit τ_w, which is given by $D\Delta p/4L$. Hence Eq. (2) gives a relationship between the flow rate Q, the pressure drop Δp, and the dimensions of the pipe D and L, the exact form of the relationship depending on the form of $f(\tau)$.

If from observed pressure drop and flow rate data in tubes, $8Q/\pi D^3$ is plotted against τ_w, that is, $D\Delta p/4L$, for different tube diameters on the same diagram, all the data should lie on a single curve for a given fluid, as shown in **Fig. 3** for several well-known types of fluid. This can form the basis of scale-up for design purposes.

If the possibility of nonlaminar flow can be ruled out and the fluid is known to be time-independent, then a separation of the curves for different tube diameters can be interpreted as evidence of anomolous flow behavior or effective slip near the tube wall due, in some cases, to the preferred orientation of particles near the wall. This effect can be shown with a suspension of paper pulp when a clear liquid annulus can be observed adjacent to the pipe wall. The fluid in this region has a lower viscosity and thus gives rise to an effective velocity of slip at the wall.

Analytical solutions of Eq. (2) for the laminar flow of time-independent fluids are possible in some cases, as follows.

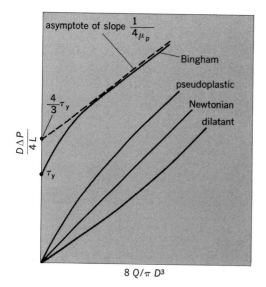

Fig. 3. Laminar flow curves illustrated for time-independent fluids in tubes.

Newtonian fluid. For a newtonian fluid one has Eq. (3), and substitution of Eq. (3) in Eq. (2) leads to the familiar Poiseuille equation for laminar newtonian flow, Eq. (4).

$$\tau = \mu\dot{\gamma} \quad f(\tau) = \tau/\mu \qquad (3) \qquad\qquad \Delta p = \frac{512\mu L Q}{\pi D^4} \qquad (4)$$

Bingham plastic. In the case of a Bingham plastic fluid one has Eq. (5), where τ_y is the yield stress and μ_p is the so-called plastic viscosity.

In pipe flow the shearing stress falls to zero at the axis, and in a region near the axis, where the local shear stress is less than the yield value of the fluid τ_y, the material does not shear but moves along the pipe as a solid plug, as illustrated in **Fig. 4**. Substitution of Eq. (5) into the general equation, Eq. (2), leads to the relationship given in Eq. (6), which is known as Bucking-

$$f(\tau) = \frac{\tau - \tau_y}{\mu_p} \qquad (5) \qquad\qquad \frac{8Q}{\pi D^3} = \frac{\tau_w}{\mu_p}\left[\frac{1}{4} - \frac{1}{3}\left(\frac{\tau_y}{\tau_w}\right) + \frac{1}{12}\left(\frac{\tau_y}{\tau_w}\right)^4\right] \qquad (6)$$

ham's equation for the laminar flow of a Bingham plastic fluid in a circular pipe.

Power-law fluids. For time-independent fluids whose behavior approximates a rheological equation, Eqs. (7) hold. For the so-called power-law fluids, one obtains by substitution of Eqs. (7) in Eq. (2) the relationship for the flow rate given in Eq. (8).

$$\tau = K\dot{\gamma}^n \quad \text{or} \quad f(\tau) = (\tau/K)^{1/n} \qquad (7) \qquad\qquad Q = \frac{\pi D^3}{8}\left(\frac{\tau_w}{K}\right)^{1/n}\left(\frac{n}{3n+1}\right) \qquad (8)$$

Turbulent flow, time-independent. The most satisfactory method available for correlating turbulent-flow data is that due to J. R. Bowen, who proposed the relationship given in Eq. (9), where τ is the shear stress at the wall (that is, $D\Delta p/4L$), U_m is the mean velocity, and a, b,

$$D^a \tau_w = b U_m^c \qquad (9)$$

and c are constants for a particular fluid which have to be determined by experiment. Equation (9) is in fact a relationship between flow rate, pressure drop, and the pipe dimensions. Other methods depend on using data obtained from steady laminar shearing experiments, and these are of doubtful utility for turbulent-flow predictions.

Turbulent suppression. It is well established that very low concentrations of some polymers, in solvents normally regarded as newtonian in tube flow at all Reynolds numbers up to the highest experimentally attained, can produce a striking reduction in the friction factor at high Reynolds numbers. The suppression of turbulence is reflected in reductions in pressure drop along the tube required to maintain a specific flow rate. Pressure drags of up to 70% have been reported.

Some materials are especially effective in this respect, and concentrations as low as 10–30 parts per million of certain polyethylene oxides produce marked reduction in the level of turbulence in water. No adequate explanation of this phenomenon is available at present.

Thixotropic fluids in pipes. If pipe flow data for thixotropic fluids are plotted on a graph in the form of **Fig. 5**, it is found that the results for all tubes do not lie on the same curve for a

Fig. 4. The Bingham plastic in pipe flow.

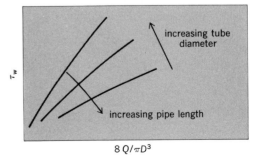

Fig. 5. Effect of thixotropy in pipe flow.

single fluid. More thixotropic breakdown would be indicated with longer pipes and smaller diameters, that is, with longer times of shear and higher rates of shear, respectively, as shown in Fig 5

Other flow problems. Only the relatively simple flow of non-newtonian fluids in tubes has been considered above. Many engineering problems are concerned with flow in more complicated geometries, for example, extrusion, flow through dies, coating operations, rolling operations, and mixing of fluids. These problems present serious analytical difficulties, and their treatments are highly empirical.

Bibliography. E. W. Billington and A. Tate, *The Physics of Deformation and Flow*, 1981; J. Harris, *Rheology and Non-Newtonian Flow*, 1978; W. R. Schowalter, *Mechanics of Non-Newtonian Fluids*, 1978.

CREEPING FLOW
SHIH I. PAI

Fluid flow in which the velocity of the flow is very small. For creeping flow, the Reynolds number $Re = Ud/\nu$ is small (less than unity). Here U is the reference velocity, d is the reference length, and ν is the kinematic viscosity of the fluid. For low Reynolds number, the inertial force is negligible and the nonlinear terms in Navier-Stokes equations are neglected. The resultant flow is known as Stokes flow. One of the important applications of creeping flow is the motion of a tiny particle in a viscous flow, which is important in two-phase flow. If the particle is a sphere, its drag coefficient is $C_D = 24/Re$ if Re is less than 1, where the reference length d is the diameter of the sphere. When Re is greater than 1, but not too large, the drag coefficient is a little larger than $24/R_e$. Such a flow is known as Oseen's flow, and may be considered as an upper limit of creeping flow. Another application of creeping flow is the lubrication problem which was initiated by O. Reynolds, who showed that two parallel or near-parallel surfaces can slide one over the other with only slight frictional resistance, even under great normal pressure, provided that a film of viscous flow is maintained. *See* FLUID FLOW; FLUID-FLOW PRINCIPLES; NAVIER-STOKES EQUATIONS; REYNOLDS NUMBER; VISCOSITY.

Bibliography. J. Happel and H. Brenner, *Low Reynolds Number Hydrodynamics*, 2d ed., 1983; W. F. Hughes, *Introduction to Viscous Flow*, 1979; S. I. Pai, *Two-Phase Flows*, 1977; S. I. Pai, *Viscous Flow Theory*, vol. 1: *Laminar Flow*, 1956; F. M. White, *Viscous Fluid Flow*, 1974.

RHEOLOGY
HERSHEL MARKOVITZ

In the broadest sense of the term, that part of mechanics which deals with the relation between force and deformation in material bodies. The nature of this relation depends on the material of which the body is constituted. It is customary to represent the deformation behavior of metals and other solids by a model called the linear or hookean elastic solid (displaying the property known as elasticity) and that of fluids by the model of the linear viscous or newtonian fluid (displaying the property known as viscosity). These classical models are, however, inadequate to depict certain nonlinear and time-dependent deformation behavior that is sometimes observed. It is these nonclassical behaviors which are the chief interest of rheologists and hence referred to as rheological behavior. *See* VISCOSITY.

Rheological behavior is particularly readily observed in materials containing polymer molecules which typically contain thousands of atoms per molecule, although such properties are also exhibited in some experiments on metals, glasses, and gases. Thus rheology is of interest not only to mathematicians and physicists, who consider it to be a part of continuum mechanics, but also to chemists and engineers who have to deal with these materials. It is of special importance in the plastics, rubber, film, and coatings industries.

This article deals with three useful nonclassical models: linear viscoelasticity, nonlinear elasticity, and nonlinear viscoelasticity. Plasticity, of interest in metals, is rarely studied by rheologists.

Models and properties. To make clear the nature of the various rheological models, the discussion will focus on some very simple types of deformation. Simple shear will be discussed in most detail; similar phenomena and properties, for the most part, also occur in extension.

Consider a block of material of height h deformed in the manner indicated in **Fig. 1**; the

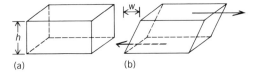

Fig. 1. Simple shear. (a) Undeformed block of height h. (b) Deformed block. The arrows indicate the net forces acting on the top and bottom faces. The forces which must be applied to left and right faces to maintain a steady state are not indicated.

bottom surface is fixed and the top moves a distance w parallel to itself. A measure of the deformation is the shear strain γ given by Eq. (1).

$$\gamma = w/h \qquad (1)$$

To achieve such a deformation if the block is a linear elastic material, it is necessary to apply uniformly distributed tangential forces on the top and bottom of the block as shown in Fig. 1b. The intensity of these forces, that is, the magnitude of the net force per unit area, is called the shear stress S. For a linear elastic material, γ is much less than unity and is related to S by Eq. (2), where the proportionality constant G is a property of the material known as the shear modulus.

$$S = G\gamma \qquad (2)$$

If the material in the block is a newtonian fluid and a similar set of forces is imposed, the result is a simple shearing flow, a deformation as pictured in Fig. 1b with the top surface moving with a velocity dw/dt. This type of motion is characterized by a rate of shear $\dot{\gamma} = (dw/dt)/h$, which is proportional to the shear stress S as given by Eq. (3), where η is a property of the material called the viscosity.

$$S = \eta\dot{\gamma} \qquad (3)$$

Linear viscoelasticity. If the imposed forces are small enough, time-dependent deformation behavior can often be described by the model of linear viscoelasticity. The material properties in this model are most easily specified in terms of simple experiments: creep, stress relaxation, and sinusoidal deformation.

Creep and relaxation. In a creep experiment a stress is suddenly applied and then held constant; the deformation is then followed as a function of time. This stress history is indicated in the solid line of **Fig. 2**a for the case of an applied constant shear stress S_0. If such an experiment is performed on a linear elastic solid, the resultant deformation is indicated by the full line in Fig. 2b and for the linear viscous fluid in Fig. 2c. In the case of elasticity, the result is an instantly achieved constant strain; in the case of the fluid, an instantly achieved constant rate of strain. In the case of viscoelastic materials, there are some which eventually attain a constant equilibrium strain (Fig. 2d) and hence are called viscoelastic solids. Others eventually achieve constant rate of strain (Fig. 2e) and are called viscoelastic fluids. If the material is linear viscoelastic, the deformation $\gamma(S_0, t)$ is a function of the time t since the stress was applied and also a linear function of S_0; that is, Eq. (4) is satisfied, where $J(t)$ is independent of S_0. The function $J(t)$ is a

$$\gamma(S_0, t) = S_0 J(t) \qquad (4)$$

property of the material known as the shear creep compliance. Similar results are obtained for creep experiments performed in extension.

The broken lines in Fig. 2 indicate the result when the stress is removed after a creep experiment. This is called a recovery experiment. The linear viscoelastic solid eventually goes back to its original undeformed shape while the fluid does not. However, the linear viscoelastic

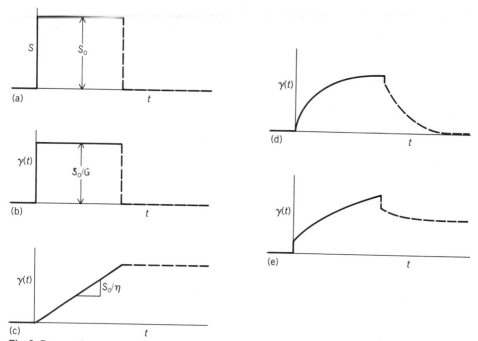

Fig. 2. Creep and recovery; solid lines indicate creep; broken lines indicate recovery. (a) Applied stress history. (b) Corresponding strain history for linear elastic solid, (c) linear viscous fluid, (d) viscoelastic solid, and (e) viscoelastic fluid.

fluid recovers somewhat. The amount of recovery is taken as a measure of the elastic character of the fluid. Some of the energy which was expended on the fluid during the creep experiment is dissipated, but some is recovered. In contrast, with the newtonian fluid, all the energy is dissipated and is converted to heat.

Stress relaxation. In a stress relaxation experiment, a deformation is suddenly imposed and then held constant; the stress is observed as a function of time. The stress decreases with time until it reaches an equilibrium value. This equilibrium value is zero for a viscoelastic fluid, and some finite value for a viscoelastic solid. If the imposed deformation is a simple shear with a constant shear strain γ_0, the required shear stress $S(t,\gamma_0)$ depends on the time t since the strain was imposed, and is proportional to γ_0; that is, Eq. (5) is satisfied, where $G(t)$ is a property of the

$$S(t,\gamma_0) = \gamma_0 G(t) \qquad (5)$$

material, called the shear relaxation modulus, which does not depend on the strain. If the experiment performed is an extension, the corresponding property is the tensile relaxation modulus $E(t)$. Both $G(t)$ and $E(t)$ are functions which decrease as t increases.

Sinusoidal deformation. Linear viscoelasticity is useful for describing experiments where the deformation is a sinusoidal function of time. For example, if a sound wave (or other mechanical vibration) is transmitted through a thick sheet of a polymeric solid, the amplitude of the oscillation decreases as the wave progresses, and the speed of propagation and the rate of damping depend on the frequency of the oscillation and on the temperature. There usually is at least one frequency at which the rate of damping has a maximum.

Behavior of linear polymers. To illustrate some of the characteristics of linear viscoelastic properties, **Fig. 3** shows a schematic logarithmic plot of $J(t)$ for a high-molecular-weight polystyrene over a range of temperatures. This set of curves is typical of high-molecular-weight linear amorphous polymers, which include the commercial transparent thermoplastics such as polymethyl methacrylate (known under the trademarks Lucite and Plexiglas). These polymers are in-

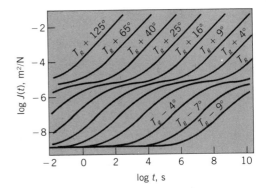

Fig. 3. Schematic shear creep compliance $J(t)$ and its temperature dependence for a high-molecular-weight polystyrene. Temperatures are in degrees Celsius (1°C = 1.8°F) above or below the glass transition temperature T_g, which is 105°C (221°F).

capable of crystallization. However, they have a characteristic temperature, the glass transition temperature T_g, where the thermal expansion coefficient and the specific heat undergo a sudden change. Below T_g it takes an appreciable time to achieve an equilibrium volume if the material is subjected to a sudden decrease of temperature or a sudden change of pressure; the material is called a glass. At temperatures appreciably below T_g the polymer behaves like a hard, brittle elastic solid.

In Fig. 3, it is seen that the main effect of temperature is to shift the curve along the log time axis. When a viscoelastic material has this property, the material is thermorheologically simple. Increasing the temperature does not change the processes which are occurring; it just makes them happen faster.

At long times, $J(t)$ is essentially proportional to time (the slope is unity in a logarithmic plot) and the polymer behaves much like a newtonian fluid. At temperatures considerably above T_g, this newtonian behavior begins at times less than a second. At such temperatures, the fluid polymer can be readily shaped by forcing it into a mold or extruding it through an orifice of the desired cross section. If the polymer is then cooled considerably below T_g, it becomes a stiff solid with very little creep over a time scale of hours or days. It behaves almost like an elastic solid with a $J(t)$ of about 10^{-9} m^2/N, which corresponds to a Young's modulus of about 3×10^9 N/m^2 (4×10^5 lb/in.^{2}). For critical and some long-term uses, even this small amount of creep may not be tolerable. Addition of solid fillers (for example, glass fibers) or replacement by a highly crystalline polymer (for example, high-density polyethylene) can reduce the amount of creep.

It is in the neighborhood of T_g that viscoelastic effects are most readily evident and where they are most sensitive to temperature for experiments in ordinary laboratory time scales.

Polymers have hundreds or thousands of atoms chemically bonded to one another in long sequences. In most synthetic polymers, the presence of single C—C (or similar) bonds makes rotation about these bonds possible. As a result, such macromolecules can have many geometric shapes, most of them coiled conformations—like exceedingly long, thin coiled-up snakes. In an amorphous polymer, these snakelike molecules are highly intertwined. When forces are applied, they cannot immediately attain the steady-state response. Increasing the temperature does not change the process; it can, however, speed it up.

Nonlinear elasticity. A strip of lightly vulcanized rubber (for example, a rubber band) can be stretched to many times its undeformed length by application of a tensile force without breaking; when the force is removed, the strip returns to its original length. The equilibrium force-deformation curve (**Fig. 4**) is clearly nonlinear. Linear elasticity is obviously not applicable, and the model of nonlinear elasticity must be employed.

If a torque is applied to the ends of a rod, the rod is simply twisted if it is a linear elastic material. However, if the rod is made of rubber, it also increases in length. This normal stress effect is a natural consequence of nonlinear elasticity.

Addition of carbon black to a rubber increases the stiffness (Fig. 4). Such a rubber compound is called filled or reinforced rubber. After stretching a strip of reinforced rubber for the first time and then removing the tensile force, the rubber is less stiff when it is stretched again. This

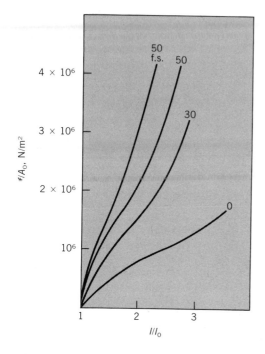

Fig. 4. Force-elongation data for a vulcanized synthetic rubber containing various amounts of carbon black (the number on the graph indicates parts of carbon black per 100 parts of rubber). The ordinate is the force f per initial cross-sectional area A_0. The abscissa is the ratio of the stretched length l to undeformed length l_0. The designation f.s. on one curve indicates that these data were obtained on the first stretching of the rubber strip; other curves show results after repeated stretching. (After E. A. Meinecke and S. Maksin, Rubber Chemistry and Technology, 54:857, 1981)

is known as the Mullins effect and is attributed to the slippage and detachment of rubber molecules from the surface of the carbon black.

Because such high recoverable deformations are possible, a great deal of energy can be stored in rubber. This property has been used not only in toys (slingshots and motors for model airplanes) but also in supports for automobile bumpers designed so that they will absorb the impact of a low-velocity collision without damage to the vehicle.

Nonlinear viscoelasticity. If stresses become too high, linear viscoelasticity is no longer an adequate model for materials which exhibit time-dependent behavior. In a creep experiment, for example, the ratio of the strain to stress, $\gamma(t,S_0)/S_0$, is no longer independent of S_0; this ratio generally decreases with increasing S_0.

An important consequence of this nonlinearity occurs in the steady flow behavior of nonlinear viscoelastic fluids. In a steady simple shearing flow, the shear rate is not proportional to the shear stress S. In analogy with the case of the newtonian fluid [Eq. (3)], the ratio $S/\dot{\gamma}$ is also used to characterize the fluid in this nonlinear behavior, as in Eq. (6), and the symbol η has been

$$S/\dot{\gamma} = \eta(\dot{\gamma}) \qquad (6)$$

adopted for this ratio, although now it is a function of $\dot{\gamma}$ as the notation indicates. It is called the viscosity function or sometimes simply the viscosity.

Shear thinning. For polymer melts, solutions, and suspensions, generally speaking, as indicated schematically in the logarithmic plot of **Fig. 5**, $\eta(\dot{\gamma})$ decreases as $\dot{\gamma}$ increases. This type of behavior, called shear thinning, is of considerable industrial significance. For example, paints are formulated to be shear-thinning. A high viscosity at low flow rates keeps the paint from dripping from the brush or roller and prevents sagging of the paint film newly applied to a vertical wall. The lower viscosity at the high deformation rates while brushing or rolling means that less energy is required, and hence the painter's arm does not become overly tired.

As $\dot{\gamma}$ goes to very small values for polymer melts and solutions, $\eta(\dot{\gamma})$ goes to a limiting value η_0, called the zero shear viscosity, as shown in curve A of Fig. 5. For high-polymer melts, η_0 is a very sensitive function of molecular weight; for a homologous series of linear polymers, η_0 is

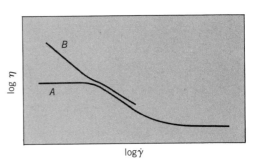

Fig. 5. Schematic logarithmic plot of the viscosity function $\eta(\dot\gamma)$ for a polymer melt or concentrated polymer solution (curve A) and for the same polymer fluid with solid particles added (curve B).

proportional to $M^{3.4}$; if the molecular weight is doubled, for example, η_0 is increased by a factor of about 10. Temperature also has a great effect on η_0, particularly in the neighborhood of T_g, where it can decrease by 75% per degree Celsius (45% per degree Fahrenheit).

On addition of solid particles (such as glass fibers or finely divided silica) to molten polymers or concentrated polymer solution, $\eta(\dot\gamma)$ changes character as indicated in curve B of Fig. 5. A much larger effect is seen at low $\dot\gamma$ than at higher γ. Instead of approaching a limit at low $\dot\gamma$, $\eta(\dot\gamma)$ keeps increasing as $\dot\gamma$ decreases. Some scientists believe that such suspensions will not flow unless the shear stress exceeds a value called the yield stress.

Normal stress effect. For newtonian fluids, steady simple shearing can be maintained by simple shear stress alone (Fig. 1). For nonlinear viscoelastic fluids, normal stresses must also be applied. If they are not, the height of the block of fluid will increase. This normal stress effect is also responsible for the observation that, when a rod is rotated in a container of some fluids (for example, some cake batters in a kitchen mixer), the fluid "climbs" around the rotating rod. With newtonian fluids, the liquid level becomes depressed near the rotating rod as the fluid moves outward due to centrifugal effects.

Die swell. When some molten plastics are forced through a tube at a slow rate, the diameter of the exiting stream can be considerably larger than that of the tube. Under similar flow conditions, newtonian fluids show a diameter increase of only a few percent. This phenomenon is known as die swell or extrudate expansion. In extrusion equipment for making plastic tubing and rods, the orifice must be made smaller than the desired size of the finished product. SEE EXTRUSION.

Drag reduction. When water is pumped through a pipe at rates high enough for the flow to be turbulent, the flow rate, for a given driving pressure, can be greatly increased by adding a very small amount (a few parts per million) of a very high-molecular-weight (greater than a million) polymer. This phenomenon has been named drag reduction. It has been used to increase the range of the water stream from a firehose while using the same pumping equipment on the firetruck. Drag reduction also was used to decrease the number of pumping stations required for the Trans-Alaskan oil pipeline. SEE TURBULENT FLOW.

Thixotropy. There are suspensions (for example, bentonite clay in water) which, after remaining at rest for a long time, act as solids; for example, they cannot be poured. However, if it is stirred, such a suspension can be poured quit freely. If the suspension is then allowed to rest, the viscosity increases with time and finally sets again. This whole process is reversibly; it can be repeated again and again. This phenomenon is called thixotropy. (The word thixotropy may be used in many different senses; a standard nomenclature for rheological phenomena and properties has not been universally adopted.) SEE NON-NEWTONIAN FLUID.

Bibliography. F. R. Eirich (ed.), *Rheology: Theory and Applications*, 5 vol., 1956–1970; H. Markovitz, The emergence of rheology, *Phys. Today*, 21(4):23–30, April, 1968; National Committee for Fluid Mechanics Films, *Illustrated Experiments in Fluid Mechanics* 1972; L. R. G. Treloar, *The Physics of Rubber Elasticity*, 1975; G. V. Vinogradov and A. Y. Malkin, *Rheology of Polymers*, 1980.

WAVES AND DISTURBANCES IN FLUIDS

Wave motion in fluids	134
Wave motion in liquids	137
Capillary wave	140
Internal wave	141
Hydraulic jump	143
Water hammer	144
Shock wave	145

WAVE MOTION IN FLUIDS
Shao-Chi Lin

Wave motion is the basic mechanism by which local disturbances are propagated from one part of a fluid to another. As characteristic of all wave motions, this mechanism allows local disturbances to be passed from one part of a fluid onto the next without net mass motion. The direction along which the local disturbances are transmitted is called the direction of propagation, and the speed of the disturbances relative to the fluid is called the wave speed, or the speed of propagation. *See* Wave motion in liquids.

Applications. Wave phenomena have widespread applications. Marine equipment such as the fathometer and sofar rely upon wave propagations. Shock waves used in sofar propagate long distances underwater, being refracted by the isothermal layers in the oceans. The study of waves is directly applicable to supersonic aircraft, wind tunnels, shock tubes, rocket combustion oscillation, nuclear-bomb blasts, controlled fusion processes in plasma, and ultrasonic processes such as cleaning and inspection. *See* Wind tunnel.

Waves in fluids are diffracted and refracted so that they can be focused to produce intense concentration of energy. For example, in shock tubes, energy is focused into a sharp pulse, the fluid there being caused to glow by the intense excitation so produced. Ultrasonic waves are focused and directed through tanks of mercury to provide short-term memory for large-scale computers.

Fluid waves caused by successive firing of charges from a vertical sounding rocket, such as the Aerobee as it climbs above the altitude in which balloons are effective, are recorded on the ground to provide data for the determination of wind velocity and air density profiles in the upper atmosphere. The technique is analogous to seismic exploration for petroleum. Waves in the upper atmosphere have been found to influence weather, and their Fourier analysis has greatly advanced the techniques of long-range weather forecasting. Shock waves are used to prepare free radicals and in forcing certain chemical processes.

Wave classification. In contrast to surface waves, which are transverse oscillations caused by gravity acting on a liquid having a free surface, wave motion within a fluid is generated by successive compression and expansion of adjacent volume elements of a compressible fluid. Because compression and expansion of an ordinary fluid can only proceed along the direction of propagation of the disturbance, waves within a fluid are mostly longitudinal waves.

Waves in a fluid can be classified as compression waves and expansion waves, according to whether the disturbance is a compression or an expansion. They can further be classified according to the amplitude of the disturbance and the chemical nature of the fluid. For example, waves of small amplitude are called acoustic (or sound) waves; compression waves propagating in chemically inert fluids are called shock waves; waves propagating in the Earth are seismic waves; and waves of large amplitude generated by rapid chemical reactions in explosive fluids are called detonation waves and can propagate much faster than sound waves. Waves in an electrically conducting fluid in the presence of strong magnetic fields are called magnetohydrodynamic waves.

Acoustic waves. The acoustical wave equation can be derived formally through linearization of the equations of motion. A more straightforward approach is to consider a plane wavefront that moves from right to left at a constant speed u_1 in a fluid which is initially at rest and of density ρ_1. If an observer fixes his attention on the wavefront by also moving from right to left at the same speed u_1, he will witness a steady flow of fluid from left to right across the wavefront (**Fig. 1**). Represent the flow velocity and the fluid density to the right of the wavefront by u_2 and ρ_2, respectively; then conservation of mass requires the validity of Eq. (1) because no fluid mass can accumulate at the wavefront in a steady state. Furthermore, an increase in fluid momentum across the wavefront can be supported only by a corresponding drop in pressure from p_1 to p_2 in the fluid. Therefore Eq. (2) holds. (All equations in this article are applicable in SI units or any

$$\rho_1 u_1 = \rho_2 u_2 \quad (1) \qquad \rho_2 u_2^2 - \rho_1 u_1^2 = p_1 - p_2 \quad (2)$$

other coherent system of units.) If the disturbance is so weak that the fractional changes in flow velocity, fluid density, and pressure across the wavefront are much smaller than unity, the changes can be written as Eqs. (3). By substituting Eqs. (3) and neglecting the product terms of

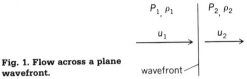

Fig. 1. Flow across a plane wavefront.

the differential quantities, Eqs. (1) and (2) become Eqs. (4) and (5). An expression for u_1 is obtained in Eq. (6) by eliminating du from these two equations. Because this derivation began with a

$$u_2 = u_1 + du \qquad \rho_2 = \rho_1 + d\rho \qquad p_2 = p_1 + dp \qquad (3)$$

$$\rho_1 du + u_1 d\rho = 0 \quad (4) \qquad 2\rho_1 u_1 du + u_1^2 d\rho = -dp \quad (5) \qquad u_1^2 = dp/d\rho \quad (6)$$

wavefront moving in a fluid initially at rest, the above result shows that any small disturbance, if propagated by wave motion at all, must propagate in relation to the fluid at the speed of sound a, defined by Eq. (7). If the disturbance is periodic in time with a fundamental frequency ν, as in a

$$a = u_1 = \sqrt{dp/d\rho} \qquad (7)$$

musical note, then there will be a wavelength λ associated with the wave motion $\lambda = a/\nu$. Furthermore, Eq. (7) shows that a depends only on the variation of pressure with density as caused by a small mechanical disturbance in the fluid, and that a is real as long as $dp/d\rho$ is positive. Historically, Isaac Newton first attempted to derive the speed of sound in air from a somewhat different approach. He arrived at a result which was equivalent to assuming an isothermal compression process. Thus, from the equation of state for a perfect gas (Boyle's law), Eq. (8), he obtained Eq. (9). However, experimental measurements of the speed of sound in air turned out to

$$p = \rho RT \qquad (8) \qquad\qquad a = \sqrt{RT} = \sqrt{p/\rho} \qquad (9)$$

be consistently higher than his prediction. This led P. S. Laplace to suspect that compression and expansion processes associated with acoustic waves should obey the adiabatic law $p\rho^{-\gamma} = $ constant ($\gamma = C_p/C_v$, being the ratio of specific heats) instead. If this pressure-density relationship is assumed, the speed of sound is then given by Eq. (10). The above result has been found to agree

$$a = \sqrt{\gamma RT} = \sqrt{\gamma p/\rho} \qquad (10)$$

so well with experimental observations under ordinary conditions that measurement of the speed of sound has become a standard method for determining the value of γ for various gases.

For liquids, and for gases at extreme temperatures and densities, or for disturbances of very high frequencies, the adiabatic law loses its usual significance, so that the speed of sound (or the compressibility $dp/d\rho$) has to be obtained from direct measurement or from more exact theories.

The speed of sound in several common fluids at room temperature is given in the **table** together with the speed of sound in some common solids for comparison.

Speed of sound in common substances	
Substance	Speed of sound, ft/s (m/s)
Air	1,130 (344)
Hydrogen	4,320 (1317)
Water	4,800 (1463)
Mercury	4,600 (1402)
Paraffin	4,300 (1311)
Lead	4,030 (1228)
Aluminum	16,700 (5090)
Iron	16,800 (5120)

At this point, it may be of value to give some numerical examples of sound amplitude. The smallest periodic pressure amplitude detectable by the human ear is in the order of 10^{-9} atm or 10^{-4} pascals (at about 2000 Hz). On the other hand, the threshold of feeling corresponds to a pressure amplitude roughly 10^6 times greater. Thus, these limits of what is ordinarily called sound correspond to pressure fluctuations of 10^{-9} to 10^{-3} atm (10^{-4} to 10^2 pascals), a fact which justifies the assumption of small disturbance in the foregoing derivation of the speed of sound.

Zone of action and zone of silence. Small disturbances can propagate in a fluid only at a finite speed. The same is true for disturbances of large amplitude. Therefore, when an object moves through a stationary fluid at a speed in excess of the speed at which disturbances can be propagated, there is a boundary that divides the fluid into two distinct regions: the zone of action that has, and the zone of silence that has not yet, been affected by the motion of the object at any given instant. To understand this behavior, consider **Fig. 2**. *See* Shock wave.

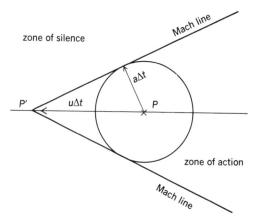

Fig. 2. Point source of disturbance in uniform rectilinear motion, depicted for supersonic motion.

Let u denote the speed of the object and a denote the speed of propagation of the fluid disturbance. To simplify the discussion, assume that the object is in uniform rectilinear motion and that the amplitude of the disturbance is so weak that a can be identified with the speed of sound in the stationary fluid. At any instant t, when the object is at a certain point P, the object will generate disturbances that will propagate away from P in all directions with velocity a. After time interval Δt, the object would have traveled a distance $u\Delta t$ and would have moved to new position P' along its trajectory, while the disturbances generated at point P would still be confined to a spherical surface of radius $a\Delta t$. Because u is greater than a, all disturbances generated by the object up to the time $t + \Delta t$ will be confined within the conical surface of the half-vertex angle given by Eq. (11), with vertex at point P' and axis along PP'. The region inside such a

$$\beta = \sin^{-1}(a/u) \qquad (11)$$

conical surface is called the zone of action at the time $t + \Delta t$; correspondingly, the region outside the conical surface, which is still out of reach of the disturbances, is called the zone of silence at the same instant.

The ratio between the speed of the object and the speed of sound u/a is known as the Mach number of the moving object, and is usually denoted by the symbol M. The motion of the object is accordingly called subsonic when $M < 1$, transonic when $M \cong 1$, supersonic when $M > 1$, and hypersonic when $M \gg 1$. From Eq. (11) and from Fig. 2, which has been depicted for supersonic motion, it can be deduced that the zone of action extends to all parts of the fluid for subsonic motion. For transonic motion, the zone of action covers approximately the rear half of the fluid, up to the plane tangent to the nose of the moving object; for hypersonic motion, the zone of action is confined to a relatively slender cone about the trajectory of the object.

Waves of larger amplitude. In physical optics, it is well known that wavelets from a distributed light source in space can be superposed according to Huygens' principle. For wave motion in fluids, however, the analogy holds only as long as the amplitude of the resultant disturbance remains small enough for the acoustic approximation to apply. For waves of larger amplitude, the nonlinear behavior of the fluid dynamic equations must be taken into account. To visualize these nonlinear effects, consider the following one-dimensional problem.

A semi-infinite tube is filled with a gas initially at rest. At one end of this tube is a movable piston. The piston is suddenly given a small velocity toward the gas. As a result, a compression wave is generated, and propagates along the tube at the speed of sound of the undisturbed gas. Suppose that, soon after this wave has started, the piston is given an additional increment in velocity. A second wave is formed and propagates along the tube behind the first. Because the pressure has increased slightly across the first wave, so has the temperature and the speed of sound. Furthermore, the gas behind the first wave is already moving along the tube at the piston velocity after the first impulse. Thus, the second wave, which propagates with respect to a moving gas ahead of itself and at a sound speed that is slightly higher than that of the first wave, will soon catch up with the first wave.

If the accelerating motion of the piston continues, the succeeding wavelets will propagate along the tube at increasingly higher velocity, so that the shape of the resultant compression pulse from the piston motion will appear steeper as it progresses. Similarly, a decelerating piston motion produces an expansion pulse that flattens out as it progresses along the tube. It is this asymmetry between the two processes that makes the occurrence of large-amplitude compression waves, called shock waves, a more noticeable phenomenon in nature than are expansion waves.

Shock waves are characterized by rapid changes in fluid density, pressure, and temperature along the direction of flow. Bomb blasts start as shock waves.

Seismic waves. Seismic waves pass through the ground; they arise either from natural readjustment of faults in the Earth's crust or by human-made explosions. According to their modes of propagation, seismic waves may be divided into two main groups, namely, body and surface waves.

Body waves, which propagate through the inside of the Earth, may further be subdivided into dilation (longitudinal) waves, which are similar to acoustic waves in compressible fluids, and shear (transverse) waves, which arise on account of the large shear resistance of most elastic solids. For any given medium, dilation waves usually have approximately twice the velocity of propagation of shear waves.

Surface waves from any distant earthquake always arrive after both the dilation waves and the shear waves, and they normally register a much larger amplitude signal on the seismogram. From the known relationship between the propagation velocities and the mechanical properties of various substances, seismologists extract valuable information about the structure of the Earth's interior from the seismograms and apply the results to such purposes as mine prospecting.

Bibliography. R. Courant and K. O. Friedrichs, *Supersonic Flow and Shock Waves*, 1948, reprint 1977; J. Lighthill, *Waves in Fluids*, 1978; Lord Rayleigh, *The Theory of Sound*, 2d ed., reprint 1945; F. W. Sears et al., *University Physics*, 6th ed., 1981; G. B. Whitham, *Linear and Nonlinear Waves*, 1974.

WAVE MOTION IN LIQUIDS
DONALD R. F. HARLEMAN

A temporal variation in the velocity of a liquid which is propagated through a liquid medium. The speed of propagation of the disturbance, or change in the velocity of the liquid relative to the initial velocity of the medium, is known as the wave celerity. The dynamic behavior of the wave depends upon the method of generation of the disturbance, the boundary conditions of the medium, and the properties of the liquid. This article treats those aspects of wave motion appropriate to liquids in which disturbances are propagated at a gas-liquid interface and are primarily dependent upon the gravitational fluid property (surface tension and viscosity being of secondary importance). Wave motions which occur in confined fluids (either liquid or gaseous) are primarily dependent upon the elastic property of the medium. SEE WATER HAMMER; WAVE MOTION IN FLUIDS.

Oscillatory waves. The term oscillatory implies a periodicity in the form of a disturbance moving past a fixed point. **Figure 1** is a definition sketch for an oscillatory wave propagating in a liquid of constant density ρ and depth h measured from the bottom to the still-water level (SWL). Wavelength L is the horizontal distance between successive crests of the wave. Wave height H is the vertical distance from crest to trough; amplitude a is the distance from the still-water level to the crest, and η is the elevation of the free surface with respect to the still-water level at any position x and instant of time t. In the linearized theory of small-amplitude waves ($H/L < 0.03$) the wave profile is sinusoidal and is given by Eq. (1), where T is the wave period. By definition, the celerity or speed of propagation $C = L/T$ and is given by Eq. (2) (in SI units or any other

$$\eta = a \sin\left[2\pi\left(\frac{t}{T} - \frac{x}{L}\right)\right] \quad (1) \qquad C = \sqrt{\left(\frac{\sigma}{\rho}\frac{2\pi}{L} + \frac{gL}{2\pi}\right)\tanh\left(2\pi\frac{h}{L}\right)} \quad (2)$$

coherent system of units). The first term on the right in Eq. (2) expresses the influence of surface tension σ and need be considered only for waves of very small length (of the order of magnitude of 1 in. or 2.5 cm). In the remaining development, only gravity waves will be considered, as expressed by the second term of the celerity equation.

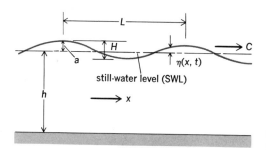

Fig. 1. Definition sketch for an oscillatory wave.

Oscillatory waves may be generated in a rectangular channel by a simple harmonic translation of a vertical wall forming one end of the flume. The wave amplitude will be determined by the displacement (stroke) of the wall, and the wavelength will be a function of the period of oscillation. The two major classes of oscillatory waves, deep-water and shallow-water waves, are determined by the magnitude of the ratio of liquid depth to wavelength h/L. An inspection of the celerity equation for gravity waves shows that as the depth becomes large in comparison with wavelength, Eqs. (3) hold. Hence, in a deep-water wave the celerity is a function only of the

$$\tanh\left(2\pi\frac{h}{L}\right) \to 1 \qquad C = \sqrt{\frac{gL}{2\pi}} \quad (3)$$

wavelength. This approximation is close if the depth is greater than one-half the wavelength. On the other hand, as the depth becomes small in comparison with wavelength, Eqs. (4) hold. Hence,

$$\tanh\left(2\pi\frac{h}{L}\right) \to 2\pi\frac{h}{L} \qquad C = \sqrt{gh} \quad (4)$$

the celerity depends only on the depth in a shallow-water wave. Somewhat arbitrarily, the limit $hL = 1/10$ is generally applied to this type of wave motion. In deep-water waves, individual fluid particles tend to move in circular orbits. The radius of the surface particle orbit is equal to the wave amplitude and the radius decreases exponentially with depth (**Fig. 2**). At a depth of one-half the wavelength, the orbital radius is about $1/20$ of the amplitude. A zone of essentially zero fluid motion is rapidly approached and the character of the wave is therefore not affected by the total depth of the liquid.

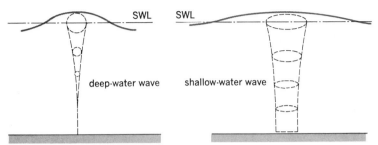

Fig. 2. Fluid particle orbital motions in deep- and shallow-water oscillatory waves.

In shallow water, no vertical particle motion can exist at the bottom; thus the wave characteristics are modified. The particle orbits are flat ellipses in which the minor axis is depressed to zero at the bottom (Fig. 2).

The energy of a wave consists of equal amounts of potential energy (due to particle position above or below the still water level) and kinetic energy (due to the motion of particles in their orbits). The rate of propagation of energy in the direction of wave travel is known as the group velocity to distinguish it from phase velocity C. In deep-water waves the group velocity is one-half the phase velocity; in shallow-water waves the two propagation velocities are equal.

Standing waves. A standing wave can be considered to be composed of two equal oscillatory wave trains traveling in opposite directions. The phase velocity of the resulting wave is zero; nevertheless the velocity of propagation of the component waves retains its usual meaning. In the notation of the previous section, the equation for the profile of a standing wave is obtained by adding the elevations of waves moving in the positive and negative x directions, given by Eqs. (5) and (6), respectively. Hence Eq. (7) holds. If the length of the basin l in which a

$$\overrightarrow{\eta_1} = a \sin\left[2\pi\left(\frac{t}{T} - \frac{x}{L}\right)\right] \quad (5) \qquad \overleftarrow{\eta_2} = a \sin\left[2\pi\left(\frac{t}{T} + \frac{x}{L}\right)\right] \quad (6)$$

disturbance occurs is an integral number n of half wavelengths, a self-perpetuating (except for frictional dissipation) standing wave will result. Therefore, if $l = nL/2$, Eq. (8) is valid. For long

$$\eta = \eta_1 + \eta_2 = H \sin\left(2\pi\frac{t}{T}\right)\cos\left(2\pi\frac{x}{L}\right) \quad (7) \qquad \eta = H \sin\left(2\pi\frac{t}{T}\right)\cos\left(\frac{\pi n x}{l}\right) \quad (8)$$

waves, as with shallow water, in a basin of uniform depth, the period of oscillation T is defined by Eq. (9). Standing waves frequently occur in canal locks as a result of filling disturbances and

$$T = 2l/(n\sqrt{gh}) \quad (9)$$

in large lakes, bays, and estuaries as a result of wind or tidal action.

Solitary waves. A solitary wave consists of a single crest above the original liquid surface which is neither preceded nor followed by another elevation or depression of the surface. Such a wave is generated by the translation of a vertical wall starting from an initial position at rest and coming to rest again some distance downstream. In practice, solitary waves are generated by a motion of barges in narrow waterways or by a sudden change in the rate of inflow into a river; they are therefore related to a form of flood wave. The amplitude of the wave is not necessarily small compared to the depth, and the wavelength is theoretically infinite because the elevation of the surface approaches the still water level asymptotically with distance as shown in **Fig. 3**. The profile of the solitary wave is given by Eq. (10) and the celerity by Eq. (11). When the solitary

$$\eta = a\,\text{sech}^2\left[\frac{x}{h}\sqrt{\frac{3}{4}\frac{a}{h}}\right] \quad (10) \qquad C = \sqrt{g(h + a)} \quad (11)$$

wave amplitude becomes approximately equal to the depth, the wave profile becomes unstable and a breaking wave results.

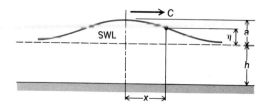

Fig. 3. Definition sketch for a solitary wave.

Surges. A surge is generated by the forward motion of a vertical wall, at a constant speed, as shown schematically in **Fig. 4**. Surges in open channels are analogous to shock waves produced in a tube by the continuous motion of a piston. A zone of violent eddy motion occurs at the wavefront and the analysis of such motions must take into account the appreciable energy dissipation in this region. The velocity of propagation of a surge is given by Eq. (12). If a velocity

$$C = \sqrt{gh_1} \left[\frac{1}{2} \frac{h_2}{h_1} \left(\frac{h_2}{h_1} + 1 \right) \right]^{1/2} \qquad (12)$$

V_1 equal and opposite to C is imposed on the fluid upstream of the disturbance, the absolute velocity of the surge front will become zero. In this form the surge is known as a hydraulic jump

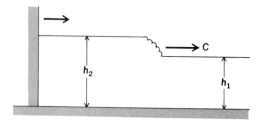

Fig. 4. Definition sketch for a surge wave.

and it is often used as a means of dissipating flow energy at the bottom of dam spillways. *See* Hydraulic jump.

Bibliography. J. Roberts, *Internal Gravity Waves in the Ocean*, 1975; R. C. H. Russell and D. H. Macmillan, *Waves and Tides*, 1953, reprint 1971; R. Silvester, *Coastal Engineering*, vol. 1: *Generation, Propagation and Influence of Waves*, 1974; J. J. Stoker, *Water Waves*, 1957.

CAPILLARY WAVE
Charles S. Cox

Capillary waves, or ripples, occur at the interface between two fluids, in which the principal restoring force is controlled by surface tension. Ripples generated by wind at the interface between air and water on oceans and lakes are of importance to the friction of air flowing over water, and to the reflection and scattering of electromagnetic and sound waves.

The formulas relating the phase velocity c and frequency f to the wavelength λ of low-amplitude sinusoidal waves, in the absence of wind forces, are shown as Eqs. (1), where Eqs. (2) apply. T is the surface tension, ρ_1 and ρ_2 are the densities of upper and lower fluids, respectively, and g is the acceleration of gravity. For air over water at 59°F (15°C), c_m is 9 in./s (23 cm/s) and λ_m is 0.67 in. (1.7 cm). The phase velocity is a minimum at $c = c_m$ when $\lambda = \lambda_m$ (see **illus.**). Shorter waves are ripples, longer waves are gravity waves.

$$c = c_m \left(\frac{\lambda}{2\lambda_m} + \frac{\lambda_m}{2\lambda}\right)^{1/2} \qquad f = \frac{c}{\lambda} \qquad (1)$$

$$c_m = \left(2\frac{\rho_2 - \rho_1}{\rho_2 + \rho_1}\right)^{1/2} \left(\frac{gT}{\rho_2 - \rho_1}\right)^{1/4} \qquad \lambda_m = 2\pi\left[\frac{T}{(\rho_2 - \rho_1)g}\right]^{1/2} \qquad (2)$$

In nature, ripples are observed to grow rapidly when the wind blows and to die away rapidly when the wind stops. When the water surface is uncontaminated, ripples die away to e^{-1} of their original amplitude in a time $t_o = \lambda^2/(8\pi^2\nu)$, where ν is the kinematic viscosity of water. For $\lambda = 0.67$ in. (1.7 cm) and $\nu = 0.0016$ in.^{2} (.01 cm^2) s, $t_o = 3.8$ s. When the water surface is contaminated, as by an oil film or other surface-active agent, ripples are damped still more rapidly because the contaminated surface acts as an inextensible film against which the water motions

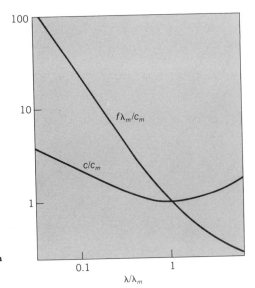

Phase velocity c and frequency f of ripples and gravity waves as functions of wavelength plotted on logarithmic scales; λ_m is wavelength at minimum phase velocity c_m.

due to the ripples must rub. For a perfectly inextensible film, t_o is equal to $(\pi^3\nu f)^{-1/2}$. For the example treated above, t_o becomes 0.86 s. For moderate winds, the increased damping of ripples almost completely inhibits their growth; the surface appears smooth and is called a slick. It has been observed that even gravity waves grow at an inappreciable rate under such conditions; the interpretation here is that the fine scale of roughness presented to the wind by a rippled surface is necessary for the formation of gravity waves. On the other hand, ripples have been observed to be formed, in the absence of wind, by momentary nonlinear interactions of steep gravity waves; consequently the formation of both gravity waves and ripples is an interconnected process.

The surface profile of ripples of large amplitude which move without change of form has been calculated. The profile changes from sinusoidal to one with sharper troughs than crests as amplitude increases.

INTERNAL WAVE
Charles S. Cox

Internal waves are wave motions of stably stratified fluids in which the maximum vertical motion takes place below the surface of the fluid. The restoring force is mainly due to gravity; when light fluid from upper layers is depressed into the heavy lower layers, buoyancy forces tend to return the layers to their equilibirium positions. Internal waves have been found in the atmosphere as lee waves (waves in the wind stream downwind from a mountain) and as waves propagated along an inversion layer (a layer of very stable air). They are also associated with wind shears at the lower boundary of the jet stream. In the oceans internal oscillations have been observed wherever suitable measurements have been made (**Fig. 1**), but it is not completely certain that all of these oscillations are manifestations of internal waves rather than turbulent eddies. The observed oscillations can be analyzed into a spectrum with periods ranging from a few minutes to many days. At a number of locations in the oceans, internal tides, or internal waves having the same periodicity as oceanic tides, are prominent.

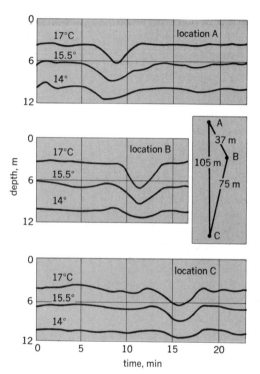

Fig. 1. Diagrams showing passage of an internal wave in the ocean off Mission Beach, California. Curves indicate depth of certain isotherms recorded at locations A, B, and C as functions of time. Prominent trough recorded at successively later times at the three locations represents a solitary internal wave trough traveling at 23 cm/s. Average depth to sea floor was 20 m. Insert shows relative location of recorders in horizontal plan. 1 m = 3.3 ft; °F = (°C × 1.8) + 32.

The vertical distribution of motions and phase velocity of internal waves depends on the vertical gradient of density in the fluid and the frequency of the generating forces. There is a simple density distribution which is illustrative: The fluid consists of two homogeneous layers, a lighter one on top of a heavier one, such as kerosine over water. The internal waves in this system are sometimes called boundary waves because the maximum vertical motion occurs at the discontinuity of density at the boundary between the two fluids. Let the thickness of the layers be h_1 and h_2, let g be the acceleration of gravity, and let $\delta\rho/\rho$ be the fractional change of density across the boundary. (In the ocean $\delta\rho/\rho$ is of the order of 0.1%, and squares of this small quantity can be neglected.) Then the phase velocity of internal waves of wavelength long compared to $h_1 + h_2$ is given by expression (1). This is to be compared with expression (2), the phase velocity

$$[(\delta\rho/\rho)(gh_1h_2)/(h_1 + h_2)]^{1/2} \qquad (1) \qquad [g(h_1 + h_2)]^{1/2} \qquad (2)$$

for surface waves of great length. Because of the factor $(\delta\rho/\rho)^{1/2}$, the internal waves move at a slow speed, of the order of a few knots in the deep oceans. The effect of the rotation of the Earth

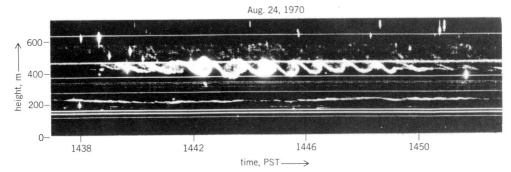

Fig. 2. Internal waves detected in the atmosphere. A narrow radar beam is pointed directly upward, and the height from which echoes are produced is recorded. The prominent waves evident just above 400 m elevation are characteristic of internal waves along a temperature inversion and show signs of breaking induced by wind shear. Radar methods can also detect internal waves at elevations above 10 km. 1 m = 3.3 ft.

is to increase the phase velocity of waves having periods long enough to approach one pendulum day.

When there is a continuous distribution of density in the fluid, as in the ocean or atmosphere, internal waves are possible only for frequencies lower than the value of expression (3),

$$(2\pi)^{-1}\left[g\frac{d}{dz}(\ln \rho) - \frac{g^2}{c^2}\right]^{1/2} \qquad (3)$$

called the Väisälä-Brunt frequency, where $d(\ln \rho)/dz$ is the maximum downward rate of increase of the logarithm of density and c is the velocity of sound in the fluid. In the ocean this maximum frequency occurs in the thermocline, where it commonly amounts to about one-fifth cycle per minute. At any frequency lower than this limit, there is an infinity of possible modes of internal waves. In the first mode, the vertical motion has a single maximum somewhere in the body of the fluid; in the second mode, there are two such maxima (180° out of phase), with a node between; and so on. The actual motion usually consists of a superposition of modes.

Internal waves in the atmosphere have been detected by a variety of instruments: microbarographs and wind recorders at ground level, and long-term recordings of the scattering of radar or sonar beams by sharp density gradients in the high atmosphere (**Fig. 2**). In the ocean, internal waves have been found by recording fluctuating currents in mid-depths by moored current meters and by studies of the fluctuations of the depths of isotherms as recorded by instruments repeatedly lowered from shipboard or by autonomous instruments floating deep in the water.

Internal waves are thought to be generated in the sea by variations of the wind pressure and stress at the sea surface, by the interaction of surface waves with each other, and by the interaction of tidal motions with the rough sea floor. The importance of internal waves is that they can transmit energy and momentum throughout the ocean, not only laterally but also vertically. They can, therefore, transmit energy from the surface to all depths. In this way the otherwise sluggishly moving water at great depths can be agitated.

HYDRAULIC JUMP
Donald R. F. Harleman

An abrupt increase of depth in a free-surface liquid flow. A hydraulic jump is characterized by rapid flow and small depths on the upstream side, and by larger depths and smaller velocities on the downstream side. A jump can form only when the upstream flow is supercritical, that is, when the fluid velocity is greater than the propagation velocity c of a small, shallow-water gravity wave ($c = \sqrt{gh}$, where h is the depth). A considerable amount of energy is dissipated in the conversion from supercritical to subcritical flow. *See* OPEN CHANNEL; WAVE MOTION IN LIQUIDS.

WATER HAMMER
VICTOR L. STREETER

A phenomenon in a piping system caused by an abrupt change in flow at some section. The disturbance, which might be caused by changing a valve opening, travels throughout the system at the speed of a sound wave in the fluid. This acoustic speed depends upon the properties of the fluid and the pipe wall material and thickness, and is given by Eq. (1), where a is the acoustic

$$a = \sqrt{\frac{K/\rho}{1 + \frac{K}{E}\frac{D}{e}}} \quad (1)$$

wave speed in ft/s or m/s, K is the bulk modulus of water in lbf/ft² or pascals, ρ is the density of water in slugs/ft³ or kg/m³, E is Young's modulus for the pipe wall material in lbf/ft² or pascals, D is the inside diameter of the pipe in feet or meters, and e is the pipe wall thickness in feet or meters.

For water at ordinary temperatures $K \cong 300{,}000$ lb/in.² (2.07 gigapascals); $\rho \cong 1.935$ slugs/ft³ (1000 kg/m³); E for steel $\cong 3 \times 10^7$ lbf/in.² (207 GPa). With pipe diameter of 3.937 in. (10 cm) and wall thickness of 0.157 in. (4 mm), the acoustic speed is 4200 ft/s (1287 m/s).

Propagation. In the simplest case, consider water flowing in a closed circuit from a reservoir with a valve at the far end. **Figure 1** shows the pressure-time relationship for the water

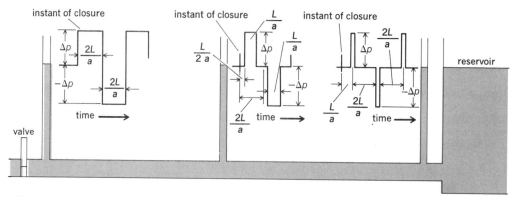

Fig. 1. Surge pressures along a conduit for instantaneous closure. (*ASME*)

hammer at three points in the conduit when the valve is closed in zero time. The magnitude of the pressure rise Δp is discussed below.

The elastic effects of the water and pipe walls affect the conditions, and all of the water will not be uniformly retarded. The closure will initially affect only the portions of the flow near the valve. These portions will be compressed, and the adjacent walls will be expanded by the pressure caused by the closure (**Fig. 2**). The increased pressure will require some time to travel along the pipe before the effect can extend to distant portions of the flow. The rate of travel is provided by the surge-wave velocity a, given by Eq. (1).

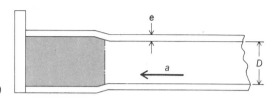

Fig. 2. Example of expansion of conduit walls and fluid compression because of water hammer. (*ASME*)

Magnitude of disturbance. There is a fixed relationship between the change in velocity in a pipe and the magnitude of the resultant pressure disturbance, given by Eq. (2), where Δp is

$$\Delta p = \rho a\, \Delta V \tag{2}$$

the change in pressure in lbf/ft² (pascals) and ΔV is change in velocity in the pipe in ft/s (m/s). Any change in flow taking place within a critical time will produce an instantaneous water hammer, given by Eq. (2). For the simple conduit discussed above, this critical time is given by Eq. (3), where μ is the critical time in seconds and L is the total length of the conduit in feet or

$$\mu = 2L/a \tag{3}$$

meters. For example, if a valve is slammed on a velocity of 3.28 ft/s (1 m/s) within the critical time, for the above case (water at ordinary temperatures, acoustic speed of 4200 ft/s or 1287 m/s), a pressure rise of 187 lbf/in.² (1.29 MPa) results on the upstream side of the valve, and a corresponding reduction downstream from the valve (unless this reduction would take the pressure below vapor pressure of the water).

Boundary conditions. The magnitude of the pressure wave travels unaltered (except for fluid friction) until it arrives at a change in pipe characteristics, such as a branch, a size change, an orifice, or perhaps a reservoir or a pump. Such a change in conditions is referred to as a boundary condition in the problem solution, and the properties and characteristics of the boundary are solved simultaneously with the pipe disturbance equation to determine how the wave proceeds. *See* HYDRODYNAMICS; PIPE FLOW.

Bibliography. V. L. Streeter and E. B. Wylie, *Fluid Mechanics*, 8th ed., 1985; E. B. Wylie and V. L. Streeter, *Fluid Transients*, 1978, reprint 1983.

SHOCK WAVE
SHAO-CHI LIN

A mechanical wave of large amplitude, propagating at supersonic velocity, across which pressure or stress, density, particle velocity, temperature, and related properties change in a nearly discontinuous manner. Unlike acoustic waves, shock waves are characterized by an amplitude-dependent wave velocity. Shock waves arise from sharp and violent disturbances generated from a lightning stroke, bomb blast, or other form of intense explosion, and from steady supersonic flow over bodies. This article discusses shock waves in gases.

The abrupt nature of a shock wave in a gas can best be visualized from a schlieren photograph or shadow graph of supersonic flow over objects. Such photographs show well-defined surfaces in the flow field across which the density changes rapidly, in contrast to waves within the range of linear dynamic behavior of the fluid. Measurements of fluid density, pressure, and temperature across the surfaces show that these quantities always increase along the direction of flow, and that the rates of change are usually so rapid as to be beyond the spatial resolution of most instruments. These surfaces of abrupt change in fluid properties are called shock waves or shock fronts. *See* SCHLIEREN PHOTOGRAPHY; SHADOWGRAPH OF FLUID FLOW; WAVE MOTION IN FLUIDS.

Shock waves in supersonic flow may be classified as normal or oblique according to whether the orientation of the surface of abrupt change is perpendicular or at an angle to the direction of flow. A schlieren photograph of a supersonic flow over a blunt object is shown in **Fig. 1**. Although this photograph was obtained from a supersonic flow over a stationary model in a shock tube, the general shape of the shock wave around the object is quite typical of those

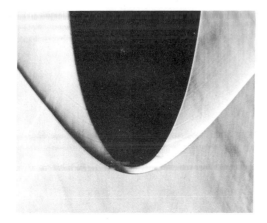

Fig. 1. Schlieren photograph of supersonic flow over blunt object. Shock wave is approximately parabolic, and detached from object. (*Avco Everett Research Laboratory, Inc.*)

observed in a supersonic wind tunnel, or of similar objects (or projectiles) flying at supersonic speeds in a stationary atmosphere. The shock wave in this case assumes an approximately parabolic shape and is clearly detached from the blunt object. The central part of the wave, just in front of the object, may be considered an approximate model of the normal shock; the outer part of the wave is an oblique shock wave of gradually changing obliqueness and strength. For a discussion of experimental techniques for supersonic flow SEE WIND TUNNEL.

Normal shock wave. The changes in thermodynamic variables and flow velocity across the shock wave are governed by the laws of conservation of mass, momentum, and energy, and also by the equation of state of the fluid. For the case of normal shock, the inset in **Fig. 2** illustrates a steady flow across a stationary wavefront (as in the case of a stationary model in a shock tube). The mass flow and momentum equations are the same as for an acoustic wave. However, in a shock wave, changes in pressure p and density ρ across the wavefront can no longer be considered small. As a consequence, the velocity of propagation of the shock wave relative to the undisturbed fluid is given by Eq. (1), where the initial state of the fluid is denoted by subscript 1

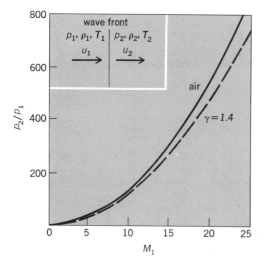

Fig. 2. Pressure ratio $p_2 p_1$ across normal shock wave in air at standard atmospheric density as a function of Mach number M_1. Inset shows variables in case of steady flow across a stationary wavefront.

and variables behind the shock front are denoted by subscript 2. In addition, conservation of thermal and kinetic energy across the shock front requires the validity of Eq. (2), where h is the

$$u_1^2 = \frac{p_2(p_2 - p_1)}{\rho_1(\rho_2 - \rho_1)} \quad (1) \qquad h_1 + 1/2 u_1^2 = h_2 + 1/2 u_2^2 \quad (2)$$

specific enthalpy (or total heat per unit mass) of the fluid, and u_1 and u_2 are fluid velocities relative to the shock wave. By eliminating u_2 and u_1 with the aid of Eq. (1) and the law of conservation of mass, the energy equation becomes Eq. (3). If the thermodynamic properties of the fluid are

$$h_2 - h_1 = 1/2 \left(\frac{1}{\rho_1} + \frac{1}{\rho_2}\right)(p_2 - p_1) \quad (3)$$

known, specific enthalpy h can be expressed as a function of pressure and density, or of any other pair of thermodynamic variables. Equations (1) and (3), together with the appropriate equation of state of the fluid, are known as the Rankine-Hugoniot equations for normal shock waves. From this set of equations, all thermodynamic variables behind the shock front (denoted by subscript 2) can be expressed as functions of the propagation velocity of the shock wave and the known initial state of the fluid (denoted by subscript 1). For example, if the fluid is a perfect gas of constant specific heats, enthalpy h can be written as Eq. (4), where C_p is the specific heat at constant

$$h = C_p T = \frac{\gamma}{\gamma - 1}\frac{p}{\rho} = \frac{a^2}{\gamma - 1} \quad (4)$$

pressure, T is the thermodynamic temperature, γ is the ratio of specific heats and a is the adiabatic speed of sound given by $(\gamma RT)^{1/2}$, where R is the gas constant. For this case, the pressure and density ratios across the shock front are given by Eqs. (5) and (6). The temperature ratio, deduced from the perfect gas law, is given by Eq. (7). These expressions show that the ratios of

$$\frac{p_2}{p_1} = \frac{2\gamma M_1^2 - (\gamma - 1)}{\gamma + 1} \quad (5) \qquad \frac{\rho_2}{\rho_1} = \frac{u_1}{u_2} = \frac{(\gamma + 1)M_1^2}{2 + (\gamma - 1)M_1^2} \quad (6)$$

$$\frac{T_2}{T_1} = \frac{[2\gamma M_1^2 - (\gamma - 1)][2 + (\gamma - 1)M_1^2]}{(\gamma + 1)^2 M_1^2} \quad (7)$$

all thermodynamic variables and flow velocities across the shock depend on only one parameter for a given gas, which is the Mach number M_1 of the flow relative to the shock front, where $M_1 = u_1/a_1$, or in other words, M_1 equals the velocity of the shock wave divided by the speed of sound for the gas into which the shock propagates. Because of this, the magnitude of M_1 is often used as a measure of the strength of the shock. For comparison with the amplitude of acoustic waves, sound waves correspond to values of $M_1 \cong 1.001$ or less. *See* Gas; Mach number.

The results of Eqs. (5), (6), and (7) have been derived for gases of constant specific heats. From molecular and atomic physics, it is well known that, when a gas is heated to high temperatures, vibrational excitation, dissociation, and ionization take place, with accompanying changes in heat capacities of the gas. Therefore, for strong shock waves, the appropriate expression for the specific enthalpy h, and the equation of state, which takes into account these phenomena, must be used in place of Eq. (4) to obtain the shock wave solution from Eqs. (1) and (3). The ratios p_2/p_1, ρ_2/ρ_1, and T_2/T_1 for normal shock waves in air at standard atmospheric density are plotted in Fig. 2 and **Figs. 3** and **4**. The approximate solutions, as given by Eqs. (6) and (7), hold only for the weaker shock waves ($M_1 < 6$), even though the pressure ratio is relatively insensitive to the changes in heat capacities of the gas.

Because the Rankine-Hugoniot equations do not impose any limit on the value of M_1, there remains the question of whether a shock wave can propagate into an undisturbed gas at a speed somewhat lower than the speed of sound of this gas, as would be the case for $M_1 < 1$. Although this question cannot be answered by the first law of thermodynamics, an examination of the

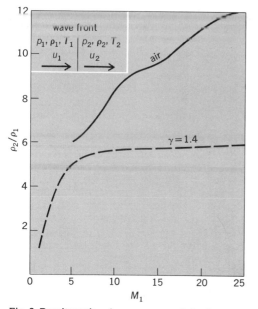

Fig. 3. Density ratio ρ_2/ρ_1 across normal shock wave in air at standard atmospheric density as a function of Mach number M_1. Inset shows variables in case of steady flow across a stationary wavefront.

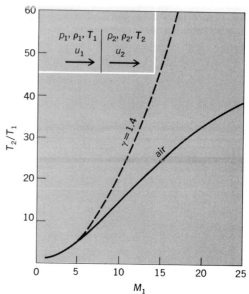

Fig. 4. Temperature ratio T_2/T_1 across normal shock wave in air at standard atmospheric density as a function of Mach number M_1. Inset shows variables in case of steady flow across a stationary wavefront.

change in specific entropies across the shock front provides an answer. Thus Eq. (8) holds, which, for gases of constant specific heats, becomes Eq. (9), showing that entropy change ΔS_{12} assumes

$$\Delta S_{12} = S_2 - S_1 = \int_1^2 \frac{dE + p\,dv}{T} \tag{8}$$

$$\frac{\Delta S_{12}}{R} = \frac{1}{\gamma - 1} \ln \frac{[2\gamma M_1^2 - (\gamma - 1)][2 + (\gamma - 1)M_1^2]^\gamma}{(\gamma + 1)^{\gamma+1} M_1^{2\gamma}} \tag{9}$$

a negative value when M_1 is noticeably less than unity. This violates the second law of thermodynamics, which states that the entropy accompanying any naturally occurring processes always tends to increase. Therefore, it follows that shock waves always travel at supersonic speeds relative to the fluids into which they propagate.

Oblique shock wave. The changes in flow variables across an oblique shock wave are also governed by the laws of conservation of mass, momentum, and energy in a coordinate system which is stationary with respect to the shock front. In this case, the problem is slightly complicated by the fact that the flow velocity will experience a sudden change of direction as well as magnitude in crossing the shock front. Thus, if β_1 and β_2 denote the acute angles between the initial and final flow velocity vectors and the shock surface (**Fig. 5**), then in crossing the oblique shock, the flow will be deflected by a finite amount $\theta = \beta_1 - \beta_2$.

The oblique shock solution can be obtained directly from the complete set of conservation equations. However, the solution already obtained for normal shock waves provides the following simplifying information.

The rate of mass flow per unit area across the shock wave is determined by the normal component of the flow velocity (Fig. 5). Thus, for conservation of mass across the shock, Eq. (10) holds. On the other hand, conservation of the parallel component of momentum across the shock

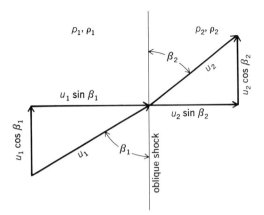

Fig. 5. Flow across an oblique shock wave.

front requires that Eq. (11) hold. Equations (10) and (11) show that Eq. (12) holds. It is equivalent

$$\rho_1 u_1 \sin \beta_1 = \rho_2 u_2 \sin \beta_2 \quad (10) \qquad \rho_1 u_1^2 \sin \beta_1 \cos \beta_1 = \rho_2 u_2^2 \sin \beta_2 \cos \beta_2 \quad (11)$$

$$u_1 \cos \beta_1 = u_2 \cos \beta_2 \quad (12)$$

to the statement that the tangential component of the flow velocity must remain unchanged in crossing the oblique shock wave. Therefore, the resultant flow across the oblique shock shown in Fig. 5 will be identical to what would be seen by an observer who moved at a uniform velocity $u_1 \cos \beta_1$ along the surface of a normal shock wave propagating at a velocity $u_1 \sin \beta_1$. Such a translation of the frame of reference in the direction parallel to the shock front should not change the strength of the shock wave; thus, the changes in thermodynamic variables across the shock should depend only on the velocity component normal to the shock wave. The substitution of M_1 in Eqs. (5), (6), (7), and (9) with $M_1 \sin \beta_1$ gives the corresponding expressions for oblique shock waves in gases of constant specific heats. Again, thermodynamic considerations show that the normal component of the flow velocity into the oblique shock wave must be at least sonic. Therefore, in a supersonic stream of Mach number $M_1 > 1$, the value of β_1 must lie within the range given by expression (13). The lower limit corresponds to the Mach angle of acoustic waves, while

$$\sin^{-1} \frac{1}{M_1} \leq \beta_1 \leq \frac{\pi}{2} \quad (13)$$

the upper limit corresponds to that of the normal shock wave. At both limits, the flow deflection angle $\theta = \beta_1 - \beta_2$ will be zero. For any intermediate value of β_1, the value of β_2 can be obtained from Eqs. (10) and (12). Thus Eq. (14) holds. The flow deflection angle θ for oblique shock waves

$$\beta_2 = \tan^{-1}\left(\frac{\rho_1}{\rho_2} \tan \beta_1\right) \quad (14)$$

in air at normal density is plotted in **Fig. 6** as a function of M_1 and β_1. For a given Mach number M_1, the flow deflection angle first increases with the wave angle β_1, reaches a maximum value θ_{max}, and then decreases toward zero again as the wave angle approaches that of the normal shock. Conversely, for any given flow deflection angle $\theta < \theta_{max}$ such as would be produced by sudden introduction of a wedge or an inclined plane surface into the initially uniform supersonic stream, there exist two possible values of β_1. The higher value of β_1 corresponds to the stronger shock. If the wedge angle or the inclination of the plane surface so introduced exceeds the value of θ_{max} for the given M_1, the shock wave either will become detached from the obstructing object or will form a more complicated pattern. The question of exactly what shock-wave pattern to

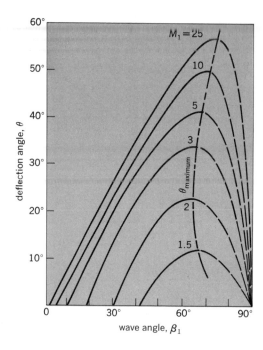

Fig. 6. Flow-detection angle for oblique shock waves in air at standard atmospheric density.

expect from a given situation is complicated by interaction of the resultant shock wave and flow pattern on the overall boundary condition as well as on the local state of the flow.

Bomb blast. When energy is suddenly released into a fluid in a concentrated form, such as by a chemical or a nuclear explosion, the local temperature and pressure may rise instantly to such high values that the fluid tends to expand at supersonic speed. When this occurs, a blast wave forms, and propagates the excess energy from the point of explosion to distant parts of the fluid. If the point of explosion is far from any fluid boundary, the blast wave assumes the form of an expanding spherical shock wave followed by a radially expanding fluid originating from the point of detonation. The changes in thermodynamic variables across the spherical shock are the same as those for a normal shock propagating at the same instantaneous velocity. However, because of the continuous expansion and the finite amount of energy available from the explosion, both the strength of the shock and the specific energy of the expanding fluid must decay with time. The decay of a blast wave goes through three principal stages. A strong shock period begins immediately after the formation of the blast wave, during which the shock strength decays rapidly with distance from the point of detonation. During this stage, the shock velocity decays with the inverse 3/2 power of the distance, and the overpressure behind the shock decays with the inverse cube of the distance. The second stage is a transition period, during which the strong spherical shock gradually changes into an acoustic wave. During the last stage or residual acoustic decay period, the acoustic wave carries the sound of explosion great distances from the point of detonation. As characteristic of sound propagation in three dimensions, the overpressure carried by the spherical acoustic wave decays inversely with distance, and velocity of propagation is constant.

Bibliography. J. D. Anderson, *Modern Compressible Flow: With Historical Perspective*, 1982; I. I. Glass, *Shock Waves and Man*, 1974; M. A. Liberman and A. L. Velikovich, *Physics of Shock Waves in Gases and Plasmas*, 1986; A. Lifshitz, (ed.), *Shock Waves in Chemistry*, 1981; S. M. Yahya, *Fundamentals of Compressible Flow*, 1982; Ya. B. Zeldovich and Yu. P. Raizer, *Physics of Shock Waves and High-Temperature Phenomena*, vols. 1 and 2, 1966.

7

SIMILITUDE

Similitude	152
Dimensional analysis	152
Dynamic similarity	157
Dimensionless groups	158
Reynolds number	172
Mach number	174
Froude number	174
Model theory	174

SIMILITUDE
FRANK H. ROCKETT

The use in scientific studies and engineering designs of the corresponding behavior between large and small objects of similar nature. Two structures behave similarly if they are geometrically, kinematically, and dynamically similar. For geometric similarity, the ratios of critical dimensions, such as ratios of diameters to lengths, must be equal. For kinematic similarity, corresponding velocities and velocity gradients must be in the same ratios at corresponding locations. For dynamic similarity, ratios of forces acting within the two structures, such as viscosity and inertia, must be equal. *See* DYNAMIC SIMILARITY

As an example, cavitation occurs in a flowing liquid when the total pressure falls below the vapor pressure. To test a hydraulic turbine for cavitation may be expensive and cumbersome, so a small model is built and tested. In scaling the model, the geometrical dimensions of the prototype are reduced and a fluid is used with a correspondingly scaled vapor pressure, or the operating pressure is scaled to preserve the relations between the characteristics that affect the behavior of the turbine and the model. *See* CAVITATION; DIMENSIONAL ANALYSIS; FLUID MECHANICS; MODEL THEORY.

Model tests in wind and water tunnels, towing tanks, dynamometers, antenna test ranges, and plasma reacting with magnetic fields are predicated on similitude relations between the model being tested and the full-sized object being studied. The use of small models greatly increases the speed with which design changes can be explored. Where explosive reactions may occur or where the structure is tested to failure, for example, the use of small models reduces hazards. Considerable power is saved and other economies are achieved in the use of small models.

DIMENSIONAL ANALYSIS
JOHN W. STEWART

A technique that involves the study of dimensions of physical quantities. Dimensional analysis is used primarily as a tool for obtaining information about physical systems too complicated for full mathematical solutions to be feasible. It enables one to predict the behavior of large systems from a study of small-scale models. It affords a convenient means of checking mathematical equations. Finally, dimensional formulas provide a useful cataloging system for physical quantities.

Theory. All the commonly used systems of units in physical science have the property that the number representing the magnitude of any quantity (other than purely numerical ratios) varies inversely with the size of the unit chosen. Thus, if the length of a given piece of land is 300 ft, its length in yards is 100. The ratio of the magnitude of 1 yd to the magnitude of 1 ft is the same as that of any length in feet to the same length in yards, that is, 3. The ratio of two different lengths measured in yards is the same as the ratio of the same two lengths measured in feet, inches, miles, or any other length units. This universal property of unit systems, often known as the absolute significance of relative magnitude, determines the structure of all dimensional formulas.

In defining a system of units for a branch of science such as mechanics or electricity, certain quantities are chosen as fundamental and others as secondary, or derived. The choice of the fundamental units is always arbitrary and is usually made on the basis of convenience in maintaining standards. In mechanics the fundamental units most often chosen are mass, length, and time. Standards of mass (the standard kilogram) and of length (the standard meter) are readily manufactured and preserved, while the rotation of the Earth gives a sufficiently reproducible standard of time. Secondary quantities such as velocity, force, and momentum are obtained from the primary set of quantities according to a definite set of rules.

Assume that there are three primary, or fundamental, quantities α, β, and γ (the following discussion, however, is not limited to there being exactly three fundamental quantities). Consider a particular secondary quantity, expressed in terms of the primaries as $F(\alpha,\beta,\gamma)$ where F represents some mathematical function. For example, if α = mass, β = length, and γ = time, the derived quantity velocity would be $F(\alpha,\beta,\gamma) = \beta/\gamma$.

Now, if it is assumed that the sizes of the units measuring α, β, and γ are changed in the proportions $1/x$, $1/y$, and $1/z$, respectively, then the numbers measuring the primary quantities become $x\alpha$, $y\beta$, and $z\gamma$, and the secondary quantity in question becomes $F(x\alpha, y\beta, z\gamma)$. Merely changing the sizes of the units must not change the rule for obtaining a particular secondary quantity.

Consider two separate values of the secondary quantity $F(\alpha_1, \beta_1, \gamma_1)$ and $F(\alpha_2, \beta_2, \gamma_2)$. Then, according to the principle of the absolute significance of relative magnitude, Eqs. (1a) and (1b) hold.

$$\frac{F(x\alpha_1, y\beta_1, z\gamma_1)}{F(x\alpha_2, y\beta_2, z\gamma_2)} = \frac{F(\alpha_1, \beta_1, \gamma_1)}{F(\alpha_2, \beta_2, \gamma_2)} \quad (1a) \qquad F(x\alpha_1, y\beta_1, z\gamma_1) = F(x\alpha_2, y\beta_2, z\gamma_2)\frac{F(\alpha_1, \beta_1, \gamma_1)}{F(\alpha_2, \beta_2, \gamma_2)} \quad (1b)$$

Differentiating partially with respect to x, holding y and z constant, gives Eq. (2), where F' represents the total derivative of the function F with respect to its first argument.

$$\alpha_1 F'(x\alpha_1, y\beta_1, z\gamma_1) = \alpha_2 F'(x\alpha_2, y\beta_2, z\gamma_2)\frac{F(\alpha_1, \beta_1, \gamma_1)}{F(\alpha_2, \beta_2, \gamma_2)} \quad (2)$$

Next set the coefficients x, y, $z = 1$. This gives Eq. (3). This relation must hold for all values

$$\frac{\alpha_1 F'(\alpha_1, \beta_1, \gamma_1)}{F(\alpha_1, \beta_1, \gamma_1)} = \frac{\alpha_2 F'(\alpha_2, \beta_2, \gamma_2)}{F(\alpha_2, \beta_2, \gamma_2)} \quad (3)$$

of the arguments α, β, and γ and hence is equal to a constant. The subscripts can now be dropped, giving Eq. (4). The general solution of this differential equation is $F = C_1 \alpha^a$, where a is

$$\frac{\alpha}{F}\frac{dF}{d\alpha} = \text{constant} \quad (4)$$

a constant and C_1 is in general a function of β and γ.

The above analysis can now be repeated for the parameters β and γ, leading to the results given in Eqs. (5a) and (5b). These solutions are consistent only if $F = C\alpha^a\beta^b\gamma^c$, where C, a, b,

$$F = C_2(\alpha, \gamma)\beta^b \quad (5a) \qquad\qquad F = C_3(\alpha, \beta)\gamma^c \quad (5b)$$

and c are constants. Thus every secondary quantity which satisfies the condition of the absolute significance of relative magnitude is expressible as a product of powers of the primary quantities. Such an expression is known as the dimensional formula of the secondary quantity. There is no requirement that the exponents a, b, c be integral.

Examples of dimensional formulas. Table 1 gives the dimensional formulas of a number of mechanical quantities in terms of their mass M, length L, and time T.

In order to extend this list to include the dimensional formulas of quantities from other branches of physics, such as electricity and magnetism, one may do either of the following:

1. Obtain the dimensional formulas in terms of a particular unit system without introducing any new fundamental quantities. Thus, if one uses Coulomb's law in the centimeter-gram-second electrostatic system in empty space (dielectric constant = 1), one obtains definitions (6a) and (6b).

$$\text{Force} = \frac{(\text{charge})^2}{(\text{distance})^2} \quad (6a) \qquad \text{Charge} = \text{distance} \times \text{square root of force} \quad (6b)$$

Thus, the dimensions of charge are $M^{1/2}L^{3/2}T^{-1}$. The dimensional formulas of other electrical quantities in the electrostatic system follow directly from definition (7) and so forth.

$$\begin{aligned} \text{Potential} &= \text{energy/charge} = M^{1/2}L^{1/2}T^{-1} \\ \text{Current} &= \text{charge/time} = M^{1/2}L^{3/2}T^{-2} \\ \text{Electric field} &= \text{force/charge} = M^{1/2}L^{-1/2}T^{-1} \end{aligned} \quad (7)$$

2. Introduce an additional fundamental quantity to take account of the fact that electricity and magnetism encompass phenomena not treated in mechanics. All electrical quantities can be defined in terms of M, L, T, and one other, without resorting to fractional exponents and without

Table 1. Dimensional formulas of common quantities

Quantity	Definition	Dimensional formula
Mass	Fundamental	M
Length	Fundamental	L
Time	Fundamental	T
Velocity	Distance/time	LT^{-1}
Acceleration	Velocity/time	LT^{-2}
Force	Mass × acceleration	MLT^{-2}
Momentum	Mass × velocity	MLT^{-1}
Energy	Force × distance	ML^2T^{-2}
Angle	Arc/radius	0
Angular velocity	Angle/time	T^{-1}
Angular acceleration	Angular velocity/time	T^{-2}
Torque	Force × lever arm	ML^2T^{-2}
Angular momentum	Momentum × lever arm	ML^2T^{-1}
Moment of inertia	Mass × radius squared	ML^2
Area	Length squared	L^2
Volume	Length cubed	L^3
Density	Mass/volume	ML^{-3}
Pressure	Force/area	$ML^{-1}T^{-2}$
Action	Energy × time	ML^2T^{-1}
Viscosity	Force per unit area per unit velocity gradient	$ML^{-1}T^{-1}$

the artificial assumption of unit (and dimensionless) dielectric constant as in the electrostatic system. The usual choice for the fourth fundamental quantity is charge A, even though charge is not a preservable electrical standard. Then definitions (8) and so forth hold.

$$\text{Potential} = \text{energy/charge} = ML^2T^{-2}Q^{-1}$$
$$\text{Current} = \text{charge/time} = QT^{-1} \qquad (8)$$

It must be realized that the choice of fundamental quantities is entirely arbitrary. For example, a system of units for mechanics has been proposed in which the velocity of light is taken as a dimensionless quantity, equal to unity in free space. All velocities are then dimensionless. All mechanical quantities can then be specified in terms of just two fundamental quantities, mass and time. This is analogous to the reduction of the number of fundamental quantities needed for electricity from four to three by taking the dielectric constant (permittivity) of empty space as unity. SEE DIMENSIONLESS GROUPS.

Furthermore, one could increase the number of fundamental quantities in mechanics from three to four by adding force F to the list and rewriting Newton's second law as $F = Kma$, where K is a constant of dimensions $M^{-1}L^{-1}T^2F$. One could then define a system of units for which K was not numerically equal to 1.

In the past there has been considerable controversy as to the absolute significance, if any, of dimensional formulas. The significant fact is that dimensional formulas always consist of products of powers of the fundamental quantities.

Applications. The important uses of the technique of dimensional analysis are considered in the following sections.

Checking of equations. It is intuitively obvious that only terms whose dimensions are the same can be equated. The equation 10 kg = 10 m/s, for example, makes no sense. A necessary condition for the correctness of any equation is that the two sides have the same dimensions. This is often a help in the verification of complicated analytic expressions. Of course, an equation can be correct dimensionally and still be wrong by a purely numerical factor.

A corollary of this is that one can add or subtract only quantities which have the same dimensions (except for the trivial case which arises when two different equations are added together). Furthermore, the arguments of trigonometric and logarithmic functions must be dimensionless; otherwise their power series expansions would involve sums of terms with different dimensions. There is no restriction on the multiplication and division of terms whose dimensions are different.

π Theorem. The application of dimensional analysis to the derivation of unknown relations depends upon the concept of completeness of equations. An expression which remains formally true no matter how the sizes of the fundamental units are changed is said to be complete. If changing the units makes the expression wrong, it is incomplete. For a body starting from rest and falling freely under gravity, $s = 16t^2$ is a correct expression only so long as the distance fallen s is measured in feet and the time t in seconds. If s is in meters and t in minutes, the equation is wrong. Thus, $s = 16t^2$ is an incomplete equation. The constant in the equation depends upon the units chosen. To make the expression complete, the numerical factor must be replaced by a dimensional constant, the acceleration of gravity g. Then $s = \frac{1}{2}gt^2$ is valid no matter how the units of length and time are changed, since the numerical value of g can be changed accordingly.

Assume a group of n physical quantities $x_1, x_2, \ldots, x_n$, for which there exists one and only one complete mathematical expression connecting them, namely, $\phi(x_1, x_2, \ldots, x_n) = 0$. Some of the quantities $x_1, x_2, \ldots, x_n$ may be dimensional constants. Assume further that the dimensional formulas of the n quantities are expressed in terms of m fundamental quantities $\alpha, \beta, \gamma, \ldots$. Then it will always be found that this single relation ϕ can be expressed in terms of some arbitrary function F of $n - m$ independent dimensionless products $\pi_1, \pi_2, \ldots, \pi_{n-m}$, made up from among the n variables, as in Eq. (9). This is known as the π theorem. It was first

$$F(\pi_1, \pi_2, \ldots, \pi_{n-m}) = 0 \qquad (9)$$

rigorously proved by E. Buckingham. The proof is straightforward but long. The only restriction on the π's is that they be independent (no one expressible as products of powers of the others). The number m of fundamental quantities chosen in a particular case is immaterial so long as they also are dimensionally independent. Increasing m by 1 always increases the number of dimensional constants (and hence n) by 1, leaving $n - m$ the same.

The main usefulness of the π theorem is in the deduction of the form of unknown relations. The successful application of the theorem to a particular problem requires a certain amount of shrewd guesswork as to which variables $x_1, x_2, \ldots, x_n$ are significant and which not. If $\phi(x_1, x_2, \ldots, x_n)$ is not known one can often still deduce the structure of $F(\pi_1, \pi_2, \ldots, \pi_{n-m}) = 0$ and so obtain useful information about the system in question.

An example is the swinging of a simple pendulum. Assume that the analytic expression for its period of vibration is unknown. Choose mass, length, and time as the fundamental quantities. Thus $m = 3$. One must make a list of all the parameters pertaining to the pendulum which might be significant. If the list is incomplete, no useful information will be obtained. If too many quantities are included, the derived information is less specific than it might otherwise be. The quantities given in **Table 2** would appear to be adequate.

The microscopic properties of the bob and string are not considered, and air resistance and the mass of the string are likewise neglected. These would be expected to have only a small

Table 2. Dimensional formulas for a simple pendulum

Quantity	Symbol	Dimensional formula
Mass of bob	m	M
Length of string	l	L
Acceleration of gravity	g	LT^{-2}
Period of swing	τ	T
Angular amplitude	θ	0

effect on the period. Thus $n = 5$, $n - m = 2$, and therefore one expects to find two independent dimensionless products. These can always be found by trial and error. There may be more than one possible set of independent π's. In this case one π is simply θ, the angular amplitude. Another is $l/\tau^2 g$. Since no other of the n variables contains M in its dimensional formula, the mass of the bob m cannot occur in any dimensionless product. The π theorem gives $F(\theta, l/\tau^2 g) = 0$. Therefore the period of vibration does not depend upon the mass of the bob.

Because τ appears in only one of the dimensionless products, this expression can be explicitly solved for τ to give $\tau = G(\theta)\sqrt{l/g}$, where $G(\theta)$ is an arbitrary function of θ. Now make the further assumption that θ is small enough to be neglected. Then $n = 4$, $n - m = 1$, and $F(l/\tau^2 g) = 0$. Thus $l/\tau^2 g = $ constant, or $\tau \propto \sqrt{l/g}$. The π theorem thus leads to the conclusion that τ varies directly as the square root of the length of the pendulum and inversely as the square root of the acceleration of gravity. The magnitude of the dimensionless constant (actually it is 2π) cannot be obtained from dimensional analysis.

In more complicated cases, where a direct solution is not feasible, this method can give information on how certain variables enter a particular problem even where $F(\pi_1, \pi_2, \ldots, \pi_{n-m}) = 0$ cannot be explicitly solved for one variable. The procedure is particularly useful in hydraulics and aeronautical engineering, where detailed solutions are often extremely complicated.

Model theory. A further application of dimensional analysis is in model design. Often the behavior of large complex systems can be deduced from studies of small-scale models at a great saving in cost. In the model each parameter is reduced in the same proportion relative to its value in the original system. SEE MODEL THEORY.

Once again the case of the simple pendulum is a good example. It was found from the π theorem that $F(\theta, l/\tau^2 g) = 0$. If the magnitudes of θ, l, τ, and g are now changed in such a way that neither argument of F is changed in numerical value, the system will behave exactly as the original system and is said to be physically similar. Evidently θ cannot be changed without altering any of the arguments of F, but l, g, and τ can be varied. Suppose that it was desired to build a very large and expensive pendulum which was to swing with finite amplitude. One could build a small model of, say, $\frac{1}{100}$ the length and time its swing for an amplitude equal to that for the desired pendulum. The acceleration of gravity g would be the same for the model as for the large pendulum. The period for the model would then be just $\frac{1}{10}$ that for the large pendulum. Thus the period of the large pendulum could be deduced before the pendulum was ever built. In practice one would never bother with the π theorem in cases as simple as this where a full analytic solution is possible. In many situations where such a solution is not feasible, models are built and extensively studied before the full-scale device is constructed. This technique is standard in wind tunnel studies of aircraft design. SEE DYNAMIC SIMILARITY.

Cataloging of physical quantities. Dimensional formulas provide a convenient shorthand notation for representing the definitions of secondary quantities. These definitions depend upon the choice of primary quantities. The π theorem is applicable no matter what the choice of primary quantities is.

Changing units. Dimensional formulas are helpful in changing units from one system to another. For example, the acceleration of gravity in the centimeter-gram-second system of units is 980 cm/s². The dimensional formula for acceleration is LT^{-2}. To find the magnitude of g in mi/h², one would proceed as in Eq. (10).

$$980 \, \frac{\text{cm}}{\text{s}^2} \times \frac{\text{(conversion factor for length)}}{\text{(conversion factor for time)}^2} = 980 \, \frac{\text{cm}}{\text{s}^2} \times \frac{\frac{1}{1.609 \times 10^5} \frac{\text{mi}}{\text{cm}}}{(1/3600)^2 (\text{h}^2/\text{s}^2)} = 7.89 \times 10^5 \, \text{mi/h}^2 \quad (10)$$

In the past the subject of dimensions has been quite controversial. For years unsuccessful attempts were made to find "ultimate rational quantities" in terms of which to express all dimensional formulas. It is now universally agreed that there is no one "absolute" set of dimensional formulas. Some systems are more symmetrical than others and for this reason are perhaps preferable. The representation of electrical quantities in terms of M, L, and T alone through the electrostatic form of Coulomb's law leads to somewhat awkward fractional exponents, but nevertheless is just as "correct" as a representation in which charge is used as a fourth fundamental unit.

A highly symmetrical pattern results if energy, linear displacement, and linear momentum are chosen as the fundamental quantities in mechanics. In electricity one can use energy, charge, and magnetic flux. The corresponding quantities for a vibrating mass on a spring and the analogous alternating-current circuit with inductance and capacitance have similar dimensional formulas. In this analogy energy is invariant, charge corresponds to displacement, and magnetic flux corresponds to linear momentum. This correspondence is not displayed in conventional dimensional formulas.

Bibliography. H. A. Becker, *Dimensionless Parameters: Theory and Methodology*, 1976; P. W. Bridgman, *Dimensional Analysis*, 1931, reprint 1976; H. L. Langhaar, *Dimensional Analysis and Theory of Models*, 1951, reprint 1980; C. I. Staicu, *Restricted and General Dimensional Analysis*, 1982.

DYNAMIC SIMILARITY
ARTHUR G. HANSEN

A relationship existing between two homologous fluid-flow systems such that corresponding parts of the systems experience similar net forces. Dynamically similar flows about geometrically similar bodies will themselves be geometrically similar (see **illus.**). Consequently, this concept is basic to the meaningful extrapolation of model results to full-scale performance. However, geometrically similar flows are not necessarily dynamically similar.

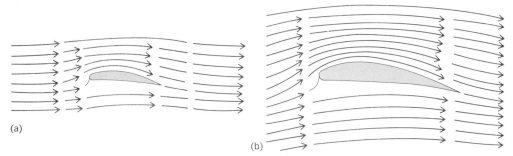

Diagrams of (a) electrical circuit and (b) mechanical system which is analogous to it.

Dynamic similarity between two flow systems will occur if certain nondimensional parameters formed from the flow variables have the same values for both systems.

One important parameter for establishing dynamically similar flows is pressure coefficient $p/\rho V^2$, where p, ρ, and V are, respectively, a reference pressure, density, and velocity. Other important nondimensional parameters are given in the **table** along with associated physical effects which characterize the parameters.

The parameters in the table can be determined analytically by dimensional analysis or by examining the invariance of the differential equations and boundary conditions for the flow systems under scalar transformation of length, time, and mass. Other useful parameters can also be defined. Often these can be obtained as ratios of the tabulated parameters.

In practice, it is often difficult to establish equality of all similarity parameters simultaneously for two flows. Equality of parameters corresponding to dominant flow properties is usually sufficient. Given low-speed viscous flows, for example, the Reynolds numbers of the two flows

Dynamic similarity parameters

Nondimensional parameter*	Name	Physical effect
$\rho VL/\mu$	Reynolds number	Viscosity
V/c	Mach number	Compressibility
V^2/Lg	Froude number	Gravity
x_m/L	Knudsen number	Pressure
$\rho V^2 L/\sigma$	Weber number	Surface tension
$c_p \mu/\kappa$	Prandtl number	Heat conduction
$\beta T g L^3 \rho^2/\mu^2$	Grashof number	Free convection

*The reference variables are defined as follows: L, length; μ, coefficient of viscosity; c, speed of sound; g, acceleration of gravity; σ, surface tension; κ, coefficient of thermal conductivity; β, coefficient of thermal expansion; T, temperature; x_m, mean free path of molecule; and c_p, specific heat at constant pressure.

would be equated, but the Mach numbers might be ignored. SEE DIMENSIONAL ANALYSIS; DIMENSIONLESS GROUPS; FROUDE NUMBER; MACH NUMBER; MODEL THEORY; REYNOLDS NUMBER.

Bibliography. H. L. Langhaar, *Dimensional Analysis and Theory of Models*, 1951, reprint 1980; S. Pai, *Viscous Flow Theory*, vol. 1, 1956; R. H. Sabersky et al., *Fluid Flow*, 2d ed., 1971; C. I. Staicu, *Restricted and General Dimensional Analysis*, 1982.

DIMENSIONLESS GROUPS
JOHN CATCHPOLE AND GEORGE D. FULFORD

A dimensionless group is any combination of dimensional or dimensionless quantities possessing zero overall dimensions. Dimensionless groups are frequently encountered in engineering studies of complicated processes or as similarity criteria in model studies. A typical dimensionless group is the Reynolds number (a dynamic similarity criterion), $N_{Re} = VD\rho/\mu$. Since the dimensions of the quantities involved are velocity V: $[L/\theta]$; characteristic dimension D: $[L]$; density ρ: $[M/L^3]$; and viscosity μ: $[M/L\theta]$ (with M, L, and θ as the fundamental units of mass, length, and time), the Reynolds number reduces to a dimensionless group and can be represented by a pure number in any coherent system of units. SEE DYNAMIC SIMILARITY.

History of named groups. The importance of dimensionless groups has long been known; as early as 1850 G. G. Stokes found the group now known as the Reynolds number and demonstrated its importance as a dynamic similarity criterion. In 1873 H. von Helmholtz showed the importance of the groups now known as the Froude and Mach numbers. The first attempt to attach names to important dimensionless groups appears to have been made in 1919 by M. G. Weber, who first named the Froude, Reynolds, and Cauchy numbers. Later a capillarity group was named after Weber, and it has become customary to name dimensionless groups after the outstanding or pioneer workers in the field. Many groups tabulated below are in this category, and are often denoted by the first two letters of the name; thus, the Reynolds number is usually denoted as N_{Re}.

No systematic procedure has been used for assigning names to dimensionless groups. Some groups have several different names, and some names have been attached to more than one group. To prevent this confusing situation from becoming worse, J. P. Catchpole and G. Fulford have compiled tables to identify unfamiliar groups that have already been named.

Sources and uses. Many important problems in applied science and engineering are too complicated to permit completely theoretical solutions to be found. However, the number of interrelated variables involved can be reduced by carrying out a dimensional analysis to group the variables as dimensionless groups. SEE DIMENSIONAL ANALYSIS.

If almost nothing is known about the problem, it is necessary to decide which physical parameters will affect the process being studied. A dimensional analysis can then be carried out by the well-known π-theorem method of E. Buckingham or by the indicial method proposed by Lord Rayleigh. The success of these methods depends on identifying all of the important parameters for the problem. On the other hand, if the differential equations describing the phenomenon are known, inspectional analysis of the equations will yield all the important dimensionless groups even when the differential equations cannot be solved. The form of inspectional analysis proposed by A. Klinkenberg and H. H. Mooy is particularly useful in studying transport phenomena. Details of the various procedures may be found in any of the usual texts on dimensional analysis, namely, those by H. L. Langhaar, A. A. Gukhman, P. W. Bridgman, and R. C. Pankhurst. All these methods give only the groups which correspond to the assumed parameters or variables. The form of the dimensionless equations linking these groups has to be determined by the use of experimental data in all except the simplest cases. Often there is no better procedure available in the case of complicated phenomena.

The advantages of using dimensionless groups in studying complicated phenomena include:

1. A significant reduction in the number of variables to be investigated; that is, each dimensionless group, containing several physical variables, may be treated as a single compound variable, thereby reducing the number of experiments needed as well as the time required to correlate and interpret the experimental data.

2. Predicting the effect of changing one of the individual variables in a process (which it may be impossible to vary much in available equipment) by determining the effect of varying the dimensionless group containing this parameter (this must be done with some caution, however).

3. Making the results independent of the scale of the system and of the system of units being used.

4. Simplifying the scaling-up or scaling-down of results obtained with models of systems by generalizing the conditions which must exist for similarity between a system and its model.

5. Deducing variation in importance of mechanisms in a process from the numerical values of the dimensionless groups involved; for instance, an increase in the Reynolds number in a flow process indicates that molecular (viscous) transfer mechanisms will be less important relative to transfer by bulk flow (inertia effects), since the Reynolds number is known to represent a measure of the ratio of inertia forces to viscous forces; the significance of each important dimensionless group is listed in the table to aid in such interpretations.

Classification. A large number of independent dimensionless groups may arise from a dimensional analysis of a complex problem; this number usually equals the number of variables in the problem minus the number of fundamental units in the system of units used. Furthermore, since any product or quotient of powers of dimensionless groups is also dimensionless, it follows that an almost limitless number of alternative dimensionless groups can be generated. Several hundred groups are known, many of them named. It is therefore useful to be able to specify the conditions for obtaining a complete statement of a problem in terms of dimensionless groups and to classify the large number of groups in some way.

Complete sets. A set of dimensionless groups of given variables is complete if each group in the set is independent of the others and if every other dimensionless combination of the variables is a product of powers of these groups. For example, a complete set of dimensionless groups for a problem involving the variables L, V, ρ, μ, g, V_s, and σ would be $N_{Re} = VL\rho/\mu$, $N_{Fr_1} = V^2/Lg$, $N_{Ma} = V/V_s$, and $N_{We_1} = V^2\rho L/\sigma$ (quantities are defined at the end of the article). None of these groups is a product of powers of the others, since μ occurs only in N_{Re}, g only in N_{Fr_1}, V_s only in N_{Ma}, and σ only in N_{We_1}. Many other dimensionless groups could be formed from these same variables, such as $V^3\rho/\mu g$, but these are not independent groups because, for instance, $V^3\rho/\mu g = N_{Re} \cdot N_{Fr_1}$. Many of these compound groups have acquired names, however, and a separate existence of their own, since any of the combined groups can be used as alternative for any one of the component groups.

F. V. A. Engel proposed that the groups forming a complete set (as defined above) for a given problem be termed true groups for that problem, and that other groups involving only independent variables in the particular problem which may be formed as combinations of the true groups be termed compound groups. He gave rules for obtaining complete sets of groups for heat transfer and flow problems, and the arguments can be readily extended to other types of problems.

Simplexes and complexes. Sometimes more than one variable of the same physical significance is important in a problem. For instance, in fluid flow through a straight rough pipe, three length parameters (pipe diameter D, pipe length L, and surface roughness ϵ) are required to define the geometry of the system. In such a case, one of these quantities (usually D in the pipe-flow problem) is used as the characteristic quantity in forming the main dimensionless groups (the Reynolds number for pipe flow), and the others occur in the form of ratios to the first (L/D, ϵ/D). These straightforward dimensionless ratios of multiple parameters arising naturally in a problem are often referred to as simplexes in the literature, while all other groups are termed complexes. Many simplexes are unnamed, being referred to simply as geometric ratios or the like, but in the case of velocity ratios a number of named simplexes exist, for example, the Alfvén, Lorentz, and Mach numbers. Simplexes and complexes have also been termed parametric and similarity criteria, respectively.

Independent and dependent groups. Groups formed entirely from parameters which are independent variables in a particular problem have been termed independent or characteristic groups, while those which also involve dependent variables for the given problem are known as dependent or unknown groups. Obviously, some groups which are independent in one problem may become dependent in another, according to which parameters are specified and which are unknown in each case.

In many practical cases, experimental data are correlated to give a relationship between a dependent group and one or more independent groups. Although there are certain philosophical problems involved, this procedure has led to many very useful engineering correlations. Examples are the Moody friction factor chart, which correlates the friction factor f for pipe flow in terms of the Reynolds number and the relative pipe roughness given by Eq. (1) and the familiar correlations for heat and mass transfer given by Eqs. (2) and (3). This type of power relationship is suggested

$$f = f(N_{Re}, \epsilon/D) \quad (1) \qquad N_{Nu} = A \cdot N_{Re}^{a} \cdot N_{Pr}^{b} \quad (2) \qquad N_{Sh} = A' \cdot N_{Re}^{a'} \cdot N_{Sc}^{b'} \quad (3)$$

by the Rayleigh indicial procedure for dimensional analysis. The constants A, a, and b are determined by fitting straight lines to suitable logarithmic plots of experimental data.

A special subclass of independent dimensionless groups contains only the parameters expressing the physical properties of the system being studied. This category includes such groups as the Prandtl numbers, the Schmidt numbers, and the Lewis numbers.

Named groups. The **table** lists all the named dimensionless groups known at present, together with a few other important groups, and includes a definition of each group, its main field of application, and its "significance," where this is important. Source references and additional information can be found in the papers of Catchpole and Fulford. It should be noted that in many cases roots of, or power functions of, these groups may be designated by the same names in the technical literature. SEE CAVITATION; FROUDE NUMBER; GAS DYNAMICS; KNUDSEN NUMBER; MACH NUMBER; REYNOLDS NUMBER.

Alphabetical list of named groups

Name	Symbol	Definition*	Field of use
Absorption no.	Ab	$k_{C_L}\sqrt{\dfrac{xL_f}{DV'_f}}$; k_{C_L} = liquid side mass transfer coeff., x = length of wetted surface, L_f = film thickness, V'_f = volume flow rate per wetted perimeter [L^2/θ]	Gas absorption in wetted wall column
Acceleration no.	$\overline{Kg}$	$E^3/\rho g^2 \mu^2 = (N_{Re} \cdot N_{Fr_1})^2/(Ho)^3$	Accelerated flow
Advance ratio	J	V/ND; V = forward speed, D = propeller diameter	Propeller studies

*For explanation of terms used in equations, see end of article.

Alphabetical list of named groups (cont.)

Name	Symbol	Definition	Field of use
Aeroelasticity parameter	—	$\equiv$ Cauchy no.	Compressible flow
Alfvén no.	N_{AL}	V_A/V (or V/V_A) (cf. Cowling no., Karman no. 2, Magnetic Mach no.)	Magneto fluid dynamics
Anonymous group 1	ϵ	$\beta^* c/r$ (see also Fedorov no. 2)	Transfer processes
Anonymous group 2	K_1	$\beta^* \Delta n/\Delta t$; Δn = concn. difference, Δt = temp. difference [T]	Transfer processes
Anonymous group 3	ϵ	$Dx/V_l L_t$ (symbols as in Absorption no.); $\epsilon = Ab^2/N_{Sh}^2$	Gas absorption in wetted wall column
Anonymous group 4	$1/\alpha$ $(1/\beta)$	$\tau_W R/V_\infty \mu$; R = cylinder radius, V_∞ = velocity outside boundary layer	Laminar boundary layer flow
Archimedes no.	N_{Ar}	$\dfrac{gL^3\rho}{\mu^2}(\rho - \rho_0)$; ρ = fluid density. ρ_0 = particle density (cf. N_{Ga_1})	Fluidization, motion of liquids due to density differences
Arrhenius group	—	$E_a/\mathcal{R}T$	Reaction rates
Bagnold no.	B	$3c_d\rho_g V^2/4dg\rho_p$; ρ_g = gas density, ρ_p = particle density	Saltation studies
Bairstow no.	—	V/V_{sw}; V_{sw} = velocity of sound at wall (cf. Mach no.)	—
Bansen no.	N_{Ba}	$h_r A_W/V_m c$; h_r = radiant heat-transfer coeff, A_W = wall area of channel (cf. N_{St})	Radiation
Batchelor no.	—	$VL\sigma_e/V_e^2 \epsilon_e$; ϵ_e = electrical permittivity $[Q^2\theta^2/L^3 M]$	Magneto fluid dynamics
Bingham no.	N_{Bm}	$\tau_y L/\mu_p V$; L = channel width	Flow of Bingham plastics
Biot no. (heat transfer)	N_{Bi_h}	hL_m/k (in French literature "Biot no." $\equiv N_{Nu}$)	Unsteady-state heat transfer
Biot no. (mass transfer)	N_{Bi_m}	$k_c L/D_{int}$; L = thickness of layer, D_{int} = diffusivity at interface	Mass transfer between fluid and solid
Blake no.	B	$V\rho/\mu S(1-e)$	Beds of particles
Bodenstein no.	N_{Bo}	$VL/D_a = N_{Pe_m}$; L = reactor length, D_a = effective axial diffusivity $[L^2/\theta]$	Diffusion in reactors
Boltzmann no.	N_{Bo}	$\equiv$ Thring radiation group	—
Bond no.	N_{Bo}	$(\rho - \rho')L^2 g/\sigma$; ρ = drop or bubble density, ρ' = medium density; if $(\rho - \rho') \simeq \rho$ (gas in liquid) $N_{Bo} = N_{We_1}/N_{Fr_1}$	Atomization, motion of bubbles and drops
Bouguer no.	N_{Bu}, B	$3C_D\lambda_r/4\rho_D \bar{R}$; C_D = wt dust/unit bed volume $[M/L^3]$, λ_r = mean path for radiation $[L]$, ρ_D = dust density, $\bar{R}$ = mean particle radius; also $N_{Bu} = kL$; L = characteristic dimension, k = adsorption coeff. of medium	Radiant heat transfer to dust-gas streams
Boussinesq no.	B	$V/(2gR_H)^{1/2}$ (cf. N_{Fr_2})	Wave behavior in open channels
Brinkmann no.	N_{Br}	$\mu V^2/k\Delta t$; Δt = temp. difference	Viscous flow
Bulygin no.	N_{Bu}	$r \cdot \dfrac{c_b}{c_q} \dfrac{P}{t_m - t_0} t_m =$ temp. of medium, t_0 = initial temp. of body	Heat transfer during evaporation
Buoyancy parameter	—	$\dfrac{\Delta T}{T}\dfrac{gL}{V^2} = N_{Gr}/N_{Re}^2 = \dfrac{\Delta T}{T \cdot N_{Fr_1}}$	Free convection
Capillarity no.	$\bar{K}_\sigma$	$\mu^2 E/\rho\sigma^2 = (N_{We_1})^2/Ho \cdot (N_{Re})^2$	Action of surface tension in flowing media
Capillarity-buoyancy no.	$\bar{K}_{\sigma g}$, K_F	$g\mu^4/\rho\sigma^3 = (\bar{K}_\sigma^3/\bar{K}_g)_{1/2} = (N_{We_1})^3/(N_{Fr_1})(N_{Re})^4$	Effects of surface tension and acceleration in flowing media (two-phase flow)
Capillary no.	Ca	$\mu V/\sigma = N_{We_1}/N_{Re}$	Atomization, two-phase flow
Carnot no.	Ca, N_{Ca}	$(T_2 - T_1)/T_2$; T_1, T_2 = abs. temp. of two heat sources or sinks	—
Cauchy no.	N_c	$\rho V^2/E_b = (N_{Ma})^2 =$ Hooke no.	Compressible flow

Alphabetical list of named groups (cont.)

Name	Symbol	Definition	Field of use
Cavitation no.	σ_c	$(p - P_V)/\tfrac{1}{2}\rho V^2$; p = local static pressure (abs), P_V = vapor pressure	Cavitation
Clausius no.	Cl, N_{Cl}	$V^3 L \rho / k \Delta T$; ΔT = temp. difference	Heat conduction in forced flows
Colburn no.		Same as Schmidt no. 1	—
Condensation no. 1	N_{Co}	$(h/k)(\mu^2/\rho^2 g)^{1/3}$	Condensation
Condensation no. 2	N_{Cv}	$L^3 \rho^2 g r / k \mu \Delta t$; r = latent heat of condensation	Condensation on vertical walls
Cowling no.	C	$(V_A/V)^2 \equiv$ (Alfvén no.)^2	Magneto fluid dynamics
Craya-Curtet no.	C_t	$V_k/(V_d^2 - V_k^2/2)^{1/2}$; V_k = kinematic mean velocity, V_d = dynamic mean velocity	Radiant heat transfer
Crispation group	N_{Cr}	$\mu\alpha/\sigma^* L$; σ^* = undisturbed surface tension, L = layer thickness	Convection currents
Crocco no.	N_{Cr}	$V/V_{max} = \left[1 + \dfrac{2}{(\gamma-1)(N_{Ma})^2}\right]^{-1/2}$ V_{max} = maximum velocity of gas expanding adiabatically	Compressible flow
Damköhler group I	Da I	UL/Vc_A	Chemical reaction, momentum, heat transfer
Damköhler group II	Da II	UL^2/Dc_A	Chemical reaction, momentum, heat transfer
Damköhler group III	Da III	$QUL/c_p \rho Vt$	Chemical reaction, momentum, heat transfer
Damköhler group IV	Da IV	QUL^2/kt	Chemical reaction, momentum, heat transfer
Damköhler group V	Da V	$= N_{Re}$	—
Darcy no. 1	Da_1	$= 4f$ (see Fanning friction factor)	Fluid friction in conduits
Darcy no. 2	Da_2	VL/D'; D' = permeability coeff. of porous medium $[L^2/\theta]$	Flow in porous media
Dean no.	N_D	$(VL\rho/\mu)(L/2R)^{1/2}$; L = pipe diameter, R = radius of curvature of bend	Flow in curved channels
Deborah no.	D	θ_r/θ_0; θ_0 = observation time	Rheology
Deborah no., generalized	N_2	$(I_e - I_w)^{1/2} \cdot \theta_n$; I_e = invariant of rate of strain tensor $[\theta^{-2}]$, I_w = invariant of vorticity tensor $[\theta^{-2}]$, θ_n = natural time $[\theta]$	Rheology
Delivery no.	ϕ	V_t/Aw; A = impeller area = $\pi d^2/4$	Flow machines
Deryagin no.	De	$L(\rho g/2\sigma)^{1/2}$; L = film thickness	Coating
Diameter group	δ	$(\pi d^2/4)^{1/2}(2H)^{1/4} V_t^{1/2}$; d = impeller diameter	Flow machines
Diffusion group	β	$D\theta/L_m^2$; D = diffusivity of solute through stationary solution contained in solid (cf. N_{Fo_m})	Mass transfer
Drag coeff.	C_d, C_D	$(F/A_p)/(\rho V^2/2)$; F = drag force, A_p = projected particle areas, ρ = medium density, V = relative velocity between particles and density	Drag forces and particle dynamics
Drew no.	N_D	$\dfrac{Z_A(M_A - M_B) + M_B}{(Z_A - Y_{AW})(M_B - M_A)} \cdot \ln\dfrac{M_V}{M_W}$; M_A, M_B = mol wt of components A and B; M_V, M_W = mol wt of mixture in vapor and at wall; Y_{AW} = mole fraction of A at wall; Z_A = mole fraction of A in diffusing stream	Boundary layer mass transfer rates, velocity profile distortion, drag coefficients for binary systems
Dufour no.	Du_2	$\Re \theta n'_{10}/c_p$; θ = thermodiffusion constant = $(D_T/D)/n_{10}n_{20}[-]$; D_T = thermal diffusion coeff $[L^2/\theta]$; $n'_{10}, n'_{20} = n'_1/n', n'_2/n'$ n' = total no. of molecules = $n'_1 + n'_2$; n'_1, n'_2 = no. of molecules of component 1 and 2 in binary mixture; also $Du_2 = (D_T/D)\rho/\rho c_p T n_{20}$	Thermodiffusion

Alphabetical list of named groups (cont.)

Name	Symbol	Definition	Field of use
Dulong no.	Du, N_{Du}	$V^2/c_p\Delta T$ = Eckert no.	—
Eckert no.	N_E	$V_\infty^2/c_p\Delta T$; V_∞ = velocity of fluid far from body; ≡ Dulong no. = 1/(Recovery factor)	Compressible flow
Einstein no.	—	$V/V\ell$; $V\ell$ = speed of light (cf. Lorentz no.)	Magneto fluid dynamics
Ekman no.	—	$(\mu/2\rho\omega L^2)^{1/2} = (N_{Ro}/N_{Re})^{1/2}$	Magneto fluid dynamics
Elasticity no. 1	N_{E_1}	$\theta_r\mu/\rho L^2$; L = pipe radius	Viscoelastic flow
Elasticity no. 2	$\overline{K}_E$	$\rho c_p/\beta E$ = Gay Lussac no. × Hooke no./Dulong no.	Effect of elasticity in flow process
Electric field parameter	R_E	$E/V\mu_e H_e$	Magnetohydrodynamics
Electrical characteristic no.	EI	$\rho\left(\dfrac{dX}{dT}\right) L^2 \Delta T \dfrac{E_1^2}{\mu^2}$; E_1 = electrical field strength $[ML/Q\theta^2]$; X = dielectric susceptibility $[Q^2\theta^2/ML^3]$	Electrical effects on transfer processes
Electrical Nusselt no.	Nu	VL/D^*; $D^* = \frac{1}{2}(D^+ + D^-)$; D^+, D^- = diffusion coeff of ions $[L^2/\theta]$[cf. Péclet no. (mass)]	Electrochemistry
Electrical Reynolds no. 1	—	$\epsilon_e V/Q' Lb'$; ϵ_e = electrical permittivity $[Q^2\theta^2/L^3M]$; Q' = space charge density $[Q/L^3]$; b' = carrier mobility $[Q\theta/M]$	Electrical effects in flow
Electrical Reynolds no. 2		Alternate name for Electrical Nusselt no.	—
Ellis no.	N_{El}	$\mu_o V/2\tau_{1/2}R$; μ_o = zero shear viscosity; $\tau_{1/2}$ = shear stress when $\mu = \mu_o/2$ $[M/L\theta^2]$; R = tube radius	Flow of non-newtonian fluids
Elsasser no.	N_{El}	$\rho/\mu\sigma_e\mu_e \equiv N_{Re}/$Magnetic Reynolds no. 1	Magneto fluid dynamics
Entry Reynolds no.	K_E	$\dfrac{X}{d} \cdot N_{Re} = XV\rho/\mu$; X = entry length	Entry or inlet processes
Eötvös no.	N_{Eo}	$(\rho - \rho')L^2 g/\sigma$ ≡ Bond no.	—
Euler no. 1	N_{Eu_1}	$\Delta P_F/\rho V^2$; ΔP_f = pressure drop due to friction	Fluid friction in conduits
Euler no. 2	N_{Eu_2}	$-(dp/dl) \cdot d/\rho V^2$; d = pipe diameter; (dp/dL) = pressure gradient	Fluid friction in conduits
Modified Euler no.	Eu'	$H_L \rho_L g/V_G^2\rho_G$; H_L = head of liquid on tray $[L]$; V_G = vapor velocity based on free area $[L/\theta]$; ρ_L, ρ_G = density or liquid and vapor	Flow of vapor across mass transfer trays
Evaporation no. 1	K_r	V^2/r; r = heat of vaporization $[L^2/\theta^2]$	Evaporation processes
Evaporation no. 2	K_r	$c_p/r\beta$; r = heat of vaporization $[L^2/\theta^2]$	Evaporation processes
Evaporation elasticity no.	$\overline{K}_{rE}$	$E/r\rho$; r = heat of vaporization $[L^2/\theta^2]$	Evaporation processes
Expansion no.	Ex	$\left(\dfrac{gd}{V^2}\right) \dfrac{(\rho_L - \rho_G)}{\rho_L}$; d = bubble diameter; V = bubble velocity; ρ_L; ρ_G = density of gas and liquid	Rise of bubbles
Fanning friction factor	f	$d\Delta p_F/2\rho V^2 L$; d = dimension of cross section; L = length (cf. Resistance coeff., Ne)	Fluid friction in conduits
Fedorov no. 1	Fe_1, N_{Fe_1}	$d_e \left\{\dfrac{4g\rho^2}{3\mu^2}\left(\dfrac{\gamma_M}{\gamma_g} - 1\right)\right\}^{1/3}$; d_e = equivalent particle diameter; γ_M = specific gravity of particles; γ_g = specific gravity of gas (cf. Archimedes no.)	Fluidized beds
Fedorov no. 2	Fe_2, N_{Fe_2}	$\delta\beta^* = K_1 Pn = \epsilon K_o Pn$	Transport processes
Fenske no.	N_F		—
Fineness coeff.	Ψ	$L/W_D^{1/3}$; W_D = volume displacement $[L^3]$	Ship modeling

Alphabetical list of named groups (cont.)

Name	Symbol	Definition	Field of use
Fliegner numbers	$F(Ma)$ $f(Ma)$ $I(Ma)$	Functions of ratio of specific heats and Mach no. $\dfrac{V_m(cT)^{1/2}}{A(P_s + P\rho V^2)} = \dfrac{\gamma Ma}{(\gamma - 1)^{1/2}}$ $a\left[1 + \dfrac{(\gamma - 1) Ma^2}{2}\right]^{1/2} = $ impulse Fliegner no.; $\gamma = $ ratio of specific heats, $Ma = $ Mach no.; $A = $ flow area	—
Flow coeff	C_Q	V_f/Nd^3; $d = $ impeller diameter	Power required by fans
Fluidization no.	—	V/V_{init}; $V_{init} = $ velocity for initial fluidization	Fluidization
Fourier no. (heat transfer)	N_{Fo_h}	$k\theta/\rho c_p L_m^2$	Unsteady-state heat transfer
Fourier no. (mass transfer)	N_{Fo_m}	$D\theta/L^2 = k_C\theta/L$ (cf. Diffusion group)	Unsteady-state mass transfer
Fourier no. flow	Fo_f	$\mu\theta/\rho L^2$	Unsteady-state flow problems
Frank-Kamenetskii no.	δ	$\dfrac{Q''}{k}\dfrac{E_a}{\mathcal{R}T^2}L^2 k_0 \exp(-E_a/\mathcal{R}T)$; $Q'' = $ heat liberated per unit mass of reacting material per unit volume $[1/L\theta^2]$; $k = $ thermal conductivity of reacting mixture $[ML/T\theta^3]$; $k_o = $ preexponential constant in Arrhenius equation $[M/\theta]$	Heat transfer in reacting systems
Frequency parameter	—	$\omega'L/V = 2\pi \times $ Strouhal no.; $2\pi/\omega' = $ period of motion $[\theta]$	Unsteady-state flow, etc.
Frequency no.	N_f	$\omega_r L/V$; $L = $ packing element diameter; $V = $ interstitial fluid velocity; $\omega_r = $ radial frequency $[\theta^{-1}]$	Flow in packed or fluidized beds
Frössling no. (heat transfer)	Fs_h	$(N_{Nu} - 2)/(N_{Re}^{1/2} N_{Pr}^{1/3})$	Heat transfer to spheres in turbulent streams
Frössling no. (mass transfer)	Fs_m	$(N_{Sh} - 2)/(N_{R2}^{1/2} N_{Sc}^{1/3})$	Mass transfer to spheres in turbulent streams
Froude no. 1	N_{Fr_1}	$V^2/gL \equiv (N_{Fr_2})^2$ (cf. Reech no., Boussinesq no., Vedernikov no.)	Wave and surface behavior
Froude no. 2	N_{Fr_2}	$V/(gL)^{1/2} \equiv (N_{Fr_1})^{1/2}$	Open channel flow, free surfaces
Froude no. rotating	Fr	DN^2/g; $D = $ impeller diameter	Agitation
Galileo no.	N_{Ga_1}	$L^3 g \rho^2/\mu^2$ (cf. Archimedes no. and Nusselt film thickness group)	Circulation of viscous liquids
Gay Lussac no.	Ga, N_{Ga_2}	$1/\beta \Delta T$	Thermal expansion processes
Geometric no.	Ge	h^*/H^*; $h^* = $ surface area of packing element/perimeter $[L]$; $H^* = $ height of packing	Mass transfer in packed beds
Goertler parameter	—	$\dfrac{V L_b \rho}{\mu}\left(\dfrac{L_b}{R_c}\right)^{1/2}$; $L_b = $ boundary layer momentum thickness; $R_c = $ radius of curvature	Boundary-layer flow on curved surfaces
Goucher no.	N_{Go}	$R(\rho g/2\sigma)^{1/2}$; $R = $ wall or wire radius	Coating
Graetz (Grätz) no.	N_{Gz}	$V_m C_p/kL$	Streamline flow
Grashof no.	N_{Gr}	$L^3 \rho^2 g\beta \Delta t/\mu^2 \equiv N_{Ga_1}/N_{Ga_2} \equiv (N_{Re})^2/(N_{Ga_2} \times N_{Fr_1})$	Free convection
Diffusional Grashof no.	Gr_{AB}	$L^3 \rho^2 g \beta'_A \Delta n'_A/\mu^2$; $n'_A = $ mass fraction of A $[-]$; $\beta'_A = $ coeff. of density change with n'_A $[-]$	Interphase transfer by free convection (density changes caused by concentration differences)
Gukhman no.	Gu, N_{Gu}	$(t_0 - t_m)/T_0$; t_0, $t_m = $ temp. of hot gas stream and moist surface (wet bulb temp) $[°C]$; $T_0 = $ temp. of gas stream $[°K]$	Convective heat transfer in evaporation

SIMILITUDE

Alphabetical list of named groups (cont.)

Name	Symbol	Definition	Field of use
Guldberg-Waage group	N_{Gw}	Given by equation relating volumes of reacting gases and reaction products	Chemical reaction in blast furnaces
Hall coeff.	N_H	$f_c J$; f_c = cyclotron frequency; J = average free path/average velocity	Magneto fluid dynamics
Hartmann no.	M_H	$(\mu_e^2 H_e^2 \sigma_e L^2/\mu^2)^{1/2}$ ≡ (Magnetic pressure no. × magnetic Reynolds no. × $N_{Re})^{1/2}$	Magneto fluid dynamics
Hatta no.	β	$\gamma/\tanh \gamma$; $\gamma = (rcD)^{1/2}/k_c$; r = reaction rate constant $[L^3/M\theta]$	Gas absorption with chemical reaction
Head coeff.	C_H	$gH'/N^2 d^2$; d = impeller diameter	Flow in pumps and fans
Heat transfer no.	K_Q	$q/V^3 L^2 \rho$	Heat transfer
Hedstrom no.	N_{He}	$\tau_y L^2 \rho/\mu_p^2 \equiv N_{Re} \times N_{Bm}$	Flow of Bingham plastics
Helmholtz resonator group	—	$(d^3/W)^{1/2}/Ma$	Pulsating combustion
Hersey no.	—	$F_b/\mu V_s$ (cf. Truncation no.)	Lubrication
Hess no.	Ge	$(KL^2/a_m)(C_0)^{n-1}$; n = order of reaction [—]; C_0 = initial concentration $[M/L^3]$; a_m = mass transfer conductivity of reaction products $[L^2/\theta]$, K = reaction rate constant $[L^{3n-3}/\theta M^{n-1}]$	Heat and mass transfer with chemical and phase changes
Hodgson no.	N_H	$Wf \Delta p_F/V_f \cdot \bar{p}_s$	Pulsating gas flow
Homochronicity no.	Ho_3	$N\theta$; N = mixer rpm; θ = mixing time	Mixing, agitation
Homochronous no.	Ho	$V\theta/L$; θ = time for liquid travel distance L	Choice of time scales
Hooke no.	Ho_2	$\rho V^2/E$ = Cauchy no.	Elasticity of flowing media
Hydraulic resistance group	Γ_e	$\Delta p_p/\rho_L gL = N_{We_3} \times$ Laplace no. = $N_{Eu_1} \times N_{Fr_1}$; Δp_p = pressure drop across liquid on tray $[M/L\theta^2]$, ρ_L = liquid density, L = depth of liquid on tray	Pressure drop in distillation columns
Ilyushin no.	I	$(Vd\rho/\mu)(4\tau_D/3V^2\rho) = (4\tau_D/3V^2\rho) \times N_{Re}$; τ_D = maximum dynamic slip resistance	Flow of viscoplastic fluids
Jakob modulus	Ja	$c_p \rho_L \Delta t/r\rho_V$; ρ_L, ρ_V = density of liquid and vapor; Δt = liquid superheat temp. difference	Boiling
J factor (heat transfer)	j_H	$(h/c_p G)(c_p \mu/k)^{2/3} \equiv N_{Nu} \times N_{Pr}^{1/3} \times N_{Re}$	Heat, mass, and momentum transfer theory
J factor (mass transfer)	j_M	$(k_c \rho/G)(\mu/\rho D)^{2/3} \equiv (k_c \rho/G)(N_{Sc})^{2/3}$	
Joule no.	J	$2\rho c_p \Delta t/\mu_e H_e^2$	Magneto fluid dynamics
Karman no. 1	N_K	$\rho d^3 \left(\dfrac{-dp}{dL}\right) / \mu^2 \equiv 2N_{Re}^2 \sqrt{f}$; d = pipe diameter; dp/dL = pressure gradient	Fluid friction in conduits
Karman no. 2	K	V/V_A (see Alfvén no.)	Magneto fluid dynamics
Kirpichev no. (heat transfer)	Ki_q, N_{Ki_q}	$q^* L/k\Delta t$ (cf. N_{Bi_h}, N_{Nu})	Heat transfer
Kirpichev no. (mass transfer)	Ki_m	$GL/D\rho n^*$ (cf. N_{Pe_m}, N_{Bi_m})	Mass transfer
Kirpitcheff no.	—	$(\rho F_R/\mu^2)^{1/3} = (N_{Re}^2 \cdot C_f)^{1/3}$	Flow around obstacles
Knudsen no. 1	N_{Kn}	λ/L	Low-pressure gas flow
Knudsen no. 2	N_{Kn_A}	$eD_{AB}/D_{KA}\zeta$	Gaseous diffusion in packed beds
Knudsen no. for diffusion	$N_{Kn_{A_2}}$	$3eD_{AB}/4\zeta K_{OA} \bar{u}_A$; K_{OA} = Knudsen flow permeability constant; $\bar{u}_A$ = equilibrium mean molecular speed of A	Gaseous diffusion in packed beds

Alphabetical list of named groups (cont.)

Name	Symbol	Definition	Field of use
Kondrat'ev no.	Kn	ΨBi_h; Bi_h = Biot no. heat transfer; Ψ = temp. field nonuniformity parameter = $(t_s - t_a)/(\bar{t} - t_a)$; t_a = temp. of surrounding medium; t_s = surface temp. of body; $\bar{t}$ = mean temp. of body	Heat transfer between fluid and body
Kossovich no.	Ko, N_{Ko}	$r_v \Delta n_m / c \Delta t$	Convective heat transfer in evaporation
Kronig no.	Kr	$4L^2 \beta \rho^2 \Delta t E_s^2 N \left(\alpha + \dfrac{2p_0^2}{3kT} \right) \Big/ \mu^2 M$; E_s = electric field at surface; N = Avogadro's no.; α = polarization coeff; p_0 = molecular dipole moment; k = Boltzmann's constant; M = molecular weight	Convective heat transfer
Kutateladze no. 1	Ku	$IEL/\rho V u'$; I = current density $[Q/L^2\theta]$; V = voltage $[ML/Q\theta^2]$; u' = enthalpy $[L^2/\theta^2]$	Electric arcs in gas streams
Kutateladze no. 2	K	$r_v/c_p(t_0 - t_w)$; t_0, t_w = stream and wall temp.	Combined heat and mass transfer in evaporation
Lagrange group 1	La_1	$\Pi/\mu L^3 N^2$; L = characteristic dimension of agitator	Agitation
Lagrange no. 2	La_2	$(D + \epsilon_D)D$	Mass transfer in turbulent systems
Lagrange no. 3	La_3	$\Delta PR/\mu V$	Magneto fluid dynamics
Laplace no.	La_4	$\Delta P_p \cdot L/\sigma = \Gamma c/N_{We_3}$; ΔP_p = pressure drop across liquid on tray $[M/L\theta^2]$; L = depth of liquid on tray	Interfacial behavior on distillation trays
Larmor no.	R_{La}	L_L/L; L_L = Larmor radius	Magneto fluid dynamics
Laval no.	La	$V \Big/ \left(\dfrac{2\gamma}{\gamma+1} \cdot \mathcal{R}T \right)^{1/2}$; γ = ratio of specific heats	Compressible flow
Lebedev no.	Le	$eb_T(t_a - t_0)/c_b P \rho_s$; b_T = intensity of vapor expansion in capillaries of body on heating $[M/L^3T]$; t_a = temp. of surrounding medium; t_0 = initial temp; ρ_s = density of solid	Drying of porous materials
Leroux no.	—	= Cavitation no.	—
Leverett function	J	$(\xi/e)^{1/2}(p_c/\sigma)$	Two-phase flow in porous media
Lewis no.	N_{Le}	$k/\rho c_p D = \alpha/D \equiv N_{Sc}/N_{Pr}$; note: sometimes defined as reciprocal of this definition	Combined heat and mass transfer
Turbulent Lewis no.	Le_T	$c_p \rho \epsilon_D/k_T = \ell_D/\ell_T = \epsilon_D/\epsilon_T$; k_T = eddy thermal conductivity $[LM/T\theta^3]$; ℓ_D, ℓ_T = mixing lengths for mass, heat transfer; ϵ_T = eddy thermal diffusivity $[L^2/\theta]$	Combined turbulent heat and mass transfer
Lewis-Semenov no.	—	= 1/Lewis no.	—
Lock no.	—	$\rho R^4 ia'/I$; ρ = fluid density; a' = rotor life curve slope $[L^2/M]$; i = blade chord $[L]$; R = rotor radius $[L]$; I = moment of inertia of blade about hinge $[L^4]$	Rotor blade dynamics
Lorentz no.	N_{Lo}	V/V_ℓ; V_ℓ = velocity of light	Magneto fluid dynamics
Luikov (Lýkov) no.	Lu	$k_c L/\alpha = k_c L \rho c_p/k$	Combined heat and mass transfer
Lukomskii no.	Lu	α/a_m; a_m = potential conductivity of mass transfer $[L^2/\theta]$	Combined heat and mass transfer
Lundquist no.	N_{Lu}	$\sigma_e H_o \mu_e^{3/2} L/\rho^{1/2}$ = Hartmann no. × (Magnetic Reynolds no./N_{Re})$^{1/2}$; L = thickness of fluid layer	Magneto fluid dynamics
Lyashchenko no.	Ly	= $(N_{Re})^3$/Archimedes no.	Fluidization

Alphabetical list of named groups (cont.)

Name	Symbol	Definition	Field of use
Lykoudis no.	N_{Ly}	$(\mu_e H_e)^2 \frac{\sigma_e}{\rho_{1/2}} (L/g\beta\Delta t)^{1/2} =$ (Hartmann no.)^2/(Grashof no.)	Magneto fluid dynamics
McAdams no.	—	$h^4 L \mu \Delta t / k^3 \rho^2 gr$	Condensation
Mach no.	Ma, N_{Ma}	V/V_s; V_s = velocity of sound in fluid $\equiv v/(E_b/\rho)^{1/2}$; E_b = bulk modulus of fluid (cf. Sarrau no.)	Compressible flow
Magnetic force parameter	N	$\mu_e^2 H_e^2 \sigma_e L/\rho V$	Magneto fluid dynamics
Magnetic Grashof no.	—	$4\pi\sigma_e\mu_e(\mu/\rho) \times$ Grashof no.	Magneto fluid dynamics
Magnetic Mach no.	M_{Ma}	V/V_A (see Alfvén no.)	Magneto fluid dynamics
Magnetic no.	R_M	$\mu_e H_e (\sigma_e L/\rho V)^{1/2} =$ (Magnetic force parameter)$^{1/2}$	Magneto fluid dynamics
Magnetic Oseen no.	k	$\frac{1}{2}(1 -$ Alfvén no.$^2) \times$ Magnetic no.	Magneto fluid dynamics
Magnetic Prandtl no.	—	$\sigma_e \mu_e \mu/\rho$ (cf. Magnetic Grashof no.)	Magneto fluid dynamics
Magnetic pressure no.	S	$\mu_e H_e^2/\rho V^2$	Magneto fluid dynamics
Magnetic Reynolds no.	R_M	$\sigma_e \mu_e LV$ (cf. Velocity no.)	Magneto fluid dynamics
Maievskii no.		$\equiv$ Mach no.	Compressible flow
Marangoni no.	N_{Ma}	$\frac{\Delta\sigma}{\Delta t} \cdot \frac{\Delta t}{\Delta L} \cdot \frac{L^2}{\mu\alpha}$; $L =$ layer thickness	Cellular convection
Margoulis no.	M	$\equiv$ Stanton no.	Forced convection
Merkel no.	N_{Me}	$GA^*W^*/(V_m)_{gas}$	Cooling towers, gas-liquid contact
Miniovich no.	Mn	SR/e; $R =$ pore radius	Drying
Mondt no.	N_{Mo}	—	Heat transfer
Naze no.	Na	$V_A/V_S \equiv$ Mach no. $\times$ Alfvén no.	Magneto fluid dynamics
Newton inertial force group	N_ℓ	$F/\rho V^2 L^2$	Agitation
Newton no.	Ne	$F_R/\rho V^2 L^2$ (cf. Fanning friction factor, Resistance coeff. 2	Friction in fluid flow
No. of velocity heads	N	$(F/L^2)/(\rho V^2/2)$	Friction in conduits
No. for similarity of physical and chemical changes	K	$r/c_p \Delta t$	Change of phase
Nusselt no.	N_{Nu}	$hL/k \equiv N_{Re} \times$ Stanton no. (cf. Biot no., heat transfer)	Forced convection
Nusselt no. for mass transfer	Nu_m, N_{Nu_m}	$k_c L/D \equiv$ Sherwood no.	Mass transfer
Nusselt film thickness group	φ, N_T	$L_f (\rho^2 g/\mu^2)^{1/3} \equiv$ (Galileo no.)$^{1/3}$; $L_f =$ film thickness	Falling films
Ocvirk no.	—	$(F_b/\mu V_S)(a/R)^2(D/b)^2$; $V_S =$ shaft surface velocity; $R =$ shaft radius; $D =$ shaft diameter	Lubrication
Ohnesorge no.	Z	$\mu/(\rho L \sigma)^{1/2} =$ (Weber no. 1)$^{1/2}/N_{Re}$	Atomization
Péclet no. (heat)	Pe, N_{Pe_h}	$LV\rho c_p/k = LV/\alpha = N_{Re} \cdot N_{Pr}$	Forced convection
Péclet no. (mass)	N_{Pe_m}	$LV/D = N_{Re} \times$ Schmidt no. 1	Mass transfer
Pipeline parameter	ρ_n	$V_w V_0/2H'_s$; $V_w =$ velocity of water hammer wave; $V_0 =$ initial velocity; $H'_s = $ static head $\times g[L^2/\theta^2]$	Water hammer
Plasticity no.	P	$\equiv$ Bingham no.	—
Poiseuille no.	—	$D^2(-dp/dL)/\mu V$; $D =$ pipe diameter; $dp/dL =$ pressure gradient	Laminar fluid friction
Pomerantsev no.	Po	$jL^2/k(t_m - t_0)$; $t_m, t_0 =$ temp. of medium and initial temp. of body (cf. Damköhler group IV)	Heat transfer with heat sources in medium
Posnov no.	Pn	$\delta\Delta t/\Delta n_m$ (cf. Federov no. 2)	Combined heat and mass transfer
Power no.	N_P	$\Pi/L^5 \rho N^3$	Power consumption by agitators, fans, pumps
Modified power no.	N'_P	$N_P \left(\frac{D'_e}{L'_e}\right) \frac{\Delta W^{-1/2}}{(N_b N_s)^{0.67}}$; $D'_e =$ effective agitator diam.; $L'_e =$ effective agitator height; $\Delta W =$ wall proximity factor; $N_b =$ no. of blades on agitator; $N_s =$ no. of blade edges	Agitation

Alphabetical list of named groups (cont.)

Name	Symbol	Definition	Field of use
Prandtl no.	N_{Pr}	$c_p \mu / k$	Forced and free convection
Total Prandtl no.	$\overline{Pr}$	$\dfrac{\epsilon_m + \mu/\rho}{\epsilon_T + \alpha}$	Heat transfer in combined turbulent and laminar flow
Turbulent Prandtl no.	Pr_T	$\epsilon_M/\epsilon_T = \ell/\ell_T$; ϵ_M, ϵ_T = eddy viscosity and eddy thermal diffusivity $[L^2/\theta]$; ℓ, ℓ_T = mixing lengths for momentum and heat transfer	Heat transfer in turbulent flow
Prandtl no. (mass transfer)	Pr_M	$\mu/\rho D$ = Schmidt no.	Diffusion in flowing systems
Prandtl velocity ratio	u^+	$V/(\tau_W/\rho)^{1/2}$	Turbulence studies
Prandtl dimensionless distance	y^+	$L(\rho\tau_W)^{1/2}/\mu$; L = distance from wall	Turbulence studies
Predvoditilev no.	Pd	$\Gamma L^2/\alpha t_0 = \left(\dfrac{dt^*}{dN_{Fo_h}}\right)_{max}$; t_0 = initial temp of body; t^* = temp of medium relative to its initial temp	Heat transfer
Predvoditilev no. (mass transfer)	Pd_m	$(\Gamma_m L^2/a_m) \cdot N_{Fo_m}$; Γ_m = rate of change of mass transfer potential of medium (mass/unit mass/time) $[1/\theta]$; a_m = mass conductivity of material $[L^2/\theta]$; N_{Fo_m} = Fourier no. (mass transfer)	Mass transfer
Pressure no. 1	Kp	$p/\{g\sigma(\rho' - \rho'')\}^{1/2}$; ρ', ρ'' = density of liquid and gas	—
Pressure no. 2	—	$H/\tfrac{1}{2}U_S^2$ ≡ Diameter no.$^{-2}$ × Speed no.$^{-2}$; U_S = circumferential velocity	Flow machines (turbines, pumps, etc.)
Psychrometric ratio	—	h_c/Gs	Wet and dry bulb thermometry
Pulsation no.	N_{Pu}	$fd_e\rho/G$; d_e = equiv diam of channel	Transfer to pulsed fluid
Radial frequency parameter 1	—	$\omega_r D/V^2 = \omega_r\alpha/V^2$; D = diffusivity or dispersion coeff of packed bed $[L^2/\theta]$; ω_r = radial frequency (radians/s) $[1/\theta]$	Packed and fluidized beds
Radial frequency parameter 2	—	$\omega_r L^2/\alpha$; L = tube radius, ω_r as above	Packed and fluidized beds
Radial frequency parameter 3	—	$\omega_r^2 DL/V^3$; quantities as above	Packed and fluidized beds
Radial frequency parameter 4	—	$L(\omega_r/2D)^{1/2} \equiv [\tfrac{1}{2}$Radial frequency parameter $2]^{1/2}$	Packed and fluidized beds
Radiation no.	$\overline{K}_S$	$kE/\eta\sigma T^3$ $= \dfrac{\text{Weber no. 1}}{\text{Hooke no.} \times \text{Stefan no.}}$	Radiant transfer
Radiation parameter	Φ	$e^+ \eta T_W^3 d_h/k$; e^+ = function of mean surface emissivity of walls $[-]$; T_W = abs wall temp	Effect of radiation on convective mass transfer
Ramzin no.	Ra	$c_b P/c_m \Delta\Omega$ = Bulygin no./Kossovich no.	Molar mass transfer
Ratio of specific heats	γ	c_p/c_v; specific heats at constant pressure and volume	Compressible flow
Rayleigh no. 1	N_{Ra_1}	$V(\rho L/\sigma)^{1/2}$ = Weber no. 2	Break up of liquid jets
Rayleigh no. 2	R_2'	$L^3\rho^2 g\beta c_p\Delta t/\mu k = L^3\rho g\beta\Delta t/\mu d$	Free convection
Rayleigh no. 3	Ra_3	$q^* L^5 \rho^2 g\beta c_p/\mu k^2 x$; L = pipe diameter	Combined free and forced convection in vertical tubes
Reaction enthalpy no.	N_H	$\Delta u_A \Delta n_A/c_P\Delta t$; Δu_A = enthalpy of reaction/unit mass of A produced $[L^2/\theta^2]$; n_A = mass fraction of A	Interphase transfer with chemical reaction
Rebinder no.	Rb	$\left(\dfrac{d\bar{t}}{d\bar{u}}\right)\dfrac{c_q}{r_v}$; $\bar{t}$ = mean temp. of body; $\bar{u}$ = mean moisture content of body $[-]$	Drying studies

Alphabetical list of named groups (cont.)

Name	Symbol	Definition	Field of use
Recovery factor	N_{rf}	$c_p(t_{aw} - t_m)/V^2$; t_{aw} = attained adiabatic wall temp.; t_m = temp. of moving medium (cf. Eckert no.)	Convective heat transfer in compressible flow
Reech no.	—	1/Froude no. 1	Wave and surface behavior
Resistance coeff 1	c_f	$F_R/\tfrac{1}{2}\rho V^2 L^2$ (cf. Drag coeff, Newton no., Fanning friction factor)	Flow resistance
Resistance coeff 2	Ψ	$\Delta p \cdot D_H / \tfrac{1}{2}\rho V^2 L$; Δ_p = pressure drop over length L (cf. Resistance coeff 1)	Fluid friction in conduits
Reynolds no.	N_{Re}	$LV\rho/\mu$	Dynamic similarity
Reynolds no. rotating	Re_r	$L^2 N \rho/\mu$; L = impeller diameter	Agitation
Richardson no.	N_{Ri}	$-(g/\rho)(d\rho/dL)/(dV/dL)_W^2$; L = height of liquid layer; $(dV/dL)_W$ = velocity gradient at wall	Stratified flow of multilayer systems
Romankov no.	Ro'	T_D/T_{PROD}	Drying
Rossby no. 1	N_{Ro}	$V/2\omega_e L \sin \Lambda$; ω_e = angular velocity of Earth's rotation [$1/\theta$]; Λ = angle between axis of Earth's rotation and direction of fluid motion	Effect of Earth's rotation on flow in pipes
Rossby no. 2	—	$V/\omega L$	—
Roughness factor	—	ϵ/L	Fluid friction
Sarrau no.	—	= Mach no.	Compressible flow
Schiller no. 1	—	$LV(\rho^2/\mu F_R)^{1/3}$	Flow around obstacles
Schiller no. 2	Sch	$V\left[\dfrac{3\rho\gamma_m}{4g\mu(\gamma_M - \gamma_m)}\right]^{1/3}$; V = velocity in fluidized bed; γ_m, γ_M = specific gravity of medium and material in bed	Fluidization
Schmidt no. 1	N_{Sc}	$\mu/\rho D$ (cf. Prandtl no., mass transfer)	Diffusion in flowing system
Schmidt no. 2	—	= Semenov no.	—
Schmidt no. 3	Sc_3	$\mu\chi/\rho\sigma_e L^2$; χ = dielectric susceptibility [$Q^2\theta^2/ML^3$]	Electrochemistry
Total Schmidt no.	$\underline{Sc}$	$\dfrac{\epsilon_M + (\mu/\rho)}{\epsilon_n + D}$; ϵ_M = eddy viscosity [L^2/θ]	Mass transfer in combined laminar and turbulent flow
Turbulent Schmidt no.	Sc_T	ϵ_M/ϵ_D; ϵ_M = eddy viscosity [L^2/θ]	Mass transfer in turbulent flow
Semenov no. 1	S_m	k_c/K; K = reaction rate constant [L/θ]	Reaction kinetics
Semenov no. 2	—	1/Lewis no.	—
Senftleben no.	Se	$NE_S^2\left(\alpha + \dfrac{2\rho_0^2}{3kT}\right) \cdot (1/4LMg)$ (cf. Kronig no.)	Convective heat transfer
Sherwood no.	N_{Sh}	$k_c L/D$ = Nusselt no. for mass transfer, also called Taylor no.	Mass transfer
Smoluchowski no.	—	L/λ = 1/Knudsen no. 1	Low-pressure gas flow
Sommerfeld no. 1	N_{S1}	$(\mu N/P_b)(D/a)^2$; D = shaft diameter (cf. Ocvirk no.)	Lubrication
Sommerfeld no. 2	N_{S2}	$(F_b/\mu V_s)(a/R)^2$; V_s = velocity of shaft surface; R = shaft radius; $N_{S2} = 4/\pi N_{S1}$	Lubrication
Soret no.	So	$\theta n'_{20}$; definitions as for Dufour no.	Coupled heat and mass transfer
Spalding function	Sp	$-\left(\dfrac{\partial \theta}{\partial \mu^+}\right)_{u^+ = 0}$; $\theta = \dfrac{T - T_\infty}{T_W - T_\infty}$; T_W = wall temperature; T_∞ = free stream temp.; u^+ = Prandtl velocity ratio	Convection
Spalding no.	B'	$c_p \Delta t/(r_v - q_r/V_m)$; q_r = radiant heat flux [ML^2/θ^3]; V_m = rate of mass transfer [M/θ]	Droplet evaporation
Specific speed	C_S	$N(V_l)^{1/2}/(gH')^{3/4}$; H' = head of liquid produced by one stage (cf. Speed no.)	Pumps and compressors

Alphabetical list of named groups (cont.)

Name	Symbol	Definition	Field of use
Speed no.	σ	$(4\pi)^{1/2}(V_t)^{1/2}N/(2H)^{3/4}$ = (Delivery no.)$^{1/2}$ × (Pressure no.)$^{-3/4}$ (cf. Specific speed)	Flow machines
Stanton no.	N_{St}	$h/c_p\rho V = h/c_p G = N_{Nu}/N_{Re} \times N_{Pr}$	Forced convection
Stark no.	Sk	$\eta T^3 L/k \equiv$ Stefan no.; L = thickness of layer	Radiant heat transfer
Stefan no.	$\overline{St}$	$\eta T^3 L/k$; L = thickness of layer	Radiant heat transfer
Stewart no.	—	$\mu_e^2 H_o^2 \sigma_e L/V\rho$	Magneto fluid dynamics
Stokes no. 1	St	$\mu\theta_v/\rho L^2$; θ_v = vibration time	Particle dynamics
Stokes no. 2	St_2	$1.042\, m_f g\rho (1 - \rho/\rho_f)\, R^{*3}/\mu^2$; ρ, μ = density, viscosity of fluid; m_f, ρ_f = mass, density of float; R^* = tube/float radius [—]	Calibration of rotameters
Strouhal no.	S_r	fL/V (cf. Thomson no.)	Vortex streets; unsteady-state flow
Suratman no.	Su	$\rho L \sigma/\mu^2 = N_{Re}^2/N_{We_1}$	Particle dynamics
Surface elasticity no.	N_{El}	$-\dfrac{\Gamma'}{D_s} L \left(\dfrac{\partial\sigma}{\partial\Gamma}\right)$; Γ' = surface concentration of surfactant in undisturbed state; D_s = surface diffusivity; L = film thickness	Convection cells
Surface tension no.	T_s	$\mu^2/h^*\sigma\rho$; h^* = surface area of packing element/perimeter [L]	Mass transfer in packed columns
Surface viscosity no.	N_{Vi}	$\mu_s/\mu L$; μ_s = surface viscosity [M/θ]; L = film thickness	Convection cells
Taylor no. 1	N_{Ta_1}	$\omega_c (R_a)^{1/2} a^{3/2} \rho/\mu$; ω_c = angular velocity of cylinder; R_a = mean radius of annulus	Stability of flow pattern in annulus with rotating cylinder
Taylor no. 2	N_{Ta_2}	$(2\omega L^2 \rho/\mu)^2$; ω = rate of spin [$1/\theta$]; L = height of fluid layer	Effect of rotation on free convection
Thiele modulus	m_T	$Q^{1/2} u^{1/2} L/(kt)^{1/2}$ = Damköhler group IV)$^{1/2}$	Diffusion in porous catalyst
Thoma no.	σ_T	$(H_a - H_s - H_v)/H$; H = total head; H_a = atm pressure head; H_s = suction head; H_v = vapor pressure head	Cavitation in pumps
Thomson no.	N_{Th}	$\theta V/L$; θ = characteristic time (cf. Strouhal no.)	Fluid flow
Thompson no.	N_{Th_2}	$-\left(\dfrac{\Delta t}{\Delta L}\right) L^2 \left(\dfrac{d\sigma}{dT}\right) \Big/ \mu\alpha$ (cf. Marangoni no.)	Cellular convection
Thring radiation group	—	$\rho c_p V/e^* \eta T^3$ (cf. Boltzmann no.)	Radiation
Thring-Newby criterion	θ	$\dfrac{(V_{ml} + V_{mo})}{V_{mo}} \times \dfrac{R}{L}$; V_{mo}, V_{ml} = mass flow rate of nozzle fluid and surrounding fluid [M/θ]; R = equivalent nozzle radius; L = furnace half width	Combustion of fuels
Thrust coeff.	T_c	$F_T/\rho V^2 D^2$; F_T = thrust force [ML/θ^2]; V = forward speed; D = tip diameter	Propeller studies
Torque coeff.	Q_c	$F'/\rho V^2 D^3$; F' = propeller torque [ML^2/θ^2]	Propeller studies
Transiency groups	K_P	$\left(1 \Big/ \dfrac{\partial P}{\partial L}\right) \cdot \dfrac{\partial(\partial P/\partial L)}{\partial(Fo_f)}$	
	K_Q	$(1/N_{Re}) \cdot \dfrac{\partial N_{Re}}{\partial(Fo_f)}$	Transient flow behavior
		$\partial P/\partial L$ = pressure gradient in flow direction [$M/L^2\theta^2$]; Fo_f = Fourier no. (flow)	
Truncation no.	r	$\mu\dot{\gamma}/\rho$ (cf. Hersey no.)	Viscous flow

Alphabetical list of named groups (cont.)

Name	Symbol	Definition	Field of use
Valensi no.	V	$\omega L^2 \rho/\mu$; ω = circular oscillation frequency when $\mu = 0$ [$1/\theta$]	Oscillation of drops and bubbles
Vedernikov no.	V	$\zeta^*\xi^*V/(V_W - V) \equiv \zeta^*\xi^*N_{Fr_2}$; ζ^* = exponent of hydraulic radius in formula [−]; ξ^* = shape factor of channel section; V_W = absolute velocity of disturbance wave	Instability of open channel flow
Velocity no.	R_V	= Magnetic Reynolds no.	—
Wave no.	k	$L(\omega_r/2\alpha)^{1/2}$; ω_r = radial frequency [$1/\theta$]	Cyclic heat transfer
Weber no. 1	N_{We_1}	$V^2\rho L/\sigma = (N_{We_2})^2$	Bubble formation
Weber no. 2	N_{We_2}	$V(\rho L/\sigma)^{1/2} = (N_{We_1})^{1/2}$	—
Weber no. rotating	We	$L^3 N^2 \rho/\sigma$; L = impeller diameter	Agitation
Weber no. 3	N_{We_3}	$\sigma/\rho_L g L^2$; ρ_L = liquid density [M/L^3]; L = depth of liquid on tray [L]	Interfacial area determination in distillation equipment
Weissenberg no.	N_{We}	$\omega_3 V/\omega_1 L$; $\omega_3 = \int_0^\infty SG(S)\,dS$; $\omega_1 = \int_0^\infty G(S)\,dS$; G = relaxation modulus of linear viscoelasticity; S = recoverable elastic strain	Viscoelastic flow
Weissenberg no., generalized	N_1	$I_e^{1/2} \times \theta_n$; definitions in Deborah no., generalized	Rheology

General nomenclature. The terms employed in this field are given in the following list. (All other quantities are defined at the point of use. Fundamental dimensions are taken to be length [L], mass [M], electrical charge [Q], temperature [T], and time [θ].)

a = annulus or clearance width, L
A = area, L^2
A^* = cooling area/unit volume, $1/L$
b = bearing breadth, L
c = specific heat, $L^2/\theta^2 T$
c_A = concentration, M/L^3
c_b = specific vapor capacity (mass/unit mass/unit pressure change), $L\theta^2/M$
c_m = mass capacity, L^3/M
c_p, c_v = specific heat at constant pressure and volume, $L^2/\theta^2 T$
c_q = heat capacity, $L^2/\theta^2 T$
d = diameter, L
d_e = equivalent diameter (of particles, etc.), L
d_h = hydraulic diameter, L
D = diffusivity (molecular, unless noted otherwise), L^2/θ
D_{AB} = binary bulk diffusion coefficient, L^2/θ
D_{KA} = Knudsen diffusion coefficient, L^2/θ
e = voidage; porosity (−)
e^* = surface emissivity (−)
E = modulus of elasticity, $M/L\theta^2$
E_a = activation energy, L^2/θ^2
E_b = bulk modulus, $M/L\theta^2$

f = frequency, $1/\theta$
F = force, ML/θ^2
F_b = force per unit length of bearing, M/θ^2
F_R = resistance force in flow, ML/θ^2
g = acceleration due to gravity, L/θ^2
G = mass velocity (mass flux density; mass transfer coefficient), $M/\theta L^2$
h = heat transfer coefficient, $M/T\theta^3$
h_c = convective heat transfer coefficient, $M/T\theta^3$
H = energy change per unit mass (= $g \times$ head), L^2/θ^2
H' = fluid head, L
H_e = field strength, $Q/L\theta$
j = heat liberated per unit volume per unit time, $M/L\theta^3$
k = thermal conductivity, $ML/T\theta^3$
k_c = mass transfer coefficient, L/θ
L = characteristic dimension (except as noted), L
L_m = distance from midpoint to surface, L
n = concentration, wt/wt (−)
n^* = specific mass content, mass/mass (−)
n_m = moisture content, wt/wt bone dry gas (−)

N = rate of rotation, $1/\theta$
P = pressure, $M/L\theta^2$
p_b = bearing pressure, $M/L\theta^2$
p_s = static pressure, $M/L\theta^2$
p_v = vapor pressure, $M/L\theta^2$
p_σ = capillary pressure, $M/L\theta^2$
Δp_σ = frictional pressure drop, $M/L\theta^2$
q = heat flux (heat flow/unit time), ML^2/θ^3
q^* = heat flux density (heat flux /unit area), M/θ^3
Q = heat liberated/unit mass, L^2/θ^2
r = latent heat of phase change, L^2/θ^2
r_v = heat of vaporization, L^2/θ^2
R = radius, L
R_H = hydraulic radius, L
$\mathcal{R}$ = gas constant, $L^2/\theta^2 T$
s = humid heat, $L^2/\theta^2 T$
S = particle area/particle volume, L^2/L^3
t = temperature, T
T = absolute temperature, T
$\left.\begin{array}{l}\Delta t \\ \Delta T\end{array}\right\}$ = temperature difference, T
U = reaction rate, $M/L^3\theta$
v_s = velocity of surface (solid), L/θ
V = fluid velocity, L/θ
V_A = velocity of Alfvén magnetic waves, L/θ
V_f = volumetric flow rate, L^3/θ
V_l = velocity of light, L/θ
V_m = mass flow rate, M/θ
V_s = velocity of sound, L/θ
w = circumferential velocity, L/θ
W = volume of system, L^3
W^* = gross volume, L^3
x = entry length; distance from entrance, L

α = thermal diffusivity L^2/θ
β = coefficient of bulk expansion, $1/T$
β^* = Dufour coefficient, T
γ = specific gravity, $(-)$
γ = rate of shear, $1/\theta$
Γ = rate of change of temperature of medium, T/θ
δ = Soret or thermogradient coefficient, $1/T$
Δ = difference in quantity
ϵ = height of roughness, L
ϵ_d = eddy mass diffusivity, L^2/θ
ζ = diffusion tortuosity $(-)$
η = radiation coefficient (Stefan-Boltzmann coefficient), $M/T^4\theta^0$
θ = time, θ
θ_r = relaxation time, θ
λ = mean free path, L
μ = dynamic viscosity, $M/L\theta$
μ_e = magnetic permeability, ML/Q^2
μ_p = rigidity coefficient, $M/L\theta$
ξ = permeability, L^2
π = 3.1416 . . .
Π = power to agitator or impeller, ML^2/θ^3
ρ = density, M/L^3
σ = surface tension, M/θ^2
σ_e = electrical conductivity, $Q^2\theta/L^3M$
τ_w = wall shear stress, $M/L\theta^2$
τ_y = yield stress, $M/L\theta^2$
ω = angular velocity (of fluid, unless noted otherwise), $1/\theta$
Ω = mass transfer potential (concentration), M/L^3
$\overline{}$ (bar over) = mean value

Bibliography. H. A. Becker, *Dimensionless Parameters: Theory and Methodology*, 1976; P. W. Bridgman, *Dimensional Analysis*, 1931, reprint 1976; J. P. Catchpole and G. Fulford, Dimensionless groups, *Ind. Eng. Chem.*, 58(3):46, 1966, and 60(3):71, 1968; A. A. Gukhman, *Introduction to the Theory of Similarity*, 1965; E. Isaacson and M. Isaacson, *Dimensional Methods in Physics*, 1975; H. L. Langhaar, *Dimensional Analysis and the Theory of Models*, 1951, reprint 1980; D. J. Schuring, *Scale Models in Engineering: Fundamentals and Applications*, 2d ed., 1977; C. I. Staicu, *Restricted and General Dimensional Analysis*, 1982; G. G. Stokes, *Mathematical and Physical Papers*, 2d ed., vol. 3, reprint 1966.

REYNOLDS NUMBER
Frank M. White

The number Re whose dimensionless parameter determines the behavior and characteristics of viscous flow patterns. It is defined by Eq. (1), where ρ is fluid density, V is stream velocity, L is a

$$\text{Re} = \frac{\rho V L}{\mu} \qquad (1)$$

characteristic length scale, and μ is fluid viscosity. This parameter, the dominant factor in viscous flow analysis, was formulated in 1883 by Osborne Reynolds and named in his honor about 40 years later.

The form of Eq. (1) is such that Re has the same value regardless of the unit system used to define its constitutents. In the International System (SI), for example, if brackets denote the dimensions of a variable, Eq. (2) is obtained. Thus the Reynolds number is not subject to any ambiguity.

$$[Re] = \frac{[kg/m^3]\,[m/s]\,[m]}{[kg/m \cdot s]} = [1] \qquad (2)$$

Laminar and turbulent flow. It had been well known since the eighteenth century that viscous flows were sometimes smooth (laminar) and sometimes disorderly (turbulent). Reynolds revealed the reason in 1883 in a classic experiment which introduced a dye streak down the centerline of water flowing in a glass tube (see **illus**.). For values of Re less than 2000 (illus. a) the dye streak remained smooth and undisturbed—that is, laminar flow resulted. For Re greater than 3000 (illus. b), the dye streak seemed to burst into turbulence and fill the whole tube with color. However, a spark-flash photograph (illus. c) showed that the turbulence structure was really a series of minute whirls and eddies, oscillating too fast for the eye to see. This classic demonstration laid the foundation for what is now a rich field of turbulent flow analysis and experimentation. SEE LAMINAR FLOW; TURBULENT FLOW.

For Re between 2000 and 3000 (not shown in the illustration), the flow pattern cycled erratically between laminar and turbulent regimes. This is now known as the transition region between the two more stable flow patterns.

The transition Reynolds number is not always equal to 2000 but rather varies somewhat with flow geometry. It is less than 2000 for unstable patterns, such as jet flow, and greater than 2000 for highly stable flows such as converging nozzles. SEE BOUNDARY-LAYER FLOW; JET FLOW.

Reynolds's original sketches of dye streak behavior in pipe flow: (*a*) laminar flow, Re > 2000; (*b*) turbulent flow, Re > 3000; (*c*) a spark-flash photograph of condition *b*. (*After O. Reynolds, An experimental investigation of the circumstances which determine whether the motion of water shall be direct or sinuous and of the law of resistance in parallel channels, Phil. Trans. Roy. Soc., London, A174:935–982, 1883*)

Model testing. In experimenting with model tests of viscous flows, the Reynolds number is the primary parameter by which the model and prototype are matched. Fulfillment of Eq. (3)

$$(\rho VL/\mu)_{\text{model}} = (\rho VL/\mu)_{\text{prototype}} \qquad (3)$$

ensures the dynamic similarity of the model test. One can vary density, velocity, length, or viscosity in any manner as long as the overall grouping $(\rho VL/\mu)$ remains the same. Of course, the model must also be geometrically similar to the prototype, that is, must have exactly the same proportions and orientation to the fluid stream. Then model test data can be used to predict the drag, pressure distribution, friction, and even the convection heat transfer from the prototype, if the data are properly interpreted. An exception is free surface flow—such as ship resistance, ocean wave motion, and buoy dynamics—where the Froude number of the flow is also extremely important. SEE DIMENSIONAL ANALYSIS; DYNAMIC SIMILARITY; FLUID FLOW; FROUDE NUMBER; MODEL THEORY.

Bibliography. H. Rouse and S. Ince, *History of Hydraulics*, 1963; H. Schlichting, *Boundary Layer Theory*, 7th ed., 1979; I. H. Shames, *Mechanics of Fluids*, 2d ed., 1982; F. M. White, *Fluid Mechanics*, 2d ed., 1986.

MACH NUMBER
Glenn Murphy

In fluid mechanics, the ratio v/c of the free stream velocity v to the velocity of sound c in the fluid at the same condition, such as temperature and pressure. Mach number is also the ratio of the inertia force of the fluid to the force of compressibility or the elastic force. In most fluid systems compressibility effects become important for values of the Mach number greater than about 0.3. A body moving through a fluid at a velocity less than sonic is preceded by a region of gradually varying density and pressure that controls the flow around the body. At Mach numbers equal to or greater than unity the gradual transition of pressure cannot exist, and shock waves, or regions of abruptly altered pressure and density, form at critical sections on or near the surface of the body and extend outward. As a result, the fluid force pattern on the moving body is markedly different at supersonic velocities. The theory for calculating the pressure patterns is well established for ideal fluids. See Shock wave.

When compressibility effects alone are significant, geometrically similar bodies will develop identical flow and shock wave patterns when operated at equal Mach numbers. However, in the hypersonic region, generally considered to mean a Mach number of 5 or more, there is appreciable interaction between the boundary layer and the shock pattern. When this occurs, the Reynolds number as well as the Mach number is significant and the similitude requirements become more difficult to satisfy. See Dynamic similarity; Model theory; Reynolds number.

Techniques of transformation of variables have been devised whereby boundary conditions may be altered to accommodate nonconformance to the Mach requirement. That is, data from incompressible flow may be used to give accurate predictions under compressible-flow conditions using distortion of geometry to balance distortion of velocity.

FROUDE NUMBER
Glenn Murphy

In fluid mechanics, the ratio v^2/gd of the inertia force v^2 to the gravitational force gd, where v is velocity, g is gravitational acceleration, and d is a characteristic length. Froude number is significant in the design of a model of any system in which the effect of gravity is important in controlling the velocities or the flow pattern. For example, it is used in the evaluation of the drag or the slipstream velocity of a ship producing surface waves. If the model is operated in the same gravitational field as the prototype, the similarity requirement based on equal Froude numbers in model and prototype leads to the equations below, in which $n(d/d_m)$ is the length scale between

$$v_m^2/d_m = v^2/d \quad \text{or} \quad v_m = v/\sqrt{n}$$

model and prototype. See Dynamic similarity; Model theory.

MODEL THEORY
Glenn Murphy

The theory underlying the design and use of models to predict the characteristics of any system. It is particularly valuable when the desired system, or prototype, is large and complex. A model can be built, tested, and modified at comparatively low cost. If the model is properly designed, the results obtained from it may be used with a high degree of confidence in predicting the performance of the prototype. Models are thus widely used in the design of bridges, dams, unusual buildings, aircraft, river and harbor works, agricultural machinery, and a variety of network systems.

If the results obtained from a model are to be applicable to the prototype, a rigorous set of conditions on the model must be satisfied or else the deviations from these conditions must be taken into account in the predictions of the prototype behavior. The similitude relationships that must exist between model and prototype may be developed conveniently by using dimensional analysis. *See* DIMENSIONAL ANALYSIS.

The first step is to identify all of the independent variables in the system which will influence the magnitude of the quantity that is to be predicted. A set of dimensionless variables is then formed from these variables, using dimensional analysis. These dimensionless variables, or pi terms, are grouped in an equation, called the characteristic equation, in the form of Eq. (1), in

$$\pi_1 = f(\pi_2, \pi_3, \pi_4, \ldots, \pi_s) \tag{1}$$

which π_1 is the dimensionless term containing the variable to be predicted, and the other π terms are dimensionless quantities based on the independent variables. A similar equation may be written for the model, which involves the same variables as the prototype; this is shown in Eq. (2). If

$$\pi_{1m} = f(\pi_{2m}, \pi_{3m}, \pi_{4m}, \ldots, \pi_{sm}) \tag{2}$$

the two systems involve the same phenomenon, the forms of the two functions will be identical. If the model is so designed that Eqs. (3) hold, it follows that Eq. (4) is true. Therefore, if the design

$$\pi_{2m} = \pi_2, \pi_{3m} = \pi_3 \qquad \pi_{4m} = \pi_4, \ldots, \pi_{sm} = \pi_s \tag{3} \qquad \pi_1 = \pi_{1m} \tag{4}$$

equations, Eqs. (3), are satisfied in the construction and operation of the model, the prediction equation, Eq. (4), may be used to predict the magnitude of π_1, and hence the desired variable in the prototype, from the measured magnitude of π_{1m} in the model.

The variables involved in a fluid-mechanics system, for example, may generally be grouped as forces and fields, geometry and dimensions, properties of the fluids, and quantities, such as velocity, that describe the behavior of the systems. *See* GAS DYNAMICS.

Typical forces include those arising from gravity, viscosity, compressibility, and surface-tension effects. Many dimensionless terms may be evolved for a given set of variables, but in general the terms arising in fluid-mechanics systems may be reduced to the combinations in Eq. (5), in which v is velocity, g is acceleration of gravity, d is a characteristic length, ρ is the density

$$\pi_1 = f\left(\frac{\lambda_i}{d}, \frac{v^2}{gd}, \frac{v}{c}, \frac{\rho v d}{\mu}, \frac{\rho v^2 d}{\sigma}\right) \tag{5}$$

of the fluid, μ is the viscosity of the fluid, c is the velocity of sound in the fluid, σ is the surface tension of the fluid, and λ_i is any significant length other than d. The terms of the form λ_i/d, of which there may be several, specify system geometry. The remaining four terms are known as the Froude number, the Mach number, the Reynolds number, and the Weber number. *See* DIMENSIONLESS GROUPS.

Similitude, as specified by Eqs. (3), requires that the model must be geometrically similar to the prototype in all significant respects, and that the Froude, Mach, Reynolds, and Weber numbers in the model be numerically equal to their counterparts in the prototype. If the performance of a given system is independent of one of the properties or forces, the corresponding term is omitted from the characteristics equation. In each of the dimensionless numbers the numerator is proportional to the inertia force of the fluid and the denominator is proportional to the corresponding significant force developed in the fluid. Thus equivalence of the respective numbers in model and prototype means that all forces are in the same ratios in model and prototype, and hence the performances of the model and prototype will be equivalent. *See* DYNAMIC SIMILARITY.

In some situations two design conditions may lead to conflicting requirements that cannot be satisfied simultaneously. For example, if the same fluid is used in model and prototype, the requirement based on the Froude number leads to Eq. (6) in which n is the length scale, or ratio of a length in the prototype to the corresponding length in the model, whereas the requirement based on the Reynolds number leads to Eq. (7). Obviously, these may be satisfied simultaneously

$$v_m = v/\sqrt{n} \tag{6} \qquad v_m = nv \tag{7}$$

only if $n = 1$, or if the model and prototype are identical in size.

In situations of this kind the design of the model is based on the predominant number, and the influence of the other number or numbers must be distorted. In order to make the model useful, either a prediction factor for π_1 must be developed from theory or test, or a compensating distortion of another π term must be introduced in the design so that Eq. (4) will be valid. Methods have been devised for both approaches.

Model theory may also be applied when the model is based on a different phenomenon from the prototype, provided that the characteristic equations are similar mathematically. The model is then known as an analog, and an analogy is said to exist between the model and prototype. The design equations and the prediction equation are Eqs. (3) and (4), respectively. Analogies have proven useful in a wide variety of situations, with the analog frequently being an electrical circuit or system because of the ease of making parametric studies in such systems.

Bibliography. A. A. Gukhman, *Introduction to the Theory of Similarity*, 1965; G. Murphy, D. J. Shippy, and H. L. Luo, *Engineering Analogies*, 1963; H. T. Schlichting and E. Truckenbrodt, *Aerodynamics of the Airplane*, 1979; E. Szucs, *Similitude and Modeling*, 1980.

PHYSICS OF FLUIDS

Fluids	178
Liquid	182
Cavitation	184
Gas	188

FLUIDS
John P. O'Connell

Substances that flow under shear stress and have a density range from essentially zero (gases) to solidlike values (liquids). Fluids comprise one of the two major forms of matter. Solids, the other form, generally deform very little when shear forces are applied, and their densities do not change significantly with pressure or temperature.

At the atomic level, the relative molecular positions in solids are well defined to very great distances, and the molecules rarely change locations. However, fluids have a uniform structure beyond a few molecular diameters. **Figure 1** schematically demonstrates the differences in molecular distribution in gases, liquids, and solids according to what x-ray and neutron scattering measurements indicate about the probability of finding molecular centers at any separation. Probabilities different from unity indicate "structure." For all phases and densities, the repulsive intermolecular forces prevent any molecules from approaching to small separations, so the probability at such separations is zero. Just beyond the repulsive barrier, the probability is higher in all phases due to attractive forces. For low-density gases, the fluid is uniform beyond the first neighbors and there is no real structure. For the denser liquids, there is only local structure, whose extent depends on the density. Fluids do not show the long-range structure observed in solids. Also, molecules in fluids change positions and orientations more frequently than do those in solids. In fact, at low densities, there is essentially no inhibition to molecular motion except for the container walls. Fluids have been conceptualized as mostly solidlike, with regions of vacuum or free individual molecules, and there even have been models, based on such ideas, that succeed in describing fluid properties. Fluid structure is not heterogeneous, as this concept implies.

The distinction between solids and fluids is most easily seen in substances and mixtures which show a well-defined melting process. For substances with large molecules, such as polymers, ceramics, and biologicals, this distinction is less clear. Instead, there is a slow evolution of structure and of resistance to flow as temperature or some other variable is changed.

Vapors and liquids. Molecular density varies greatly in fluids and is their most important characteristic. The distinction between vapors (or gases) and liquids is most clear for substances and mixtures that show well-defined vaporizing (boiling) and condensing processes. The high-density liquid boils to make a low-density gaseous vapor. **Figure 2** shows the pressures and temperatures for which pure substances are single phases. At the conditions of the lines between the single-phase regions, two phases can be observed to coexist. At the state of intersection of the lines (the triple point), three phases can coexist. For most substances, the triple-point pressure is well below atmospheric. However, for carbon dioxide, it is very high, so that dry ice sublimes rather than melts, as water ice does. Beyond the end of the liquid-vapor (saturation or vapor-pressure) line, vaporization and condensation cannot be observed. The state at the end of this line is called the critical point, where all the properties of the vapor and liquid become the same. There is no such end to the solid-liquid (or melting) line, because the solid and fluid structures cannot become the same.

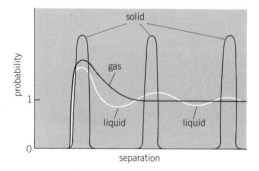

Fig. 1. Probability of finding molecular centers at various separations in gases, liquids, and solids.

PHYSICS OF FLUIDS 179

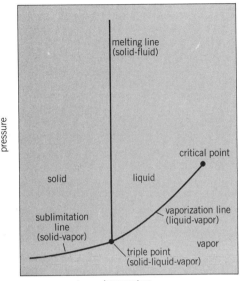

Fig. 2. Conditions of pure-component phase behavior.

Figure 3 shows the relationship of pressure and density of fluids at temperatures above the melting point and at pressures above the triple point. As the pressure on a vapor is increased at fixed temperature (T_1), the density changes rapidly until the condensation point is reached. During condensation, the pressure on the two phases is constant until all the vapor is gone. Then the fluid is relatively incompressible, and the density changes very little with even large changes in pressure. The region of conditions where two phases are observed is capped by the critical point. At a temperature just above the critical (T_2), the variation of density with pressure is very great, even though the system is one phase. At the highest temperatures (T_3), the pressure-density variation is more gradual everywhere. The relation between fluid-state variables of pressure P, density ρ, and temperature T is often represented by a quantity called the compressibility factor, $z = P/\rho RT$, where R is a universal constant equal to 1.987 calories/mol K, 1.987 Btu/lb mole °R, or 8.3143 J/mol K and where density is in units of moles per unit volume. The behavior of z is

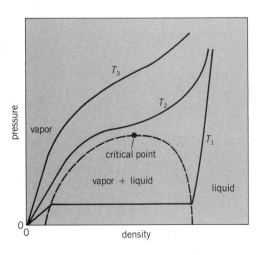

Fig. 3. Behavior of pure-component pressures with density at various temperatures.

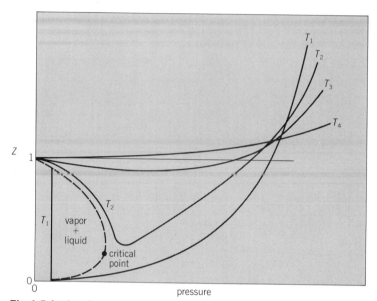

Fig. 4. Behavior of pure-component compressibility factors with pressure at various temperatures.

shown in **Fig. 4**. At the lowest pressures and highest temperatures (T_3 and T_4), z approaches 1, where the fluid is called an ideal gas and the molecules behave as if they do not interact with each other. At most temperatures (T_1 and T_2), z decreases as the pressure is raised, becoming very small during the condensation process. However, because of the relative incompressibility of the liquid, z becomes large at the highest pressures. Mathematical representations of the variation of z with T, P, or ρ and composition are called equations of state, a famous form being that due to J. D. van der Waals. Many equations of state of much higher accuracy have been developed, though the most complete expressions are sufficiently complex that a computer is necessary to actually calculate the properties they can give. SEE GAS.

Mixtures. Mixtures of fluids show the same general density and multiphase behavior as pure fluids, but the composition is an extra variable to be considered. For example, the density differences between the vapor and the liquid phases cause them to have different relative amounts of the components. This difference in composition is the basis of the separation process of distillation, where the vapor will be richer in some components while the liquid will be richer in others. In the distillation refining of petroleum, the vapor is enriched with the components of gasoline, while the liquid is richer in heavier oils. It is also possible for mixtures of liquids to be partially or nearly wholly immiscible, as are water and oil. The separation process of liquid extraction, used in some metal-purification systems and chemical-pollution-abatement processes, depends on different preferences of chemical solutes for one liquid phase or the other. Both vapor-liquid and liquid-liquid systems can show critical points where the densities and compositions become the same. In the single-phase conditions near the critical point, the great variation of density with small changes in pressure can be used to selectively remove components from sensitive substances such as foods and pharmaceuticals. This process (called supercritical extraction) is used to extract caffeine from coffee with carbon dioxide near its critical point. In some systems, it is possible to observe many liquid phases. A well-known example has nine liquid phases made from mixing nine liquid substances.

Fluid interfaces. The usual observation of the presence of more than one fluid phase is the appearance of the boundary or interface between them. This is seen because the density or composition (or both) changes over a distance of a few molecular diameters, and this variation bends or scatters light in a detectable way. At the interface, the molecules feel different forces than in the bulk phases and thus have special properties. Energy is always required to create interface from bulk, the amount per unit area being called the interfacial tension. Water is a fluid with an extremely high vapor-liquid (or surface) tension; this surface tension allows insects to crawl on ponds and causes sprinkler streams to break up into sprays of droplets. One way to measure the surface tension of a fluid is to measure its capillary rise in small tubes (**Fig. 5**). The height of the fluid rise decreases with the tube diameter and increases with the surface tension. Another effect of interfacial tension is the increase of vapor pressure of a component in a very small droplet. Liquid mercury will not evaporate in an open container with an essentially flat interface. However, if it is spilled into little drops, those too small to be easily seen will have their vapor pressure so enhanced that the mercury will vaporize. (This can be a health hazard in chemical and other laboratories.) As the conditions of temperature and composition are changed toward a critical point, the interfacial tension decreases to zero.

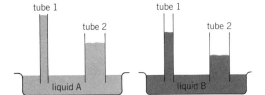

Fig. 5. Capillary rise of liquids due to surface tension; liquid A has higher surface tension than liquid B.

In mixtures, the molecules respond differently to the interfacial forces, so the interfacial composition is generally different from that of the bulk. This has also been the basis of a separation process. If the difference of composition is great enough and it varies with time and position because of evaporation of one or more of the components, the interfacial forces can push the fluid into motion, as can be observed on the walls of a glass of brandy (the Marangoni effect). Some substances strongly adsorb at the interface because their chemical structure has one part that prefers to be in one phase, such as water, and another part that prefers the other phase, such as oil or air. Such surfactants or detergents help solubilize dirt into wash water, keep cosmetics and other immiscible mixtures together, and form foams when air and soapy water are whipped together.

Transport properties. Besides the relations among pressure, density, temperature, and composition of static or equilibrium fluids, there are also characteristics associated with fluid flow, heat transfer, and material transport. For example, when a liquid or gas flows through a tube, energy must be supplied by a pump, and there is a drop in pressure from the beginning to the end of the tube that matches the rise in pressure in the pump. The pump work and pressure drop depend on the flow rate, the tube size and shape, the density, and a property of the molecules called the viscosity. The effect arises because the fluid molecules at the solid tube wall do not move and there are velocity gradients and shear in the flow. The molecules that collide with one another transfer momentum to the wall and work against one another, in a sort of friction which dissipates mechanical energy into internal energy or heat. The greater the viscosity, the greater the amount of energy dissipated by the collisions and the greater the pressure drop. The viscosity generally increases with the density. In gases, viscosity increases with temperature and decreases with molecular size and attractive forces. In liquids, the effects are opposite; the main effect is due to changes in density. If only chemical constitution and physical state are needed to characterize the viscosity, and if shear stress is directly proportional to velocity gradient, the fluid is called newtonian and the relation for pressure drop is relatively simple. If the molecules are large or the attractive forces are very strong over long ranges, as in polymers, gels, and foods such as bread dough and cornstarch, the resistance to flow can also depend on the rate of flow and even the recent deformations of the substance. These fluids are called non-newtonian, and the relationship of flow resistance to the applied forces can be very complex. S<small>EE</small> F<small>LUID FLOW</small>; N<small>EWTONIAN FLUID</small>; N<small>ON-NEWTONIAN FLUID</small>; V<small>ISCOSITY</small>.

Another fluid-transport property, thermal conductivity, indicates the ability of a static fluid to pass heat from higher to lower temperature. This characteristic is a function of chemical constitution and physical state in a similar way as is the viscosity. In mixtures, these properties may involve simple or complex dependence on composition, the variation becoming extreme if the unlike species strongly attract each other. The values of both properties increase rapidly near a critical point.

Finally, the ability of molecules to change their relative position in a static fluid is called the diffusivity. This is a particularly important characteristic for separation processes whose efficiency depends on molecular motion from one phase to another through a relatively static interface, or on the ability of some molecules to move faster than others in a static fluid under an applied force. The diffusivity is highest for the smallest molecules with the weakest attractive intermolecular forces, decreasing as the size and forces increase. The diffusivity in gases increases with temperature and decreases with density; often, the product of the diffusivity and the density divided by the product of the viscosity and the absolute temperature is nearly constant. In liquids the diffusivity tends to be inversely proportional to viscosity of solvent. SEE GAS; LIQUID.

Bibliography. A. W. Adamson, *A Textbook in Physical Chemistry*, 3d ed., 1986; A. W. Adamson, *Understanding Physical Chemistry*, 3d ed., 1980; G. M. Barrow, *Physical Chemistry*, 4th ed., 1979; R. H. Perry and D. Green (eds.), *Perry's Chemical Engineers' Handbook*, 6th ed., 1984; R. C. Reid, J. M. Prausnitz, and B. E. Poling, *The Properties of Gases and Liquids*, 4th ed., 1987.

LIQUID
NORMAN H. NACHTRIEB

A state of matter intermediate between that of crystalline solids and gases. Macroscopically, liquids are distinguished from crystalline solids in their capacity to flow under the action of extremely small shear stresses and to conform to the shape of a confining vessel. Liquids differ from gases in possessing a free surface and in lacking the capacity to expand without limit. On the scale of molecular dimensions liquids lack the long-range order that characterizes the crystalline state, but nevertheless they possess a degree of structural regularity that extends over distances of a few molecular diameters. In this respect, liquids are wholly unlike gases, whose molecular organization is completely random.

Thermodynamic relations. The thermodynamic conditions under which a substance may exist indefinitely in the liquid state are described by its phase diagram, shown schematically in the **illustration.** The area designated by L depicts those pressures and temperatures for which the liquid state is energetically the lowest and therefore the stable state. The areas denoted by S and V similarly indicate those pressures and temperatures for which only the solid or vapor phase may exist. The connecting lines OC, OB, and OA define pressures and temperatures for which the liquid and its vapor, the solid and its liquid, and the solid and its vapor, respectively, may coexist in equilibrium. They are usually termed phase boundary or phase coexistence lines. The intersection of the three lines at O defines a triple point which, for the three states of matter under discussion, is the unique pressure and temperature at which they may coexist at equilibrium. Other triple points exist in the phase diagram of a substance that possesses two or more crystalline modifications, but the one depicted in the figure is the only triple point for the coexistence of the vapor, liquid, and solid. Line OA has its origin at the absolute zero of temperature and OB, the melting line, has no upper limit. The liquid-vapor pressure line OC is different from OB, however, in that it terminates at a precisely reproducible point C, called the critical point. Above the critical temperature no pressure, however large, will liquefy a gas.

Along any of the coexistence curves the relationship between pressure and temperature is given by the Clausius-Clapeyron equation: $dP/dT = \Delta H/T\Delta V$, where ΔV is the difference in molar volume of the corresponding phases (gas-liquid, gas-solid, or liquid-solid) and ΔH is the molar heat of transition at the temperature in question. By means of this equation the change in the melting point of the solid or the boiling point of the liquid as a function of pressure may be calculated. When a liquid in equilibrium with its vapor is heated in a closed vessel, its vapor pressure and temperature increase along the line OC. ΔH and ΔV both decrease and become zero at the critical point, where all distinction between the two phases vanishes.

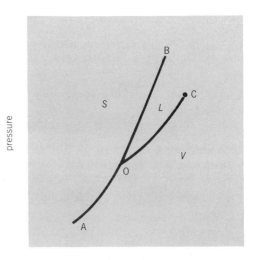

Phase diagram of a pure substance.

Transport properties. Liquids possess important transport properties, notably their capacity to transmit heat (thermal conductivity), to tranfer momentum under shear stresses (viscosity), and to attain a state of homogeneous composition when mixed with other miscible liquids (diffusion). These nonequilibrium properties of liquids are well understood in macroscopic terms and are exploited in large-scale engineering and chemical-process operations. Thus, the rate of flow of heat across a layer of liquid is given by $\dot{Q} = \kappa \, dT/dx$, where $\dot{Q}$ is the heat flux, dT/dx is the thermal gradient, and κ is the coefficient of thermal conductivity. Similarly, the shearing of one liquid layer against another is resisted by a force equal to the momentum transfer: $F = \dot{p} = \eta \, dv/dx$, where dv/dx is the velocity gradient and η is the coefficient of viscosity. Likewise, the rate of transport of matter under nonconvective conditions is governed by the gradient of concentration of the diffusing species: $J = -D \, dC/dx$, where J is the matter flux and D is the coefficient of diffusion. Each of these transport coefficients depends upon temperature, pressure, and composition and may be determined experimentally. An a priori calculation of κ, η, and D is a very difficult problem, however, and only approximate theories exist.

Theoretical explanations. In fact, although a great deal of effort has been expended, there still exists no satisfactory theory of the liquid state. Even so commonplace a phenomenon as the melting of ice has no adequate theoretical explanation. The reason for this state of affairs lies in the tremendous structural and dynamical complexity of the liquid state. To understand this, it is useful to compare the structural and kinetic properties of liquids with those of crystalline solids on the one hand and with those of gases on the other.

In crystals, atoms or molecules occupy well-defined positions on a three-dimensional lattice, oscillating about them with small amplitudes; their kinetic energy is entirely distributed among these quantized vibrational states up to the melting point. This nearly perfect spatial order is revealed by diffraction techniques, which utilize the coherent scattering of x-rays or particles have wavelengths comparable with interatomic spacings. The structural and dynamical properties are sufficiently tractable mathematically so that the theory of solids is quite well understood.

The theory of gases is also simple, but for quite a different reason. No vestige of positional regularity of atoms remains in gases, and their energy resides entirely in high-speed translational motion. Except for collisions, which deflect their motions into new straight-line trajectories, atoms in gases do not interact with one another; vibrational modes in monatomic gases are absent.

Liquids, by contrast, lie intermediate between gases and crystals from both a structural and dynamic point of view. Kinetic energy is partitioned among translational and vibrational modes, and diffraction studies reveal a degree of short-range order that extends over several molecular diameters. Moreover, this "structure" is continually changing under the influence of vibrational and translational displacements of atoms. Physical reality may be attributed to this short-range structure, nevertheless, in the sense that a time average over the huge number of possible configurations of atoms may show that a fairly definite number of neighboring atoms lie close to any arbitrary atom. At a somewhat greater distance from this reference atom, the density of neighbors oscillates above and below the average density of atoms in the liquid as a whole.

Information about the degree of local order is contained in the radial distribution function, a mathematical property which may be deduced from diffraction measurements. This is the starting point for a theory of the liquid state, and although research efforts have yielded partial successes, prodigious mathematical difficulties lie in the path of an entirely satisfactory solution. SEE VISCOSITY.

Bibliography. J. P. Hansen and I. R. McDonald, *The Theory of Simple Liquids*, 1977; P. Kruus, *Liquids and Solutions: Structure and Dynamics*, 1977; W. G. Rothschild, *Dynamics of Molecular Liquids*, 1984; D. Tabor, *Gases, Liquids, and Solids*, 2d ed., 1980; H. N. Temperley and D. H. Trevena, *Liquids and Their Properties: A Molecular and Microscopic Treatise*, 1978.

CAVITATION
K. E. SCHOENHERR AND JACQUES B. HADLER

The formation of vapor- or gas-filled cavities in liquids by mechanical forces. If understood in this broad sense, cavitation includes the familiar phenomenon of bubble formation when water is brought to a boil and the effervescence of champagne wines and carbonated soft drinks. In engineering terminology, the term cavitation is used in a narrower sense, namely, to describe the formation of vapor-filled cavities in the interior or on the solid boundaries of vaporizable liquids in motion when the pressure is reduced to a critical value without change in ambient temperature. Cavitation in the engineering sense occurs at suitable combinations of low pressure and high speed in pipelines; in hydraulic machines such as turbines, pumps, and propellers; on submerged hydrofoils; behind blunt submerged bodies; and in the cores of vortices. This type of cavitation has great practical significance because it restricts the speed at which hydraulic machines may be operated and, when severe, lowers efficiency, produces noise and vibrations, and causes rapid erosion of the boundary surfaces, even though these surfaces consist of concrete, cast iron, bronze, or other hard and normally durable material. The subsequent discussion will be limited to cavitation in the narrower sense as understood in engineering terminology.

As mentioned, cavitation occurs when the pressure is in a liquid is reduced to a critical value. For the present, it will be assumed that this critical value is the vapor pressure p_v of the liquid. For clean, fresh water at 70°F (21°C) and at sea level, p_v has a value of about 52 lb/ft^2 (360 kilopascals); hence, when a body is moving with velocity V through water at ordinary temperature and the pressure on its surface is reduced to or near 52 lb/ft^2 (360 kPa), cavitation may be expected to occur. The condition for the onset of cavitation is therefore given by relation (1), where p_m is the minimum pressure at any point on the surface of a moving body and p_v is the vapor pressure of the liquid at the prevailing temperature. Inversely, the condition for avoidance of cavitation is given by relation (2).

$$p_m \lessapprox p_v \qquad (1) \qquad\qquad p_m > p_v \qquad (2)$$

Bernoulli's principle. A more practical relation than (2) is obtained when the pressure p_m is expressed in terms of easily measurable reference values. This can be done by use of Bernoulli's principle. According to this principle, the sum of pressure head and velocity head in a frictionless incompressible medium remains constant along a streamline. To make this clear, assume a sphere to be held stationary in a stream flowing from right to left with constant velocity V as shown in **Fig. 1**. Consider two points A and B lying on the same streamline, and let point B be the point where the local pressure is minimum. Writing Bernoulli's equation for these points yields Eq. (3), where p_m = absolute local pressure at point B, v = velocity at point B, V = velocity

$$p_m = p_a + p_s - \frac{\rho}{2}V^2\left[\left(\frac{v}{V}\right)^2 - 1\right] \qquad (3)$$

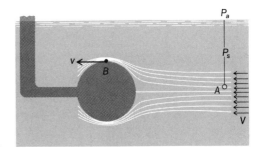

Fig. 1. Flow around a sphere in a frictionless, incompressible fluid; symbols defined in text.

of undisturbed flow at point A, p_a = pressure on free surface, p_s = hydrostatic pressure at point A, and ρ = density of liquid. The term in the bracket is independent of the absolute speed V, because v changes proportionally to V. It follows that for constant values of p_a and p_s the pressure p_m may be reduced to the vapor pressure p_v, to zero, or even to negative values by increasing the speed V. Negative values for p_m have been obtained in carefully conducted laboratory tests, proving that a pure liquid, such as clean air-free water, may sustain tension. However, in most practical cases, negative values are not obtained; instead, the flow is disrupted and cavities are formed at small positive values as previously explained. Combining (2) and (3), the condition for avoidance of cavitation becomes relation (4). SEE BERNOULLI'S THEOREM.

$$\frac{p_a + p_s - p_v}{(\rho/2)V^2} > \left[\left(\frac{v}{V}\right)^2 - 1\right] \qquad (4)$$

Cavitation number. The term on the left of inequality (4) contains easily measurable values and is usually denoted by σ. It is called the cavitation number. The magnitude of the term on the right can be calculated for the sphere and other simple bodies, but for more complex configurations it cannot be calculated and must be obtained by experiment. Denoting this value by σ_c, the condition for avoidance of cavitation has the form of relation (5).

$$\sigma = \frac{p_a + p_s - p_v}{(\rho/2)V^2} > \sigma_c \qquad (5)$$

The cavitation number σ is used for flow through pipes, flow around submerged bodies, and in the design of marine propellers. In pump and hydraulic turbine work, slightly different expressions are used. The simplest one is the expression first introduced by D. Thoma which has form of relation (6), where H_{sv} is the net positive suction head at the pump inlet, or just below the

$$\sigma_T = \frac{H_{sv}}{H} > (\sigma_T)_c \qquad (6)$$

turbine runner, and H is the total head under which the turbine or pump operates. The value of $(\sigma_T)_c$, like the value of σ_c, is a fixed number for a given design of pump or turbine which in general must be found by experiment.

Experiments to determine σ_c and $(\sigma_T)_c$ are usually made on models geometrically similar to, but smaller than, the prototype installations. This has the advantage of less cost and more precise control of the experiments and permits correction of undesirable characteristics of a design before the prototype machine is actually constructed. For instance, should it be found in a model test that a given propeller design cavitates heavily at the design operating value, different combinations of diameter and pitch, revolutions, blade width and outline, or section shape may be tried to eliminate or alleviate the observed cavitation. The same procedure is followed in the case of pumps and turbines. SEE DYNAMIC SIMILARITY.

Types. It has been found convenient to differentiate between a type of cavitation in which small bubbles suddenly appear on the solid boundary, grow in extent, and disappear and another type in which cavities form on the boundary and remain attached as long as the conditions that

Fig. 2. Bubble or transient cavitation on screw propeller in David Taylor Model Basin water tunnel at Carderock, Maryland. (*U.S. Navy*)

led to their formation remain unaltered. The former type is known as bubble or transient cavitation (**Fig. 2**) and the latter as steady-state or sheet cavitation (**Figs. 3** and **4**).

Physical causes. The exact mechanism of the disruption of a liquid when pressure is reduced is not fully understood. Thus, experiments and calculations show that pure, air-free, still water may sustain tension up to several hundreds of atmospheres before disruption takes place; on the other hand, numerous experiments have shown that in ordinary flowing water cavitation commences as the pressure approaches or reaches the vapor pressure. This discrepancy may be accounted for by the fact that ordinary water contains dissolved and entrained air or gas as well as solid particles and therefore is not pure in the foregoing sense. By way of explanation it is postulated that the presence of these foreign substances together with pressure fluctuations in turbulent flow interrupts the continuity of the liquid and lowers its surface tension. Microscopic cavities formed at these discontinuities or nuclei fill with vapor and small amounts of air or gas drawn from solution as the ambient pressure is reduced. This process continues until equilibrium

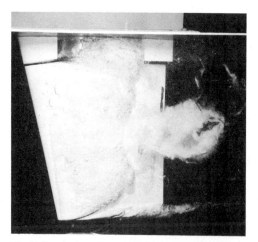

Fig. 3. Steady-state, intermittent, and vortex cavitation on model of ship's rudder observed in 24-in. (61-cm) water tunnel at David Taylor Model Basin. (*U.S. Navy*)

Fig. 4. Sheet cavitation on screw propeller in David Taylor Model Basin water tunnel. (*U.S. Navy*)

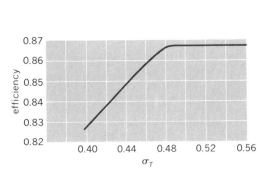

Fig. 5. Drop in efficiency of hydraulic turbine caused by cavitation.

Fig. 6. Thrust loss of screw propeller from cavitation.

between the various forces acting on the surface of these cavities is established. This postulate fits the observed facts that bubbles appear at discrete spots in low-pressure regions, grow quickly to relatively large size, and suddenly collapse as they are swept into regions of higher pressure.

Effects. It was mentioned initially that cavitation, when severe, lowers the efficiency of a machine, produces noise and vibrations, and causes rapid erosion of boundary surfaces. An example of the drop in efficiency of a hydraulic turbine after cavitation is fully developed is shown in **Fig. 5**.

An example of the deleterious effect of cavitation on the thrust of a marine propeller with decreasing σ is shown in **Fig. 6**. Examples of the destructive effect of cavitation on solid surfaces are shown in **Figs. 7** and **8**. This effect may be explained as follows. When cavitation bubbles

Fig. 7. Cavitation erosion on concrete specimens. (*U.S. Bureau of Reclamation*)

Fig. 8. Section of marine propeller surface showing erosion caused by cavitation. (*U.S. Navy*)

Fig. 9. Supercavitating propeller in 24-in. (61-cm) water tunnel at David Taylor Model Basin. (*U.S. Navy*)

form, they grow to full size in a very short time—about 2 microseconds. This time interval is too short for much air or gas to come out of solution, so that the bubbles are highly evacuated. On subsequent collapse in an equally short time interval, the liquid particles rush toward the center of the bubble virtually unimpeded and impinge on the surface with very high velocities. It is estimated that the surface stress caused by the impingement is of the order of 1000 atm (100 megapascals), which is sufficiently high to cause fatigue failure of the material in relatively short time. Some investigators hold that the explosive formation of the bubbles, intercrystalline electrolytic action, and the collapse of the bubbles all are factors contributing to the observed destruction.

Supercavitating propellers. The limitation on ship speed caused by loss of thrust when cavitation is severe has been overcome by a radical departure from conventional propeller designs. In this new design, cavitation on the backs of the blades (forward side) is induced by special blade sections at relatively low forward speed so that, when revolutions and engine power are increased, the whole back of each blade becomes enveloped by a sheet of cavitation. When this is completed, further increase in thrust at still higher engine power and rpm is obtained by the increase in positive pressure on the blade face (rear side); erosion is avoided because the collapse of the cavitation bubbles occurs some distance behind the trailing edges of the blades. The blade sections are usually wedge-shaped with a sharp leading edge to initiate cavitation at this point, a blunt trailing edge, and a concave face. Such a supercavitating propeller in action is shown in **Fig. 9**. The supercavitating propeller is no replacement for the conventional propeller, being suitable only for very high ship and engine speeds.

Bibliography. F. G. Hammitt, *Cavitation and Multiphase Flow Phenomena*, 1980.

GAS

C. F. Curtiss and J. O. Hirschfelder

A phase of matter characterized by relatively low density, high fluidity, and lack of rigidity. A gas expands readily to fill any containing vessel. Usually a small change of pressure or temperature produces a large change in the volume of the gas. The equation of state describes the relation between the pressure, volume, and temperature of the gas. In contrast to a crystal, the molecules in a gas have no long-range order.

At sufficiently high temperatures and sufficiently low pressures, all substances obey the ideal gas, or perfect gas, equation of state, shown as Eq. (1), where p is the pressure, T is the

$$pv = RT \tag{1}$$

absolute temperature, v is the molar volume, and R is the gas constant. Absolute temperature T expressed on the Kelvin scale is related to temperature t expressed on the Celsius scale as in Eq. (2).

$$T = t + 273.16 \tag{2}$$

The gas constant is

$$R = 82.0567 \text{ cm}^3\text{-atm/(mole)(K)} = 82.0544 \text{ ml-atm/(mole)(K)}$$

The molar volume is the molecular weight divided by the gas density.

Empirical equations of state. At lower temperatures and higher pressures, the equation of state of a real gas deviates from that of a perfect gas. Various empirical relations have been proposed to explain the behavior of real gases. The equations of J. van der Waals (1899), Eq. (3), of P. E. M. Berthelot (1907), Eq. (4), and F. Dieterici (1899), Eq. (5), are frequently used. In these

$$\left(p + \frac{a}{v^2}\right)(v - b) = RT \tag{3}$$

$$\left(p + \frac{a}{Tv_2}\right)(v - b) = RT \tag{4}$$

$$pe^{a/vRT}(v - b) = RT \tag{5}$$

equations, a and b are constants characteristic of the particular substance under considerations. In a qualitative sense, b is the excluded volume due to the finite size of the molecules and roughly equal to four times the volume of 1 mole of molecules. The constant a represents the effect of the forces of attraction between the molecules. In particular, the internal energy of a van der Waals gas is $-a/v$. None of these relations gives a good representation of the compressibility of real gases over a wide range of temperature and pressure. However, they reproduce qualitatively the leading features of experimental pressure-volume-temperature surfaces.

Schematic isotherms of a real gas, or curves showing the pressure as a function of the volume for fixed values of the temperature, are shown in **Fig. 1**. Here T_1 is a very high temperature and its isotherm deviates only slightly from that of a perfect gas; T_2 is a somewhat lower temperature where the deviations from the perfect gas equation are quite large; and T_c is the critical temperature. The critical temperature is the highest temperature at which a liquid can exist. That is, at temperatures equal to or greater than the critical temperature, the gas phase is the only phase that can exist (at equilibrium) regardless of the pressure. Along the isotherm for T_c lies the critical point, C, which is characterized by zero first and second partial derivatives of the pressure with respect to the volume. This is expressed as Eq. (6). At temperatures lower than the

$$(\partial p/\partial v)_c = (\partial^2 p/\partial v^2)_c = 0 \tag{6}$$

critical, such as T_3 or T_4, the equilibrium isotherms have a discontinuous slope at the vapor pressure. At pressures less than the vapor pressure, the substance is gaseous; at pressures greater than the vapor pressure, the substance is liquid; at the vapor pressure, the gas and liquid phases (separated by an interface) exist in equilibrium.

Along one of the isotherms of the empirical equations of state discussed above, the first and second derivatives of the pressure with respect to the volume are zero. The location of this

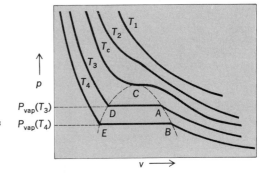

Fig. 1. Schematic isotherms of a real gas. C is the critical point. Points A and B give the volume of gas in equilibrium with the liquid phase at their respective vapor pressures. Similarly, D and E are the volumes of liquid in equilibrium with the gas phase.

critical point in terms of the constants a and b is shown below; p_c and v_c are the pressure and volume at the critical temperature.

	van der Waals	Berthelot	Dieterici
p_c	$\dfrac{a}{27b^2}$	$\left(\dfrac{aR}{216b^3}\right)^{1/2}$	$\dfrac{a}{4e^2 b^2}$
v_c	$3b$	$3b$	$2b$
T_c	$\dfrac{8a}{27Rb}$	$\left(\dfrac{8a}{27Rb}\right)^{1/2}$	$\dfrac{a}{4Rb}$
$\dfrac{p_c v_c}{RT_c}$	0.3750	0.3750	0.2706

Some typical values of $p_c v_c / RT_c$ for real gases are as follows: 0.30 for the noble gases, 0.27 for most of the hydrocarbons, 0.243 for ammonia, and 0.232 for water. The van der Waals and Berthelot equations of state (3) and (4) cannot quantitatively reproduce the critical behavior of real gases because no substance has a value of $p_c v_c / RT_c$ as large as 0.375. The Dieterici equation (5) gives a good representation of the critical region for the light hydrocarbons but does not represent well the noble gases or water.

At temperatures lower than the critical point, the analytical equations of state, such as the van der Waals, Berthelot, or Dieterici equations, give S-shaped isotherms as shown in **Fig. 2**. From thermodynamic considerations, the vapor pressure is determined by the requirement that

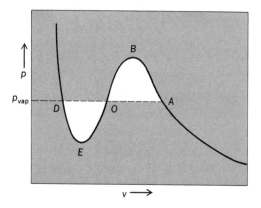

Fig. 2. Schematic low-temperature isotherm as given by van der Waals, Berthelot, or Dieterici equations of state. Here the line *DOA* corresponds to the vapor pressure. The point *A* gives the volume of the gas in equilibrium with the liquid phase, and *D* gives the volume of the liquid. The segment of the curve *DE* represents overexpansion of the liquid. The segment *AB* corresponds to supersaturation of the vapor. However, the segment *EOB* could not be attained experimentally.

the cross-hatched area *DEO* be equal to the cross-hatched area *AOB*. Under equilibrium conditions, the portion of this isotherm lying between *A* and *D* cannot occur. However, if a gas is suddenly compressed, points along the segment *AB* may be realized for a short period until enough condensation nuclei form to create the liquid phase. Similarly, if a liquid is suddenly overexpanded, points along *DE* may occur for a short time. For low temperatures, the point *E* may represent a negative pressure corresponding to the tensile strength of the liquid. However, the simple analytical equations of state cannot be used for a quantitative estimate of these transient phenomena. Actually, it is easy to show that the van der Waals, Berthelot, and Dieterici equations give poor representations of the liquid phase since the volume of most liquids (near their freezing point) is considerably less than the constant b.

Principle of corresponding states. In the early studies, it was observed that the equations of state of many substances are qualitatively similar and can be correlated by a simple scaling of the variables. To describe this result, the reduced or dimensionless variables, indicated by a subscript r, are defined by dividing each variable by its value at the critical point. These variables are given in Eqs. (7)–(9).

$$p_r = p/p_c \quad (7) \qquad T_r = T/T_c \quad (8) \qquad v_r = v/v_c \quad (9)$$

In its most elementary form, the principle of corresponding states asserts that the reduced pressure, p_r, is the same function of the reduced volume and temperature, v_r and T_r, for all substances. An immediate consequence of this statement is the statement that the compressibility factor, expressed as Eq. (10), is a universal function of the reduced pressure and the reduced

$$z = pv/RT \tag{10}$$

temperature. This principle is the basis of the generalized compressibility chart shown in **Fig. 3**. This chart was derived from data on the equation-of-state behavior of a number of common gases.

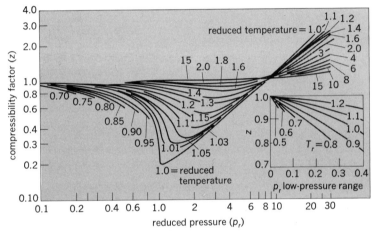

Fig. 3. The compressibility factor pv/RT as a function of the reduced pressure $p_r = p/p_c$ and reduced temperature $T_r = T/T_c$. (After O. A. Hougen, K. M. Watson, and R. A. Ragatz, *Chemical Process Principles*, pt. 2, Wiley, 1959)

It follows directly from the principle of corresponding states that the compressibility factor at the critical point z_c should be a universal constant. It is found experimentally that this constant varies somewhat from one substance to another. On this account, empirical tables were developed of the compressibility factor and other thermodynamic properties of gases as functions of the reduced pressure and reduced temperature for a range of values of z_c. Such generalized corresponding-states treatments are very useful in predicting the behavior of a substance on the basis of scant experimental data.

Theoretical considerations. The equation-of-state behavior of a substance is closely related to the manner in which the constituent molecules interact. Through statistical mechanical considerations, it is possible to obtain some information about this relationship. If the molecules are spherically symmetrical, the force acting between a pair of molecules depends only on r, the distance between them. It is then convenient to describe this interaction by means of the intermolecular potential $\varphi(r)$ defined so that the force is the negative of the derivative of $\varphi(r)$ with respect to r.

Two theoretical approaches to the equation of state have been developed. In one of these approaches, the pressure is expressed in terms of the partition function Z and the total volume V of the container in the manner of Eq. (11). Here k is the Boltzmann constant or the gas constant

$$p = kT \, (\partial \ln Z/\partial V) \tag{11}$$

divided by the Avogadro number N_0, $k = R/N_0$. For a gas made up of spherical molecules or atoms with no internal structure, the partition function is given as Eq. (12). In this expression, φ_{ij}

$$Z = \frac{1}{N!}\left(\frac{2\pi m k T}{h^2}\right)^{3N/2} \times \int \exp\left(-\sum_{i>j}\frac{\varphi_{ij}}{kT}\right) dv_1\, dv_2 \cdots dv_N \qquad (12)$$

is the energy of interaction of molecules i and j and the summation is over all pairs of molecules, h is Planck's constant, N is the total number of molecules, and the integration is over the three cartesian coordinates of each of the N molecules. The expression for the partition function may easily be generalized to include the effects of the structure of the molecules and the effects of quantum mechanics.

In another theoretical approach to the equation of state, the pressure may be written as Eq. (13), where $g(r)$ is the radial distribution function. This function is defined by the statement

$$p = \frac{NkT}{V} - \frac{2\pi N^2}{3V^2}\int g(r)\frac{d\varphi}{dr}r^3\, dr \qquad (13)$$

that $2\pi(N^2/V)g(r)r^2\, dr$ is the number of pairs of molecules in the gas for which the separation distance lies between r and $r + dr$. The radial distribution function may be determined experimentally by the scattering of x-rays. Theoretical expressions for $g(r)$ are being developed.

The compressibility factor $z = pV/NkT$ may be considered as a function of the temperature, T, and the molar volume, v. In the virial form of the equation of state, z is expressed as a series expansion in inverse powers of v, as in Eq. (14). The coefficients $B(T)$, $C(T)$, . . . , which

$$z = 1 + B(T)/v + C(T)/v^2 + \cdots \qquad (14)$$

are functions of the temperature, are referred to as the second, third, . . . , virial coefficients. This expansion is an important method of representing the deviations from ideal gas behavior. From statistical mechanics, the virial coefficients can be expressed in terms of the intermolecular potential. In particular, the second virial coefficient is Eq. (15). If the intermolecular potential is

$$B(T) = 2\pi N_0 \int (1 - e^{-\varphi/kT}) r^2\, dr \qquad (15)$$

known, Eq. (15) provides a convenient method of predicting the first-order deviation of the gas from perfect gas behavior. The relation has often been used in the reverse manner to obtain information about the intermolecular potential. Often $\varphi(r)$ is expressed in the Lennard-Jones (6–12) form, Eq. (16), where ϵ and σ are constants characteristic of a particular substance. Values of

$$\varphi(r) = 4\epsilon\left[\left(\frac{\sigma}{r}\right)^{12} - \left(\frac{\sigma}{r}\right)^6\right] \qquad (16)$$

these constants for many substances have been tabulated. In terms of these constants, the second virial coefficient has the form of Eq. (17), where $B^*(kT/\epsilon)$ is a universal function. If all substances

$$B(T) = (2/3)\pi N_0 \sigma^3 B^*(kT/\epsilon) \qquad (17)$$

obeyed this Lennard-Jones (6–12) potential, the simple form of the law of corresponding states would be rigorously correct.

Bibliography. I. B. Cohen (ed.), *Laws of Gases*, 1981; J. O. Hirschfelder et al., *Molecular Theory of Gases and Liquids*, 1964; R. Holub and P. Vonka, *The Chemical Equilibrium of Gaseous Systems*, 1976; R. Mohilla and B. Ferencz, *Chemical Process Dynamics*, 1982.

MEASUREMENT AND DISPLAY OF PROPERTIES

Pressure measurement	194
Manometer	196
Barometer	198
Bourdon-spring pressure gage	198
Pressure tranducer	200
Flow measurement	203
Torricelli's theorem	221
Metering orifice	221
Venturi tube	223
Borda mouthpiece	224
Pitot tube	225
Air-velocity measurement	226
Anemometer	228
Ripple tank	230
Towing tank	232
Water tunnel	235
Wind tunnel	238
Shock-wave display	254
Schlieren photography	256
Shadowgraph of fluid flow	258

PRESSURE MEASUREMENT
John H. Zifcak

The determination of the magnitude of a fluid force applied to a unit area. Pressure measurements are generally classified as gage pressure, absolute pressure, or differential pressure. Gage pressure is the difference between a given pressure and the pressure of the atmosphere. Absolute pressure is the total pressure, including that of the atmosphere. Atmospheric pressure was the first pressure that was really measured. Differential pressure is the difference between any two pressures, neither of which is atmospheric. Pressures less than atmospheric are called vacuum.

The **table** compares nine common units of pressure measurement. To avoid confusion, gage, absolute, or differential is often suffixed; for example, 100 pounds per square inch gage pressure (100 psig), 115 pounds per square inch absolute pressure (115 psia), or 30 pounds per square inch differential (30 psid).

In the laboratory, pressure is an important measurement, since the pressure level has a significant effect on most physical, chemical, and biological processes.

In industry—particularly in the process industries—pressure is measured and controlled to maintain uniformity of product, to guide in safe plant operation, to determine pumping head for fluid transfer, and to measure other variables indirectly, including weight, liquid level, temperature, flow and density of fluids, and hydraulic forces.

Pressure gages generally fall in one of three categories, based on the principle of operation: liquid columns, expansible-element gages, and electrical pressure transducers.

Liquid-column gage. This type of pressure gage includes barometers and manometers. It consists of a U-shaped tube partly filled with a nonvolatile liquid. Water and mercury are the two most common liquids used in this type of gage. *See Barometer; Manometer.*

If one leg is left open to the atmosphere, the difference in level is a direct measure of gage pressure (see **Fig. 1**). If the differential of two pressures is desired, each is applied to a leg of the gage. The level rises in the low-pressure leg and drops in the high-pressure leg. The differential pressure is the difference in level (head) multiplied by the density of the instrument liquid.

Expansible metallic-element gages. These are in wide use throughout industry, due to their low cost and freedom from the operational limitations of liquid gages. There are three classes: bourdon, diaphragm, and bellows. All forms—as single elements—are affected by variations in external (atmospheric) pressures and hence are generally used as gage elements. Accuracies vary depending on materials, design, and precision of components.

These elements may be designed to produce either motion or force under applied pressure.

Pressure equivalents

Unit	kPa	bar	kg/cm²	psi	atm	in. Hg (0°C)	mmHg (0°C)	in. H₂O (4°C)	mmH₂O (4°C)
1 kilopascal	= 1	0.01	0.01020	0.1450	0.009870	0.2953	7.501	4.015	102.0
1 bar	= 100	1	1.020	14.50	0.9870	29.53	750.1	401.5	10,200
1 kilogram per square centimeter	= 98.07	0.9807	1	14.22	0.9678	28.96	735.6	393.7	10,000
1 pound per square inch	= 6.895	0.06895	0.07031	1	0.06805	2.036	51.72	27.68	703.1
1 atmosphere	= 101.3	1,013	1.033	14.70	1	29.92	760.0	406.8	10,330
1 inch of mercury*	= 3.386	0.03386	0.03453	0.4912	0.03342	1	25.40	13.60	345.3
1 millimeter of mercury*	= 0.1333	0.001333	0.001360	0.01934	0.001316	0.03937	1	0.5352	13.60
1 inch of water†	= 0.2491	0.002491	0.002540	0.03613	0.002458	0.07356	1.868	1	25.40
1 millimeter of water†	= 0.009807	0.9807 × 10⁻⁴	0.0001	0.001422	0.9678 × 10⁻⁴	0.002896	0.07356	0.03937	1

*Closed column, in vacuum at 0°C. †Closed column, in vacuum at 4°C.

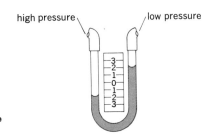

Fig. 1. Liquid-column gage (U-tube manometer).

The more common motion type may directly position the pointer of a concentric indicating gage; position a linkage to operate a recording pen, or pneumatic relaying system to convert the measurement into a pneumatic signal; or position an electrical transducer to convert to an electrical signal.

Bourdon-spring gages (**Fig. 2**), in which pressure acts on a shaped, flattened, elastic tube, are by far the most widely used type of instrument. These gages are simple, rugged, and inexpensive. Common shapes are the C-tube, spiral-, helix-, and twisted-tube element. While most bourdons are metal, improved accuracy is sometimes accomplished through the use of quartz bourdons. SEE BOURDON-SPRING PRESSURE GAGE.

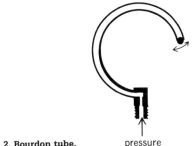

Fig. 2. Bourdon tube.

In diaphragm-element gages, pressure applied to one or more contoured diaphragm disks acts against a spring or against the spring rate of the diaphragms, producing a measurable motion (**Fig. 3**). The size, number, and thickness of the disks determine the range.

For lower pressures, slack membrane diaphragms are used, opposed by a calibration spring; these instruments can detect differential pressures as low as 0.01 in. (0.25 mm) of water (2.5 pascals).

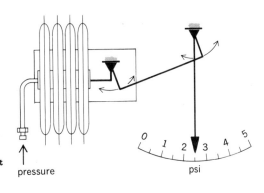

Fig. 3. Diaphragm-element gage.

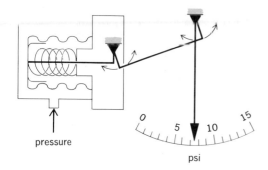

Fig. 4. Bellows element gage.

In bellows-element gages, pressure in or around the bellows moves the end plate of the bellows against a calibrated spring, producing a measurable motion (**Fig. 4**). Advances in the quality of metals and in the manufacture of metallic-element gages have led to a corresponding improvement in the quality of this type of gage. This is also true of the bourdon and diaphragm elements.

Electrical pressure transducers. Pressure measurement using these transducers has become more common. These devices convert a pressure to an electrical signal which may be used to indicate a pressure or to control a process. The electrical signal may be analog where the output is proportional to the measured pressure, or digital where the pressure is converted to a numerical value directly equivalent to its written value.

Such devices as strain gages, resistive, magnetic, crystal, and capacitive pressure transducers are commonly used to convert the measured pressure to an electrical signal. SEE PRESSURE TRANSDUCER.

Measurement standards. For pressures below 20 psig (138 kilopascals gage), the universally accepted standard of pressure measurement—both in the laboratory and in the industrial plant—is the classic manometer, using mercury or water.

For higher pressures, the standard is the deadweight tester. The principle is the balance of the force exerted by a precisely known weight on a piston of precisely measured area against a variable hydraulic pressure.

The pressure gage which is to be checked or calibrated is connected to the hydraulic reservoir. Weights corresponding to the desired pressure are placed on the piston. The piston and weights are rotated to reduce the effect of friction. Hydraulic pressure is increased. When the pressure is reached, the measuring piston floats freely.

The deadweight tester is often used as a primary standard for laboratory use, and secondary standards (test gages which have been calibrated against the deadweight tester) are used as field or plant test instruments. Deadweight equipment is produced having accuracy to within 0.025% of reading.

Bibliography. R. P. Benedict, *Fundamentals of Temperature, Pressure and Flow Measurements*, 3d ed., 1984.

MANOMETER
JOHN H. ZIFCAK

A double-leg liquid-column gage that is used to measure the difference between two fluid pressures. Micromanometers are precision instruments which typically measure from very low pressures to 50 mmHg (6.7 kilopascals). The barometer is a special case of manometer with one pressure at zero absolute. SEE BAROMETER.

The various types of manometers have much in common with the U-tube manometer, which consists of a hollow tube, usually glass, a liquid partially filling the tube, and a scale to measure the height of one liquid surface with respect to the other.

One end of the hollow tube of a manometer is the high-pressure end and the other the low-pressure. If the low-pressure end is open to atmospheric pressure, the remaining end may be used to measure vacuum or gage pressure.

U-tube manometer. If the legs of this manometer are connected to separate sources of pressure, the liquid will rise in the leg with the lower pressure and drop in the other leg (**Fig. 1**). The difference between the levels is a function of the applied pressure and the specific gravity of the pressurizing and fill fluids. The cross-sectional area of the tubes does not affect the difference between the levels. A scale graduated in inches or centimeters is commonly affixed between the legs of the manometer.

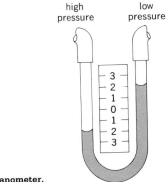

Fig. 1. U-tube manometer.

Well-type manometer. One leg of this manometer has a relatively small diameter, and the second leg is a reservoir. The cross-sectional area of the reservoir may be as much as 1500 times that of the vertical leg, so that the level of the reservoir does not change appreciably with a change of pressure. Small adjustments to the scale of the vertical leg compensate for the little reservoir level change that does occur. Readings of differential or gage pressure may then be made directly on the vertical scale. Mercurial barometers are commonly made as well-type manometers (**Fig. 2**).

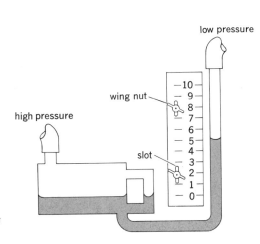

Fig. 2. Well-type manometer with zeroing adjustment.

Inclined-tube manometer. This is used for gage pressures below 10 in. (250 mm) of water differential. The leg of the well-type manometer is inclined from the vertical to elongate the scale (**Fig. 3**). Inclined double-leg U-tube manometers are also used to measure very low differential pressures. Water or some other low-specific-gravity liquid is employed for this application.

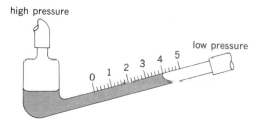

Fig. 3. Inclined-tube manometer.

Micromanometer. Micromanometer U-tubes have been made using precision-bore glass tubing, a metallic float in one leg, and an inductive coil to sense the position of the float. A null-balance electronic indicator can detect pressure changes as minute as 0.005 in. (13 micrometers) of water. Such devices are normally used as laboratory standards. *See* Pressure measurement.

BAROMETER
John H. Zifcak

An absolute pressure gage which is specifically designed to measure atmospheric pressure. This instrument is a type of manometer with one leg at zero pressure absolute. *See* Manometer.

The common meteorological barometer is a liquid-column gage with mercury. The top of the column is sealed, and the bottom is open and submerged below the surface of a reservoir of mercury. The atmospheric pressure on the reservoir keeps the mercury at a height proportional to that pressure. An adjustable scale, with a vernier scale, allows a reading of column height. Before each reading, the scale must be adjusted to correct for mercury level in the reservoir. Often a peg (the zero reference point) attached to the scale is adjusted by a micrometer to just touch the mercury surface. The apparent mercury height must be corrected for temperature and gravity, with standard conditions being 0°C and 980.665 cm/s^2.

Typically, the barometer scale is in millimeters or inches of mercury, with sea-level pressures in the order of 760 mm or 29.9 in. of mercury.

Aneroid barometers using metallic diaphragm elements are usually less accurate, though often more sensitive, devices, and not only indicate pressure but may be used to record. *See* Pressure measurement.

BOURDON-SPRING PRESSURE GAGE
H. G. Sell and P. J. Walsh

A mechanical pressure-measuring instrument employing as its sensing element a curved or twisted metallic tube, flattened in cross section. One end of the tube is closed, and the fluid pressure to be measured is applied through the other end. As the pressure is increased, the tube becomes more nearly circular in cross section and tends to straighten. The motion of the free (closed) end of the tube is a measure of the internal pressure.

The fundamental principle that motion of a bourdon tube is proportional to applied pressure is credited to Eugene Bourdon, nineteenth-century French inventor.

Fig. 1. C-tube bourdon element with mechanism for industrial pressure gage. (*Foxboro Co.*)

Pressure-measuring instruments using bourdon tubes are used for pressure ranges from 0–10 pounds per square inch, gage (psig) to 0–100,000 psig (0–70 kilopascals to 0–700 megapascals gage), as well as for vacuum. The accuracy may be from 0.1 to 2.0% of full-scale reading, depending on the materials, design, and precision of components.

The bourdon pressure element may be made in any one of a number of shapes. The commonest forms are the C tube, the spiral, and the helical element.

The C tube (**Fig. 1**) is simple and rugged, and is used in concentric dial-indicating pressure gages. The C tube is also used in some electrical pressure transducers where small tip motion is permissible or desirable. Rangers are to 20,000 psi (140 MPa).

The spiral and helical forms (**Figs. 2** and **3**) are used in recording and controlling instru-

Fig. 2. An example of a spiral-type bourdon element. (*Foxboro Co.*)

Fig. 3. An example of a helical-type bourdon element. (*Foxboro Co.*)

Fig. 4. Twisted bourdon element. (*Norwood Controls*)

ments for greater tip motion or lower wall stress. Spiral bourdon elements are available for ranges as high as 0–4000 psi (0–28 MPa), and helical elements are made for pressures as high as 100,000 psi (700 MPa).

The twisted tube (**Fig. 4**), rugged and compact, is often preferred, especially for use with electrical pressure transducers. SEE PRESSURE TRANSDUCER.

Bourdon tubes are available in a variety of copper alloys and AISI 300 series stainless steels. In some respects, the copper alloys give better performance, but the stainless steels offer greater corrosion resistance. Nickel-iron alloy bourdons, which have a nearly constant modulus, are also available, so that the pressure reading is not affected by the temperature of the instrument.

Mechanical and pneumatic instruments with bourdon elements allow measurement accuracies to within 0.5% of full-scale pressure. Electrical transducers are available if greater accuracy is needed. Bourdon gages measure the difference between internal and external pressure on the tube. Since the external pressure is almost always atmospheric pressure, the gage reads the difference between the measured pressure and atmospheric pressure, hence the term gage pressure.

The bourdon gage is a popular industrial pressure-measuring instrument because of its relatively low cost, reasonable accuracy, and durability. SEE PRESSURE MEASUREMENT.

PRESSURE TRANDUCER
JOHN H. ZIFCAK

An instrument component which detects a fluid pressure and produces an electrical, mechanical, or pneumatic signal related to the pressure. SEE TRANSDUCER.

In general, the complete instrument system comprises a pressure-sensing element such as a bourdon tube, bellows, or diaphragm element: a device which converts motion or force produced by the sensing element to a change of an electrical, mechanical, or pneumatic parameter; and an indicating or recording instrument. Frequently the instrument is used in an autocontrol loop to maintain a desired pressure.

Although pneumatic and mechanical transducers are commonly used, electrical measurement of pressure is often preferred because of a need for long-distance transmission, higher accuracy requirements, more favorable economics, or quicker response. Especially for control applications, pneumatic pressure signal transmission may be desirable over electrical where flammable materials are present.

Electrical pressure transducers may be classified by the operating principle as resistive transducers, strain gages, magnetic transducers, crystal transducers, capacitive transducers, and resonant transducers.

Resistive pressure transducers. Pressure is measured in these transducers by an element that changes its electrical resistance as a function of pressure.

Many types of resistive pressure transducers use a movable contact, positioned by the pressure-sensing element. One form is a contact sliding along a continuous resistor, which may be straight-wire, wire-wound, or nonmetallic such as carbon. If the cross section of the resistor is constant, the change in resistance will be proportional to the motion of the contact. The cross section may be made nonuniform to give a nonlinear relation between motion and change of resistance.

The resistance element may be curved or part of an arc for convenience in measuring angular motion.

Figure 1 shows one type of resistive pressure transducer. A bellows opposed by a precisely designed spring senses the pressure and converts the pressure to a linear motion of the plate between the bellows and the spring. The plate bears a contact which wipes the surface of the precision wire-wound resistor. If a constant potential (ac or dc) is maintained across the resistor and if the resistance of the voltmeter is high with respect to the resistor, the measured voltage is a precise measure of the pressure.

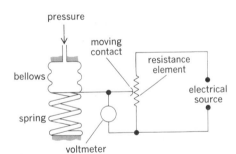

Fig. 1. Resistive pressure transducer.

Strain-gage pressure transducers. These might be considered to be resistive transducers, but are usually classified separately. Strain gage pressure transducers convert a physical displacement into an electrical signal. When a wire is placed in tension, its electrical resistance increases. The change in resistance is a measure of the displacement, hence of the pressure. Its advantages include infinite resolution and small size. The strain gage is usually used in conjunction with a bridge circuit.

Another variety of strain gage transducer uses integrated circuit technology. Resistors are diffused onto the surface of a silicon crystal within the boundaries of an area which is etched to form a thin diaphragm. Conductive pads at the ends of the resistors and fine wires provide the electrical connection to external electronic circuitry. Often four resistors are connected in a bridge circuit so that an applied pressure leaves one pair of resistors in compression and the other pair in tension, yielding the maximum output change.

Magnetic pressure transducers. In this type, a change of pressure is converted into change of magnetic reluctance or inductance when one part of a magnetic circuit is moved by a pressure-sensing element—bourdon tube, bellows, or diaphragm.

Reluctance-type pressure transducer. This type produces in a magnetic circuit a change of magnetic reluctance which is directly related to pressure. The change of reluctance is

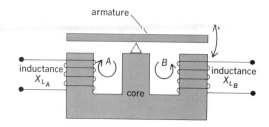

Fig. 2. Reluctance-type pressure transducer.

usually within one or two coils, wound intimately about the magnetic material in the magnetic circuit.

A representative reluctance-changing device is shown in **Fig. 2**. A bourdon-type or other pressure-sensing device rotates the armature. The reluctances in the magnetic paths A and B are determined chiefly by the lengths of the air gaps between the armature and the core. The inductance and inductive reactance of each winding depend on the reluctance in its magnetic path.

If the armature is at a neutral symmetrical position, the air gaps are equal, and the inductive reactances X_{LA} and X_{LB} are equal. Change of pressure decreases one air gap and increases the other, thus changing the ratio of the inductive reactances X_{LA} and X_{LB}. These changes can be used in a variety of circuits to produce an electrical signal which is a measure of pressure. The signal is transmitted to a measuring or controlling instrument.

Inductive-type pressure transducer. A change in inductance and inductive reactance of one or more windings is produced by the movement of a magnetic core that is positioned by a bourdon tube or other pressure-sensing element. Unlike the action of a reluctance-type transducer, the inductance change is caused by a change in air gap within the winding, rather than in a relatively remote portion of the magnetic circuit.

Figure 3 shows a representative ratio-type inductive device. The pressure-sensing element moves the core in response to changes of pressure. When the core is in a central position, the inductances of the two coils are equal. When a pressure change moves the core, the ratio of the two inductances is changed. Energy is supplied to the coils by the same bridge circuit that measures the ratio of inductances.

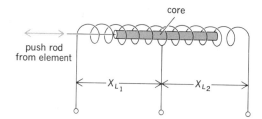

Fig. 3. Inductive pressure transducer.

Another form of inductive pressure transducer is the linear variable differential transformer (LVDT; **Fig. 4**). When the core is centered, equal voltages are induced in two oppositely wound secondary windings and the output voltage is zero. A change of pressure moves the core, increasing the voltage induced in one secondary and decreasing the voltage induced in the other. The change in output (differential) voltage is thus a measure of the pressure.

Piezoelectric pressure transducers. Some crystals produce an electric potential when placed under stress by a pressure-sensing element. The stress must be carefully oriented with respect to a specific axis of the crystal.

Suitable crystals include naturally occurring quartz and tourmaline, and synthetic crystals

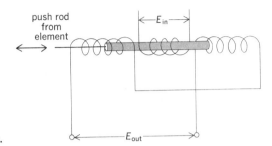

Fig. 4. Differential transformer.

such as Rochelle salts and barium titanate. The natural crystals are more rugged and less subject to drift. Although the synthetic crystals offer much higher voltage output, an amplifier is usually required for both types. Crystal transducers offer a high speed of response and are widely used for dynamic pressure measurements in such applications as ballistics and engine pressures.

Capacitive pressure transducers. Almost invariably, these sense pressure by means of a metallic diaphragm, which is also used as one plate of a capacitor. Any variation in pressure changes the distance between the diaphragm and the other plate or plates, thereby changing the electrical capacitance of the system. The change in capacitance can be used to modify the amplitude of an electrical signal.

Resonant transducers. This transducer consists of a wire or tube fixed at one end and attached at the other (under tension) to a pressure-sensing element. The wire is placed in a magnetic field and allowed to oscillate at its resonant frequency by means of an electronic circuit. As the pressure is increased, the element increases the tension in the wire or tube, thus raising its resonant frequency. This frequency may be used directly as the transducer output or converted to a dc voltage or digital output through the use of intermediate electronics. SEE PRESSURE MEASUREMENT.

Bibliography. R. P. Benedict, *Fundamentals of Temperature, Pressure, and Flow Measurements*, 3d ed., 1984.

FLOW MEASUREMENT
MEAD BRADNER AND LEWIS P. EMERSON

The determination of the quantity of a fluid, either a liquid, vapor, or gas, that passes through a pipe, duct, or open channel. Flow may be expressed as a rate of volumetric flow (such as liters per second, gallons per minute, cubic meters per second, cubic feet per minute), mass rate of flow (such as kilograms per second, pounds per hour), or in terms of a total volume or mass flow (integrated rate of flow for a given period of time).

Flow measurement, though centuries old, has become a science in the industrial age. This is because of the need for controlled process flows, stricter accounting methods, and more efficient operations, and because of the realization that most heating, cooling, and materials transport in the process industries is in the form of fluids, the flow rates of which are simple and convenient to control with valves or variable speed pumps.

Measurement is accomplished by a variety of means, depending upon the quantities, flow rates, and types of fluids involved. Many industrial process flow measurements consist of a combination of two devices: a primary device that is placed in intimate contact with the fluid and generates a signal, and a secondary device that translates this signal into a motion or a secondary signal for indicating, recording, controlling, or totalizing the flow. Other devices indicate or totalize the flow directly through the interaction of the flowing fluid and the measuring device that is placed directly or indirectly in contact with the fluid stream. **Table 1** shows several types of metering devices, the principles upon which they operate, and comparative characteristics.

Table 1. Examples of fluid meters

Name or type	Operating principle	Useful flow range
Orifice	Differential pressure (ΔP) over a restriction in the flow	3.5:1
Venturi		4:1
Nozzle		4:1
Flow tubes		4:1
Pitot-static	Compares impact pressure with the static pressure	3.5:1
Averaging pitot		3.5:1
Pipe elbow	Pressure from change in it	3.5:1
Target	Force on circular obstruction	3.5:1
Weir	Variable head, variable area	75:1
Flume	Variable head, variable area	up to 100:1
Rotameter (free float)	Automatic variable area	10:1
Piston		10:1
Plug		10:1
Magnetic	Electromagnetic induction	20:1
Tracer	Transit time	10:1
Nutating disk	Positive displacement; separation into and counting of discrete fluid quantities (most of these are available with high accuracy for custody transfer, with a less expensive, less accurate one for monitoring uses)	20:1
Rotary vane		25:1
Oval-shaped gear		50:1
Lobed impeller		10:1
Abutment rotor		25:1
Drum gas meter		40:1
Gas turbine meter	Converts velocity to proportional rotational speed	10:1 to 25:1
Liquid turbine		10:1 to 25:1
Bearingless		10:1 to 50:1
Insertion turbine		10:1
Propeller		25:1
Helical impeller		15:1
Vortex cage		15:1
Axial flow mass	Momentum effect due to change in direction of fluid path	15:1
Gyroscopic/Coriolis		10:1
Gyroscopic		>10:1
Thermal loss	Heat extracted by fluid	10:1
Vortex flowmeter	Counting vortices shedding from an obstruction	10:1
Ultrasonic detection of vortex		8:1 to 100:1
Swirl	Precessing rotating vortex	8:1 to 100:1
		10:1 to 100:1
Fluidic oscillator	Coanda wall attachment	30:1
Ultrasonic contrapropagating: clamp-on built-in spool piece	Time difference upstream/downstream	25:1
Ultrasonic contrapropagating: multiple path in spool piece		
Ultrasonic contrapropagating: axial in spool piece		
Ultrasonic correlation	Transit time of turbulence	10:1
Ultrasonic correlation: deflection (drift)	Deflection of beam across pipe	5:1
Ultrasonic contrapropagating: Doppler	Doppler shift–reflection from particles	10:1
Noise measurement	Spectral distribution of noise	5:1
Nuclear magnetic resonance	Transit time of nuclear disturbance	10:1
Laser Doppler velocimeter	Frequency shift of light reflected from particles	>1000:1

*Best accuracy can normally be achieved by having the calibration installed, or in a section of pipe to be installed with it; urv = upper range value.

VOLUME FLOW RATE METERS

Flow rate meters may modify the flow pattern by insertion of some obstruction or by modification of the shape of the walls of the conduit or channel carrying the fluid or they may measure the velocity without affecting the flow pattern.

Differential-producing primary devices. Differential-producing primary devices (sometimes called head meters) produce a difference in pressure caused by a modification of the flow pattern. The pressure difference is based on the laws of conservation of energy and conservation of mass. The law of conservation of energy states that the total energy at any given point in the stream is equal to the total energy at a second point in the stream, neglecting frictional and turbulent losses between the points. It is possible, however, to convert pressure (potential energy) to a velocity (kinetic energy), and vice versa. By use of a restriction in the pipe, such as an orifice plate (**Fig. 1**), venturi tube (**Fig. 2**), flow nozzle (**Fig. 3**), or flow tube (**Fig. 4**), a portion of the potential energy of the stream is temporarily converted to kinetic energy as the fluid speeds up to pass through these primary devices. *See* BERNOULLI'S THEOREM.

The secondary device measures this change in energy as a differential pressure ΔP, which is related to the flow by Eq. (1), where V is the velocity of flow through the restriction and ρ_f is

$$V = K_1\sqrt{\Delta P/\rho_f} \tag{1}$$

the mass density of the flowing fluid. The constant K_1 includes the orifice or throat diameter and the necessary dimension units, sizes, local gravity, and so forth. It also contains the discharge coefficient which has been experimentally determined and reduced to equation form for certain standard differential producers [namely, the square-edged orifice with specified pressure taps, ASME (American Society of Mechanical Engineers) nozzles, and classical venturis].

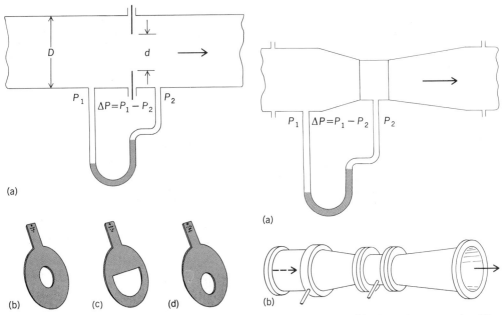

Fig. 1. Orifice plate primary device. (a) Schematic cross section. (b) Concentric plate. (c) Segmental plate. (d) Eccentric plate.

Fig. 2. Venturi tube. (a) Schematic cross section. (b) Perspective view.

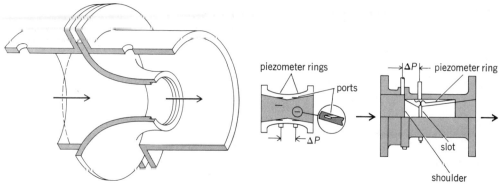

Fig. 3. Cutaway view of a flow nozzle.

Fig. 4. Two types of flow tube. (*After 1969 Guide to process instrument elements, Chem. Eng., 76(12):137–164, 1969*)

Mass flow for liquids, q_m, is given by Eq. (2), where K_2 is given by Eq. (3), and the quantities in Eqs. (2) and (3) are given in **Table 2**. The approximate value of the sizing factor S, upon which K_2 depends, can be found from **Fig. 5**.

$$q_m = K_2 \sqrt{\Delta P \rho_f} \qquad (2)$$

$$K_2 = NSD^2 \qquad (3)$$

Beyond the restriction, the fluid particles slow down until the velocity is once more the same as that in the upstream conduit section. However, the static pressure is less due to the effect of turbulent and frictional losses associated with the acceleration and deceleration at the restriction. The magnitude of the loss depends upon the configuration of the primary device. SEE METERING ORIFICE.

Orifice plate. The common orifice plate (Fig. 1) is a thin plate inserted between pipe flanges, usually having a round, concentric hole with a sharp, square upstream edge. The extensive empirical data available and its simplicity make it the most common of the primary devices. Segmental and eccentric orifices are useful for measurement of liquids with solids, and of vapors or gases with entrained liquids.

Venturi tube. This device (Fig. 2) has the advantage of introducing less permanent loss of pressure than the orifice plate. The converging inlet cone permits solids and dirt to be flushed through it, and the outlet cone reduces the turbulent losses. SEE VENTURI TUBE.

Table 2. Quantities used in liquid-flow calculation

Symbol	Quantity	Unit or value		
		SI units	U.S. Customary units	
N	Constant that depends upon the system of units used	1.11	0.099	0.52
D	Inside diameter of pipe	meter	inch	inch
ΔP	Pressure difference	pascal	in. H_2O*	$lbf/in.^2$ (psi)
ρ_f	Density of the flowing fluid	kg/m^3	lb_m/ft^3	lb_m/ft^3
β	Ratio, (d/D)	dimensionless		
C	Discharge coefficient	dimensionless		
S	Sizing factor = $C\beta^2/\sqrt{1-\beta^4}$	dimensionless		
d	Orifice or throat diameter	meter	inch	inch
q_m	Mass flow rate	kg/s	lb_m/s	lb_m/s

*The pressure of a column of water 1 in. (25.4 mm) high at 68°F (20°C) and acceleration of gravity of 32.174 ft/s² (9.80665 m/s²).

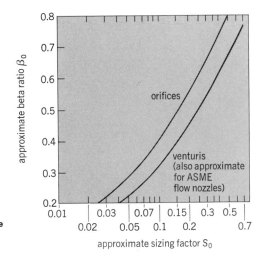

Fig. 5. Sizing factors for differential-primary-producing elements, for liquid flows, based on $C = 0.6 + 0.06\ S$ for orifices, $C = 0.995$ for venturis. The factors do not include the Reynolds number correction. (Foxboro Co.)

Flow nozzle. The flow nozzle (Fig. 3) also has a converging inlet, but it has no diverging outlet section. The pressure loss is similar to that with an orifice plate for the same differential pressure. It is used where solids are entrained in the liquid and also where fluid velocities are particularly high (it will pass more fluid than an orifice of the same diameter).

Flow tubes. Several designs of primary devices have been developed to give lower permanent pressure loss than orifices and shorter overall length than standard venturis, and they sometimes incorporate piezometer chamber averaging of the pressure tap. Figure 4 shows two such designs.

Pitot tubes. In industrial flow measurement, pitot tubes are used in spot checks and for comparative measurements to determine trends. A pitot tube measures the difference between total impact pressure and static pressure. In the device shown in **Fig. 6** the central tube receives the impact pressure, while the holes in the outer tube are subjected to the static pressure. A secondary device senses the difference in these pressures and converts it into an indication of fluid velocity at the location of the tip. There are a wide variety of other tube configurations

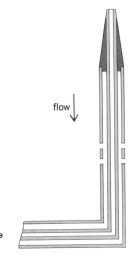

Fig. 6. Laboratory-type pitot tube.

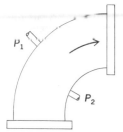

Fig. 7. Pipe elbow. Difference between outside pressure P_1 and inside pressure P_2 indicates volume flow rate.

available for sensing both impact and static pressure or combinations thereof. Several readings should be taken across the pipe along a diameter (preferably two diameters) of the pipe and the average velocity determined. The pitot tube is used extensively as a laboratory device for the point velocity measurement and also for speed measurement of vehicles and in wind tunnel work.

The averaging pitot tube stimulates an array of pitot tubes across the pipe. It consists of a rod extending across the pipe with a number of interconnected upstream holes and a downstream hole for the static pressure reference. In one design, the diamond shape of the rod is claimed to present a very low flow obstruction and still, with the interconnected ports, to provide a differential pressure that is representative of the flow relatively unaffected by profile differences. SEE AIR-VELOCITY MEASUREMENT; PITOT TUBE.

Pipe elbow. Fluid flow around the bend in a pipe elbow (**Fig. 7**) produces a centrifugal force greater on the outside than on the inside, and thereby produces a difference in pressure between pressure connections on the inside and outside of the bend. This differential pressure is used as an indication of the volume flow rate. This device is often used for rough flow measure-

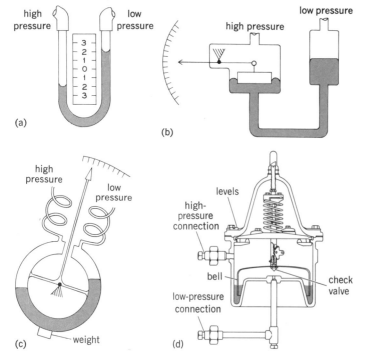

Fig. 8. Differential-pressure secondary devices. (a) U-tube manometer. (b) Mercury float mechanism. (c) Weight-balanced ring-type meter. (d) Bell type (*Foxboro Co.*).

ment because it is not necessary to disturb existing piping. Carefully installed and calibrated, the device can provide an accurate measurement.

Differential-pressure secondary devices. All the primary devices discussed above create a difference in pressure between two points. This differential pressure must be measured and converted by the equations given to obtain flow. Differential pressure may be measured accurately by simple liquid-filled (usually mercury) U-tube manometers or by more refined types of meters, such as the float-operated mechanism or the weight-balanced, ring-type meter, to provide controlling, recording, and totalizing functions. These devices (**Fig. 8**) are ordinarily connected directly to the pressure connections on the primary device and are therefore exposed to the process fluids. Sometimes an inert-liquid-seal fluid is used to isolate the secondary device from corrosive process fluids. In either case, the manometers are best located adjacent to the primary device. These manometers are accurate if the mercury is clean and the fluid flow rate is not changing too rapidly. SEE MANOMETER.

Dry-type sensors. Diaphragm- and bellows-type differential-pressure measuring devices were developed to eliminate manometer fluids and to provide faster response and easier installations. **Figure 9** shows a representative bellows-type differential-pressure sensor. The range of these instruments is changed by removing one cover and adjusting the range spring or replacing it with one of a different spring rate. Thermal expansion of the fill fluid contained in the two bellows is compensated for by the bimetal-supported additional convolutions shown at the left of the figure. These are very stable and reliable instruments.

Many direct-connected meters have been modified to provide a pneumatic or electric signal to remote instrumentation. Serviceable transmitters have been developed. Two types of meters, the force-balance and deflection types, are suited to either pneumatic or electric transmission of the flow signal.

The force-balance type (**Fig. 10**) causes the variation of an air pressure in a small bellows to oppose the net force created by the differential pressure on opposite sides of a diaphragm or diaphragm capsule. This air pressure is used as a signal representing the flow rate. An electronic force-balance design utilizes a restoring force due to a force motor and uses an inductive detector for unbalance; the current in the motor is the transmitted (4–20 mA) signal. One electronic version

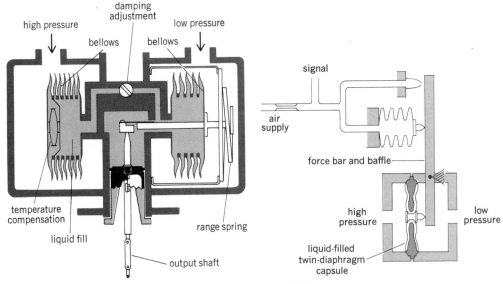

Fig. 9. Bellows-type differential-pressure-measuring element. (*Foxboro Co.*)

Fig. 10. Diaphragm meter, force balance type.

provides an output signal proportional to flow by making the feedback force the interaction between two motors, with the current flowing through them (the force is squared).

Many types of deflection differential-pressure cells are offered commercially. Several apply the differential pressure across a diaphragm containing or connected to a strain gage (bonded-semiconductor, diffused strain gage in a silicon diaphragm, and so forth; **Fig. 11**a). A representative two-wire circuit is shown in Fig. 11b. The signal and power are carried by the same two wires.

Some differential-pressure cells use variable capacity or differential capacity between a diaphragm and an adjacent wall (Fig. 11c). Some use the moving diaphragm to cause a variable reluctance or variable inductance. In most cases the measuring elements are protected from the fluid being measured by a pair of seal-out diaphragms or bellows, and the elements are submerged in silicon oil. In one type the measuring diaphragm provides a variable tension on a wire that is placed in a magnetic field and connected to a self-oscillating circuit that provides a frequency that varies with the differential pressure (Fig. 11d).

Readout scales. Since these secondary devices measure the differential pressure developed, their outputs are proportional to the square of the flow rate. In the case of direct-reading devices, they must be fitted with square root scales and charts for reading flow rate. These scales are particularly difficult to read at low values, thereby limiting the useful rangeability. Uniform

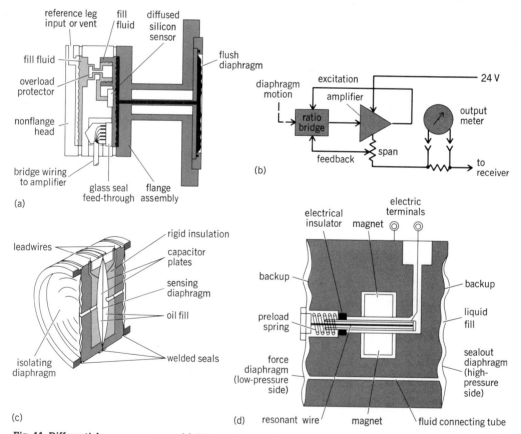

Fig. 11. Differential-pressure sensors. (a) Silicon-chip type (*Honeywell Industrial Division*) with (b) 4- to 20- mA two-wire transmitter circuit (*after 1969 Guide to process instrument elements, Chem. Eng., 76(12):137–164, 1969*). (c) Capacitance type (*Rosemount Engineering Corp.*). (d) Resonant-wire type (*Foxboro Co.*).

flow-rate scales are obtained with manometers with specially shaped chambers or with shaped plugs (Ledoux bell) or with special cams in the linkage between the manometer and the indicating pointers or pens. Some differential-pressure transmitters provide linear-with-flow transmitted signals by incorporating squaring functions in the feedback, or mechanical or electronic square rooting can be done either at the transmitter or at the receiver location.

Target meter. The target-type meter (**Fig. 12**), besides being particularly suitable for dirty fluids, has the important advantage of a built-in secondary device and is without the disadvantage of fluid-filled connection pipes to the secondary device.

Open-channel meters. Flow in open channels is determined by measuring the height of liquid passing through a weir or flume. Consistent results require that the measuring device be protected from direct impact of the flowing stream. Because the height-flow velocity relationship is exponential, the secondary device frequently uses cams to obtain a uniform flow scale.

Variable-area meters. These meters work on the principle of a variable restrictor in the flowing stream being forced by the fluid to a position to allow the required flow-through. Those that depend upon gravity provide essentially a constant differential pressure, while those that supplement the gravity with a spring will have a variable differential pressure.

In the tapered-tube rotameter (**Fig. 13**a), the fluid flows upward through a tapered tube,

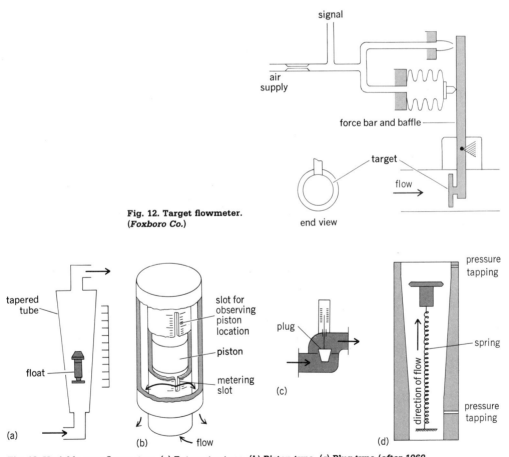

Fig. 12. Target flowmeter. (*Foxboro Co.*)

Fig. 13. Variable-area flowmeters. (a) Rotameter type. (b) Piston type. (c) Plug type (*after 1969 Guide to process instrument elements, Chem. Eng., 76(12):137–164, 1969*). (d) Spring-loaded differential-pressure producer (*after Institute of Mechanical Engineers, How to Choose a Flowmeter, 1975*).

lifting a shaped weight (possibly misnamed a float) to a position where the upward fluid force just balances the float weight, thereby giving a float position that indicates the flow rate. For remote indication, and also where the pressure requires a metal tube, a center rod attached to the float carries the core of an inductance pickup for transmission of the float position through a nonmagnetic tube seal. In other designs the rod carries one side of a magnetic coupling to a follower mechanism outside of the nonmagnetic tube.

In a position type (Fig. 13b) the buoyant force of the liquid carries the piston upward until a sufficient area has been uncovered in a slot in the side of the vertical tube to allow the liquid to flow through the slot. The position of the piston indicates the flow rate and, by properly shaping the slot, this type of meter may be made with a uniform flow scale. A spring is used in some designs to supplement the gravity of the plug and further characterize the indication.

In the tapered-plug design (Fig. 13c), the plug, located in an orifice, is raised until the opening is sufficient to handle the fluid flow. Readouts can be the same as in the tapered-tube design.

A spring-loaded variable-area meter (Fig. 13d) provides a differential pressure for a secondary device. By combining the characteristics of a spring-loaded variable-area meter and a standard differential-pressure measurement across the float, it is claimed that the rangeability of the measurement can be raised to as high as 50:1.

Electromagnetic flowmeter. A magnetic field is applied across a metering tube carrying a flowing liquid, and the voltage generated between two perpendicularly located electrodes contacting the liquid is measured by a secondary or transmitter instrument. Since this voltage is strictly proportional to the average velocity, a linear volume flow signal results. Practical flow measurements can be made on any liquid that is at all conductive. The linear scale allows accurate flow measurements over a greater range than is possible in the differential-pressure meters. The lack of obstruction to fluid flow makes it possible to measure thick slurries and gummy liquids. In order to avoid errors due to electrodes becoming coated with an insulating material, some electromagnetic flow meters are equipped with ultrasonic electrode cleaners.

Classical electromagnetic flow tubes power the field coils with a continuous ac mains voltage and measure the induced ac voltage on the electrodes. After an approximately 90° phase shift, a ratio is made of the two voltages to correct for mains voltage change. Some electromagnetic flow meters apply a periodic dc voltage pulse to the coils and measure the dc voltage across the electrodes at a carefully selected time relative to the field excitation. These flowmeters are available with nonmagnetic stainless steel pipe and various types of insulating lining, and with all-plastic construction. When the field coils are mounted inside the pipe, a magnetic iron may be used for the pipe.

Tracer method. This classic method consists of injecting a small quantity of salt into the flowing stream and measuring the transit time of passage between two fixed points by measuring electrical conductivity at those points. Inaccuracies are introduced by variations in flow cross section in the pipe or channel being measured. Because of the convenience for a quick check of flow under comparatively adverse conditions, standards have been written around the use of salt, radioactive material, and even pulses of heat. Appropriate sensors are offered with correlation techniques for improving the accuracy.

QUANTITY METERS

Quantity meters are those that measure or pass fixed amounts of fluid. By introducing a timing element, all quantity meters can be converted to volume flow-rate meters. Quantity meters may be classified as positive-displacement meters or rotating impeller-type meters. These are illustrated in **Figs. 14** and **15**.

Positive-displacement meters. The flowing stream is separated into discrete quantities by these meters, which capture definite volumes one after another and pass them downstream. Some representative examples will be discussed.

Nutating-disk meter. A circular disk attached to a spherical center is restrained from rotation by a vertical partition through a slot in the disk (Fig. 14a). It is allowed to nutate so that its shaft is in contact with a conical guide. The liquid entering the inlet port moves the disk in a nutating motion until the liquid discharges from the outlet port. The double conical shape of the measuring chamber makes a seal with the disk as it goes through the nutating motion. Normally

MEASUREMENT AND DISPLAY OF PROPERTIES 213

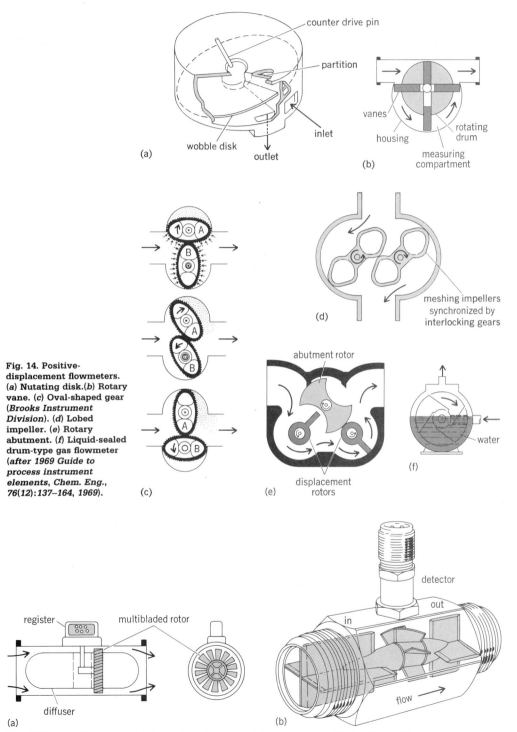

Fig. 14. Positive-displacement flowmeters. (a) Nutating disk. (b) Rotary vane. (c) Oval-shaped gear (*Brooks Instrument Division*). (d) Lobed impeller. (e) Rotary abutment. (f) Liquid-sealed drum-type gas flowmeter (*after 1969 Guide to process instrument elements, Chem. Eng., 76(12):137–164, 1969*).

Fig. 15. Turbine flowmeters. (a) Gas turbine flowmeter (*after Gas meters, Factory Mag., 9(1):39–45, January 1976*). (b) Liquid turbine flowmeter (*Foxboro Co.*).

a mechanical counter indicates the number of cycles as the fluid flows through smoothly with no pulsations.

Rotary-vane meter. An eccentrically mounted drum carries radial spring-loaded vanes (Fig. 14b). These vanes slide in and out, and seal against the meter casing to form pockets that carry a measured amount of fluid on each cycle.

Oval-shaped gear meter. Two carefully formed gears provide their own synchronization, and with close-fitting teeth to the outer circular-shaped chambers they deliver a fixed quantity of liquid for each rotation (Fig. 14c). Rotation of the output shaft can be recorded smoothly by a contact on each revolution. If a higher-frequency signal is desired for flow control or closer monitoring, a compensating gear or optical disk can be used to make the output uniform for a constant liquid flow

Lobed impeller. The lobed impeller (Fig. 14d) is similar to the oval gear except that the synchronization is done by external circular gears and therefore the rotation is smooth.

Rotary-abutment meter. Two displacement rotating vanes interleave with cavities on an abutment rotor (Fig. 14e). The three elements are geared together, and the meter depends upon close clearances rather than rubbing surfaces to provide the seals.

Liquid-sealed drum-type gas meter. A cylindrical chamber is filled more than half full with water and divided into four rotating compartments formed by trailing vanes (Fig. 14f). Gas entering through the center shaft from one compartment to another forces rotation that allows the gas then to exhaust out the top as it is displaced by the water.

Remotely indicating meters may incorporate a simple electronic contact, or a magnetic or optical sensor which can transmit one or more pulses per revolution.

Rotating impeller meters. These convert velocity to a proportional rotational speed. They include various types of turbine meters, propellers, helical impellers, and vortex cage meters.

Turbine meters. The gas turbine meter (Fig. 15a) is characterized by a large central hub that increases the velocity of the gas and also puts it through the tips of the rotor to increase the torque. Readout can be by a directly connected mechanical register or by a proximity sensor counting the blade tips as they pass by. Low-friction ball bearings require special lubricants and procedures.

Most of the available types of liquid-turbine meters (Fig. 15b) are designed to cause a fluid flow or hydraulic lift to reduce end thrust, and several depend upon the flowing fluid to lubricate the bearings.

One bearingless-turbine meter is designed to float a double-ended turbine on the flowing fluid and thereby avoid the problem of wearing bearings. The turbine rotor can be made of various materials to suit the fluid, making it an attractive flow measurement device for nonlubricating and corrosive fluids.

In one insertion meter a small propeller or turbine rotor is mounted at right angles to the end of a support rod. Readout is normally through an inductive pickup and appropriate electronics for obtaining rate of flow. The propeller can be inserted into a flowing stream or into a closed pipe with a direct readout of the flow rate at the location of the propeller. Proper location can give an average-flow-rate either locally or remotely.

Propeller, helical impeller, and vortex cage meters. Since the elements in these meters (**Fig. 16**) rotate at a speed that is linear with fluid velocity, the useful rangeability is large and limited mostly by the bearings. For local indication, revolutions can be counted by coupling to a local-mounted counter. Proximity detectors, giving pulses from 1 to as many as 100 pulses per revolution, can be used for transmitting to remote integrators or flow controllers.

Flow correction factors. The accuracy of quantity meters is dependent to varying degrees on the physical properties of the fluid being measured. Changes from the reference (calibration) conditions of viscosity and density must be considered if accurate measurement is required. It is usually the practice to take pressure and temperature measurements of vapors and gases, or temperature measurement of liquids, at the flow-metering device so that corrections either may be applied by the mechanism of a recording or totalizing device or introduced manually. Some meters are available in which these corrections are introduced automatically with great accuracy, while others are made in which only a first-order approximation is introduced. Proper selection is determined by the requirements of the installation. A meter compensated for fluid conditions can become a mass flowmeter.

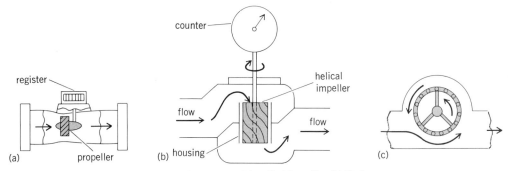

Fig. 16. Types of quantity flowmeters. (a) Propeller. (b) Helical impeller. (c) Vortex cage.

MASS FLOW RATE METERS

There have been many attempts to develop practical industrial mass flowmeters. The most consistently satisfactory method of determining mass flow has been through calculation using volume flow and density determined either by measurement or inference from pressure and temperatures (**Fig. 17**). Only a few of the mechanical type will be mentioned here.

In one type, the fluid in the pipe is made to rotate at a constant speed by a motor-driven impeller. The torque required by a second, stationary impeller to straighten the flow again is a direct measurement of the mass flow (**Fig. 18**a).

A second type, described as gyroscopic/Coriolis mass flowmeter, employs a C-shaped pipe and a T-shaped leaf spring as opposite legs of a tuning fork (Fig. 18b). An electromagnetic forcer excites the tuning fork, thereby subjecting each moving particle within the pipe to a Coriolis-type acceleration. The resulting forces angularly deflect the C-shaped pipe an amount that is inversely proportional to the stiffness of the pipe and proportional to the mass flow rate within the pipe. The angular deflection (twisting) of the C-shaped pipe is optically measured twice during each cycle of the tuning-fork oscillation (oscillating at natural frequency). A digital logic circuit converts the timing of the signals into a mass-flow-rate signal.

In another configuration (using the Coriolis effect) a ribbed disk is fastened by a torque-sensing member within a housing, and both are rotated at constant speed. The fluid enters at the center of the disk and is accelerated radially; the torque on the disk is a direct measure of the mass flow rate. Unfortunately, the practical problem of maintaining the rotating seals in the device has hindered its commercial development.

Another type is based on the principle of the gyroscope (Fig. 18c). The flowing stream is sent through a pipe of suitable shape so that the mass of material flowing corresponds to a gyro wheel rotating about axis C. The entire gyrolike assembly is rotating about axis A, and a torque

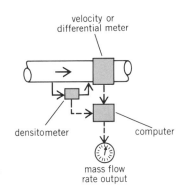

Fig. 17. Inferential mass-flowmeter which computes real-time flow rate in terms of volume rate times gas density. (*After Gas meters* , Factory Mag., *9(1):39–45, January 1976*)

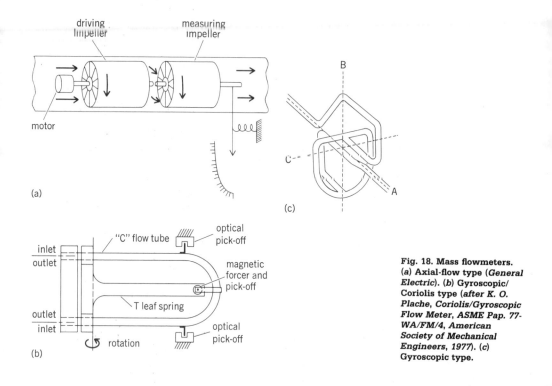

Fig. 18. Mass flowmeters. (a) Axial-flow type (*General Electric*). (b) Gyroscopic/Coriolis type (after K. O. Plache, Coriolis/Gyroscopic Flow Meter, ASME Pap. 77-WA/FM/4, American Society of Mechanical Engineers, 1977). (c) Gyroscopic type.

is produced about axis B proportional to the angular momentum of the simulated gyro wheel. This torque is therefore directly proportional to the mass rate of flow. The pretzellike configuration of pipe is introduced to eliminate centrifugal and other extraneous effects.

THERMAL FLOW MEASUREMENT DEVICES

Several heated or self-heating types of sensor are used for gas-velocity or gas-mass flowmeters.

Hot-wire anemometer. A thermopile heated by a constant ac voltage is cooled by the gas flow. The degree of cooling, proportional to local flow velocity, is indicated by the thermopile millivoltage output.

Heat-loss flowmeter. A constant current is supplied to a thermistor bead. The temperature of the bead depends upon the cooling effect of the mass flow of the fluid, and the measure of the voltage drop across the bead is therefore an indication of the flow. A second thermistor bead is used to compensate for the temperature of the fluid.

Another heat-loss flowmeter measures the amount of electrical heating that is required to keep the temperature of a platinum resistor constant since the amount of heat carried away changes with the velocity of the fluid. A second sensor is used to compensate for the temperature of the fluid.

VORTEX FLOWMETERS

Flowmeters have been developed based on the phenomena of vortex shedding and vortex precession. SEE VORTEX.

Vortex-shedding flowmeters. One of the newest successful methods of measuring flow is based on one of the oldest observed phenomena relating to flow—the eddies (vortices) forming behind an obstruction in a flowing stream. Close observation has shown that each vortex builds up and breaks loose, followed by a repeat in the opposite rotation on the other side of the obstruction, and so alternately the vortices shed and go downstream in what is called a vortex street.

Further observation shows that the distance between successive vortices is directly proportional to the blocking width of the obstruction. The distance is also dependent of the fluid velocity over a large range, and thus the frequency of the shedding is directly proportional to the velocity of the fluid. SEE KARMAN VORTEX STREET.

Application of vortex-shedding flowmeters is normally limited to the turbulent-flow regime, and various shedding-element shapes and internal constructions have been developed in order to improve the shedding characteristics and the detection of the shedding. The common characteristic is sharp edges around which the vortices can form.

Sensing of the shedding vortices is done by: (1) measuring the cyclical change in flow across the front surface of the shedder (self-heated resistance elements); (2) sensing the change in pressure on the sides of the "tail" of the shedder (inductive sensing of motion of a "shuttlecock" in a cross passage, or piezoelectric or capacitive sensing of the pressure on diaphragms sealing a cross passage); (3) sensing the forces on the shedder or an extended tail (piezoelectric sensing of the bending of the shedder, or strain-gage sensing of tail motion); (4) sensing of the vortices after they have left the shedder (by an ultrasonic beam across the wake); (5) sensors mounted outside of the pipe at the junction between two tubes connected to openings in the sides of the tail of the shedder.

One design, which uses a rod or a taut wire as the obstruction and as the ultrasonic detector, is offered for open-channel measurement as well as ship velocity.

Vortex flowmeters are available for measuring the flow of liquids, vapors, and gases. The pulse rate output is particularly convenient for input to integrators for recording total flow and for digital computers. The linear-with-flow output makes the measurement particularly attractive for widely varying flows. However, the relatively low frequency (as low as 1 Hz) of the large blockage designs must be considered when using this measurement in a flow-control loop.

Vortex-precession flowmeter. In this instrument, used to measure gas flows, a swirl is imparted to the flowing fluid by a fixed set of radial vanes. As this swirl goes into an expanding tube, it forms a precessing vortex that is detected by a heated thermistor sensor. The frequency of the precession is claimed to be proportional to the volumetric flow rate within an accuracy of 0.75%.

FLUIDIC-FLOW MEASUREMENT

Two types of flowmeters based on fluidic phenomena have been developed: the fluidic oscillator meter and the fluidic-flow sensor.

Fluidic-oscillator meter. This type of meter works on the principle of the Coanda effect, the tendency of fluid coming out of a jet to follow a wall contour. The fluid entering the meter will attach to one of two opposing diverging side walls (**Fig. 19**a). A small portion of the stream is split off and channeled back through the feedback passage on that side to force the incoming stream to attach to the other side wall (Fig. 19b). The frequency of the oscillation back and forth is directly proportional to the volume flow through the meter. A sensor detects the oscillations and transmits the signal. This principle is used in both liquid and gas flowmeters.

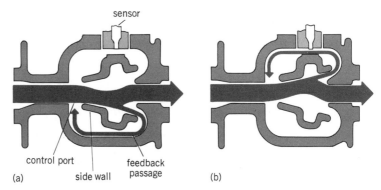

Fig. 19. Fluidic oscillator meter in which feedback stream attaches to (a) first one wall, and (b) then the other. (*After S. J. Bailey, Tradeoffs complicate decisions in selecting flowmeters, Control Eng., 27(4):75–79, 1980*)

Fluidic-flow sensor. Figure 20 shows this device (also known as a deflected-jet fluidic flowmeter) as used in gas flows, particularly in dirty environments. It consists of a jet of air or other selected gas directed back from the outer nozzle onto two adjacent small openings. The flow of the gas being measured will deflect the jet to change the relative pressure on the two ports and thereby give a signal corresponding to the gas velocity.

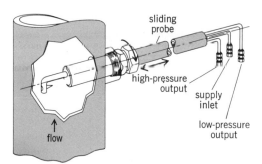

Fig. 20. Fluidic flow sensor.
(*Fluid Dynamics Devices*)

ULTRASONIC FLOW MEASUREMENT

Ultrasonic flowmeters are available in many physical arrangements and are based on a great variety of fundamental principles. Most are found in one of the following categories.

Contrapropagating meter. Contrapropagating diagonal-beam designs (**Fig. 21***a*) have transducers mounted in sockets machined in the pipe or welded on wedges, and also as clamp-on elements. The difference between upstream and downstream "flight" times combined with the speed of sound in the fluid will give fluid velocity. Since the flow near the middle of the pipe is faster, a multiple parallel-beam design (Fig. 21*b*) will give a better average. In a symmetrical profile a beam offset from the center of the pipe may also give a better average.

Axial meter. For small pipes, the short path length across the fluid stream makes measurement difficult. Figure 21*c* shows a contrapropagating arrangement for measuring lengthwise over a relatively longer path where timing measurements are more practical.

Correlation meter. Two pairs of elements with beams across the pipe can be used with a correlation circuit to detect the time it requires for discontinuities in the fluid stream to pass between the two detectors (Fig. 21*d*).

Deflection (drift) meter. An ultrasonic beam directed across the pipe will be deflected by an amount depending on the relative velocity of sound in the fluid and the flowing rate of the fluid Fig. 21*e*). One arrangement detects this deflection with two receivers mounted on the other side of the pipe. The relative intensity of the two receivers is an indication of the flowing-fluid velocity. A change in the velocity of sound in the fluid will introduce an error.

Doppler meter. Flow rate can be detected by the Doppler shift of a reflection from particles or discontinuities in the flowing fluid. A transceiver on one side of the pipe may have an error due to the reflections from the slower particles near the wall. Figure 21*f* shows the measurement being made by opposing elements.

Noise-type meter. One flow measurement device is available where the noise that is generated by the flowing fluid is measured in a selected frequency band, with a useful but relatively inaccurate readout (Fig. 21*g*).

Open channels. A different application of ultrasonics to flow measurement can be found in flumes and weirs where the level of the flowing fluid, measured by ultrasonic reflection, is used to calculate the flow rate. In unrestricted open channels the level can be measured with one reflecting transducer (either above or below the surface) and the flowing rate with a pair of submerged detectors. An electronic computation incorporating the cross-section shape of the channel will give the flow rate.

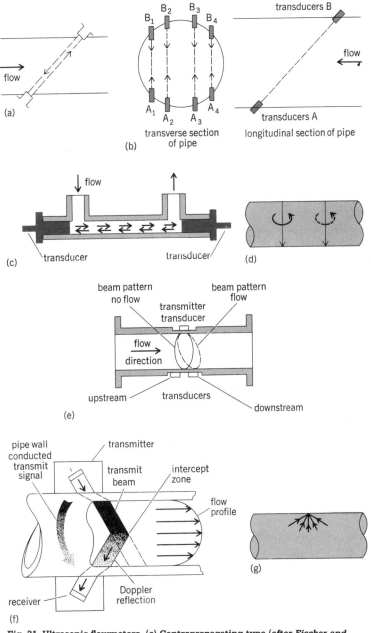

Fig. 21. Ultrasonic flowmeters. (a) Contrapropagating type (*after Fischer and Porter Co., Publ. 12257, 1957*). (b) Multiple-beam contrapropagating type (*after Institute of Mechanical Engineers, How To Choose a Flowmeter, 1975*). (c) Axial-transmission contrapropagating type (*after Dupont Co., Industrial Products*). (d) Correlation type (*after W. P. Mason and R. Thurston, eds., Physical Acoustics, vol. 14, Academic Press*). (e) Deflection (drift) type (*after H. E. Dalke and W. Walkowitz, A new ultrasonic flowmeter for industry, ISA J., 7(10):60–63, 1960*). (f) Doppler type (*after H. M. Morris, Ultrasonic flowmeter uses wide beam technique to measure flow, Control Eng., 27(7):99–101, 1980*). (g) Noise type (*after W. P. Mason and R. Thurston, eds., Physical Acoustics, vol. 14, Academic Press*)

NUCLEAR MAGNETIC RESONANCE FLOWMETER

In this device (**Fig. 22**), nuclei of the flowing fluid are resonated by a radio-frequency field superimposed on an intense permanent magnetic field. A detector downstream measures the amount of decay of the resonance and thereby senses the velocity of the fluid. The most effective fluids are hydrocarbons, fluorocarbons, and water-bearing liquids because this resonance is most pronounced in fluorine and hydrogen nuclei.

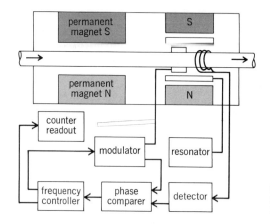

Fig. 22. Nuclear magnetic resonance type. (*After 1969 Guide to process instrument elements, Chem. Eng., 76(12):137-164, 1969*)

LASER DOPPLER VELOCIMETER

When light is scattered from a moving object, a stationary observer will see a change in the frequency of the scattered light (Doppler shift) proportional to the velocity of the object. In the laser Doppler velocimeter (**Fig. 23**) this Doppler shift is used to measure the velocity of particles in a fluid. From the particle velocity the velocity of the fluid is inferred. A laser is used as the light source because it is easily focused and is coherent and thereby enables the measurement of a frequency shift that is due to a Doppler effect.

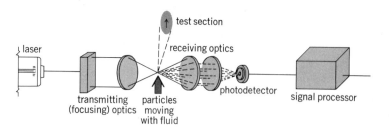

Fig. 23. Laser Doppler velocimeter. (*Thermo-Systems, Inc.*)

ACCURACY OF MEASUREMENT

The practical accuracy statement must be for the total flow-measurement system (combined primary and secondary devices). This has been customarily expressed as a percentage of the upper-range value ("full scale") in the case of rate meters used for industrial process control, and as a percentage of total flow rate for quantity meters. However, linear meters have often been specified in percentage of flow regardless of the use. Well-established standards for terminology, test procedures, and statements of operating environments have made it relatively easy to know what to expect for performance of secondary devices. Theory does not yet allow an accurate enough prediction of the actual flow of fluid inside of a pipe to predict the performance of a flowmeter. Square-edged orifices, venturis, and nozzles are the only types for which sufficient laboratory data have been collected and analyzed to enable a user to purchase and install a meter without individual calibration. One can expect a 1% maximum uncertainty in the primary if care is taken to follow the installation requirements of standards.

The readability or resolution of the readout display is important. This is poor at the low end of the square-root scale of differential-pressure meters. Rangeability of a flowmeter, sometimes termed turndown ratio in process applications, refers to the ratio of full-scale flow to a practical minimum flow rate. A figure of 3.5:1 is normally used for square-root scale flowmeters, while linear flowmeters, such as magnetic flowmeters, vortex-shedding flowmeters, and ultrasonic flowmeters normally have figures of 10:1 or higher. Flowmeters with rotating parts in the flowing stream, although linear, may have a smaller useful range, depending on the effect of viscosity and friction at low flow rates. *See Fluid flow.*

Bibliography. American National Standards Institute, *Standard ANSI/API 2530*, 1979; H. S. Bean (ed.), *Fluid Meters, Their Theory and Application,* American Society of Mechanical Engineers, 6th ed., 1970; D. M. Considine (ed.), *Process Instruments and Controls Handbook,* 3d ed., 1985; International Standards Organization, *Standard ISO 5167,* 1980; G. P. Katys, *Continuous Measurement of Unsteady Flow,* 1964; W. P. Mason and R. Thurston (eds.), *Physical Acoustics,* vol. 14, 1979; R. W. Miller, *Flow Measurement Engineering Handbook,* 1983; R. H. Perry and D. Green (eds.), *Perry's Chemical Engineers' Handbook,* 6th ed., 1984; K. O. Plache, *Coriolis/Gyroscopic Flow Meter,* ASME Pap. 77-WA/FM/4, 1977; A. E. Rodely, Vortex shedding flow meters, *Meas. Control,* 14(1):134–138, February 1980.

TORRICELLI'S THEOREM
Victor L. Streeter

The proposition that the speed of efflux of a liquid from an opening in a reservoir equals the speed that the liquid would acquire if allowed to fall from rest from the surface of the reservoir to the opening.

Torricelli, a student of Galileo, observed this relationship in 1643. In equation form, $v^2 = 2gh$, in which v is the speed of efflux, h the head (or elevation difference between reservoir surface and center line of opening if in a vertical plane), and g the acceleration due to gravity. (The equation is the same as that for a solid particle dropped a distance h in a vacuum.) The relationship can be derived from the energy equation for flow along a streamline, if energy losses are neglected.

An orifice (opening in the wall or bottom of a reservoir) is used as a flow-measuring device. From Torricelli's theorem, by solving for v and multiplying by the flow area, an expression for discharge Q, in volume per unit time, is obtained. In equation form, $Q = C_d A \sqrt{2gh}$, in which A is the area of opening and C_d is a dimensionless coefficient, determined experimentally, that corrects for contraction of the jet as it leaves the orifice and for energy loss due to viscosity. When h is measured, Q may be determined from the formula. *See Flow measurement; Metering orifice.*

METERING ORIFICE
Raymond E. Sprenkle

A wall opening with sharp inlet edges, through which fluid flows, the wall thickness being less than an eighth of the area of the opening. An orifice serves basically to control or to meter the rate of fluid flow. It may be used as an instrument for the measurement of fluid flow, including oils, air, gases, steam, and other vapors (**Fig. 1**), or as an element in a machine to limit the flow of fluid. Common shapes are sharp-edge, with bevel facing out, and square. Often, fuel enters a chamber through an orifice, as in a carburetor. *See Flow measurement.*

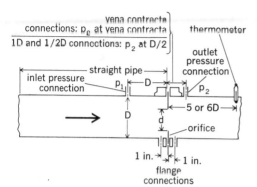

Fig. 1. Orifice installation in pipe of diameter D. 1 in. = 2.5 cm.

In both liquid and gas devices, an orifice acts to convert potential energy to kinetic energy. For liquids, the potential energy is usually the pressure head; for gases, the potential energy includes the pressure differential and the temperature available for conversion into velocity. Theoretical velocity v of discharge of liquid initially at rest from an orifice is $v = (2gh)^{1/2}$, where g is gravitational constant and h is pressure head. SEE TORRICELLI'S THEOREM.

For incompressible liquid density ρ, in lb/ft^3 or kg/m^3, flowing through an orifice of area a,

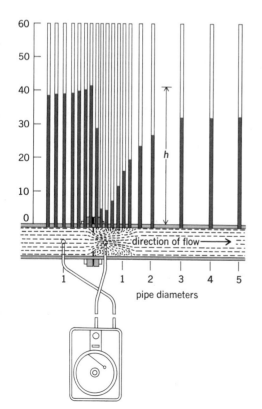

Fig. 2. Fluid flow through a thin-plate orifice.

in ft² or m², at atmospheric temperature whose discharge coefficient, including velocity of approach, is K, the rate of flow w_s in lb/s or kg/s is given by Eq. (1). Here g is the acceleration of

$$w_s = Ka\rho\sqrt{2gh} \qquad (1)$$

gravity in ft/s² or m/s² and h is the effective differential head or change in pressure in feet or meters measured across the orifice.

For gas, air, steam, and other vapors, the following corrections must be made to Eq. (1): an expansion factor Y to compensate for the expansion of the fluid as it passes through the orifice; and a thermal expansion factor F_a to compensate for the increased physical size of the orifice opening when the fluid is at a temperature substantially higher than that of atmosphere. Thus the rate of flow w_s in lb/s is given by Eq. (2).

$$w_s = Ka\rho\sqrt{2gh}YF_a \qquad (2)$$

Equation (2) also applies for liquids at substantially higher than atmospheric temperatures, except that the factor Y is taken as unity.

For liquid measurement, an orifice is arranged as in **Fig. 2**. The liquid passes through the orifice, experiencing a pressure drop which is transmitted to a liquid-filled manometer, differential pressure gage, or a recording instrument. The flow rate is calculated from the change in pressure, the area of the orifice, and the factor k. Figure 2 also shows, as the rise of liquid in manometers along the pipe, the effect of the orifice on pressure. The region where the pressure is lowest, with the streamlines closest together, is called the vena contracta.

Orifices for flow instrumentation usually are clamped between pipe flanges. They may have a concentric hole, although for self-cleaning the hole may be off center with one edge flush with the bottom of the pipe or with a segmented hole equal to the inside diameter of the pipe (**Fig. 3**).

Fig. 3. Commonly used orifice shapes. (*a*) Concentric. (*b*) Eccentric. (*c*) Segmental.

As a differential pressure producer, the orifice is the most widely used device. More test data are available on the thin-plate orifice than on other type flow meters; thus it is chosen despite its small pressure recovery. S*EE* V*ENTURI TUBE*.

Bibliography. H. S. Bean (ed.), *Fluid Meters, Their Theory and Application*, American Society of Mechanical Engineers, 6th ed., 1970; D. M. Considine (ed.), *Process Instruments and Controls Handbook*, 3d ed., 1985; R. W. Miller, *Flow Measurement Engineering Handbook*, 1983; R. H. Perry and D. Green, *Perry's Chemical Engineers' Handbook*, 6th ed., 1984.

VENTURI TUBE
R*AYMOND* E. S*PRENKLE*

A device that causes a drop in pressure as a fluid flows through it. Essentially, a venturi tube is a short straight pipe section, or throat, between two tapered sections. Local pressure varies in the vicinity of the constriction; thus, by attaching at the throat a manometer or recording instrument, the drop in pressure can be measured and the flow rate calculated from it, or, by attaching a fuel source, fuel can be drawn into the main flow stream. S*EE* M*ANOMETER*.

Proportions of the venturi tube for flow measurement, as established by its inventor Cle-

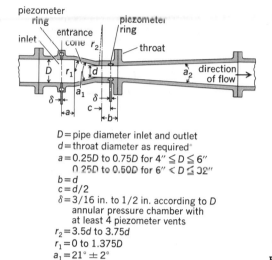

D = pipe diameter inlet and outlet
d = throat diameter as required
a = 0.25D to 0.75D for $4'' \leq D \leq 6''$
      0.25D to 0.50D for $6'' < D \leq 32''$
$b = d$
$c = d/2$
δ = 3/16 in. to 1/2 in. according to D
  annular pressure chamber with
  at least 4 piezometer vents
r_2 = 3.5d to 3.75d
r_1 = 0 to 1.375D
a_1 = 21° ± 2°
a_2 = 5° to 15°

Proportions of Herschel-type venturi tube for standard fluid-flow measurement. 1 in. = 2.5 cm.

mens Herschel, are generally as shown in the **illustration**. The inlet is a short cylindrical section of the same diameter as the pipe to which it is attached. The entrance cone, having an included angle a_1, leads by an easy tangential curve into the short throat section of diameter d. A long diverging or pressure-restoring cone, having an included angle a_2, expands the fluid again to the full pipe diameter. Throat diameter ranges from one-third to three-fourths of the pipe diameter.

The pressure preceding the inlet taper is transferred through multiple openings into an annular opening, called a piezometer ring. In similar fashion, the pressure in the throat is transferred through multiple openings into another piezometer ring. A single pressure line from each ring leads to the manometer or recording meter. In some designs, the piezometer rings are replaced with single pressure connections into the inlet section and into the throat.

The principal advantage of the venturi tube is that not more than 10–20% of the difference in pressure between the inlet and the throat is permanently lost. This is accomplished by the discharge cone gradually decelerating the flow with minimum turbulence. SEE FLOW MEASUREMENT.

Bibliography. H. S. Bean (ed.), *Fluid Meters, Their Theory and Application*, American Society of Mechanical Engineers, 6th ed., 1970; D. M. Considine (ed.), *Process Instruments and Controls Handbook*, 3d ed., 1985; R. W. Miller, *Flow Measurement Engineering Handbook*, 1983; R. H. Perry and D. Green, *Perry's Chemical Engineers' Handbook*, 6th ed., 1984.

BORDA MOUTHPIECE
FRANK H. ROCKETT

A reentrant tube in a hydraulic reservoir, as shown in the **illustration**. The contraction coefficient of a Borda mouthpiece can be calculated more simply than for other discharge openings.

As a liquid accelerates toward a discharge opening, the streamlines converge. This convergence may cause the issuing jet to contract to a smaller cross section than that of the opening, the region of the jet of least cross section being called the vena contracta. The ratio of vena contracta area to opening area is the contraction coefficient. The frictional forces in the vicinity of a simple opening alter the accelerations in the liquid, but with a Borda mouthpiece, friction is so reduced near the opening that the contraction coefficient can be calculated from hydrostatic

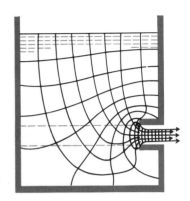

Streamlines around a Borda mouthpiece.

considerations along with close agreement to measurements. The coefficient thus calculated is 0.5. *See* METERING ORIFICE; VENTURI TUBE.

Bibliography. T. Baumeister, E. A. Avallone, and T. Baumeister III (eds.), *Marks' Standard Handbook for Mechanical Engineers*, 9th ed., 1987.

PITOT TUBE
LLOYD N. KRAUSE

An instrument that measures the stagnation pressure of a flowing fluid; also called an impact tube. Stagnation (also called impact or total) pressure is the pressure that would be obtained if the fluid were brought to rest isentropically. When total pressure is being measured, the system consists of a primary sensing element mounted on a suitable support, pressure connecting lines, and a pressure-indicating device. Normally, the connecting lines and indicating devices are considered secondary elements and are not treated as part of the pitot tube. *See* MANOMETER.

Application. The pitot tube is used primarily to obtain fluid velocity, total energy as measured being composed of impact or stagnation pressure P_2, static pressure P_1, and velocity v_1 of the fluid of density ρ. Then for incompressible (low-speed) flow, the equation shown below

$$\frac{P_2}{\rho} = \frac{P_1}{\rho} + \frac{v_1^2}{2}$$

holds (in SI units or any other coherent system of units). For incompressible flow, the total energy is expressed by Bernoulli's theorem. For compressible flow, total energy can be expressed in terms of impact pressure, static pressure, and Mach number, which is related to velocity. *See* BERNOULLI'S THEOREM.

Pitot tubes of many shapes and sizes are used in a wide variety of applications. A square-ended circular tube pointing upstream will measure true total pressure at subsonic speeds and will measure true total pressure existing behind a normal shock wave across its nose at supersonic speeds.

Accuracy. Depending on design, tubes can be made insensitive to flow misalignment up to 45° (see **illus**.). Another error arises when a pitot tube is in a total pressure gradient. The effective center of the tube is then displaced from the geometric center toward the region of higher total pressure. Other errors will arise when dealing with turbulent or pulsating flow because of the pressure-averaging effect of the tube.

In certain flow regions, conventional pitot-tube response must be corrected to obtain velocity accurately. One such region is that of low Reynolds numbers, where the viscous effect of the fluid predominates; another such region is that of high Knudsen number (slip and free molecular

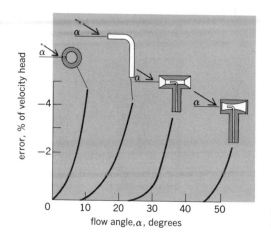

Effect of the flow alignment of a pitot tube with streamline on accuracy of measurement.

flow), which is associated with measurements in a rarefied gas. SEE AIR-VELOCITY MEASUREMENT; FLOW MEASUREMENT; REYNOLDS NUMBER.

Bibliography. R. C. Folsom, Review of the pitot tube, *Trans. ASME*, 78:1447–1460, 1956; R. W. Miller, *Flow Measurement Engineering Handbook*, 1983; V. L. Streeter (ed.), *Handbook of Fluid Dynamics*, 1961; V. L. Streeter and E. B. Wylie, *Fluid Mechanics*, 8th ed., 1985.

AIR-VELOCITY MEASUREMENT
HERBERT FOX

The measurement of the rate of displacement of air or gas at a specific location, for example, wind speed, air velocity in the test section of a wind tunnel, airspeed of an aircraft, or air velocity produced by a fan or blower in air conditioning. Velocity indicates the magnitude and direction of the flow rate; the latter is usually measured separately. To obtain the total flow velocity as required in pipes or conduits, integration over the cross section is required, or another method of measurement is used. SEE FLOW MEASUREMENT.

Three primary methods are used in measuring air velocity. These methods and the instruments which make use of each are listed in the **table**. For discussion of the three anemometers SEE ANEMOMETER.

Pitot-static tube. This, coupled with temperature-measuring devices, is widely used in wind tunnels and in aircraft to measure airspeed. For all velocity ranges (subsonic or supersonic), measurements of total (pitot) or stagnation pressure and temperature, p_0 and T_0, respectively, and

Methods of measuring air velocity	
Method	Instruments
Pressure on an element in the airstream	Pitot-static tube Venturi tube Bridled pressure plate
Speed or number of revolutions of a rotating element in the airstream	Cup anemometer Vane anemometer
Effect of velocity on a heated element in the airstream	Hot-wire anemometer Kata thermometer

static pressure p are required. For any gas the relation between these and velocity v is shown in Eq. (1), where T_0 is the absolute temperature, M the molecular weight, R the universal gas con-

$$v = \sqrt{\left(\frac{2k}{k-1}\right)\left(\frac{RT_0}{M}\right)\left[1 - \left(\frac{p}{p_0}\right)^{(k-1)/k}\right]} \quad (1)$$

stant (8.314 joules/g-mole K for SI units), and k the ratio of specific heats. For air $k = 1.4$ and $M = 29$ g/g-mole, Eq. (2) holds, where v is in meters/second and T_0 in kelvins.

$$v = 4.480 \, T_0^{1/2} \, [1 - (p/p_0)^{0.286}]^{1/2} \quad (2)$$

The methods of measurement of p_0 and p are different for subsonic and supersonic flow. For subsonic flow a typical configuration is shown in **Fig. 1**. The airstream enters the open end of the tube and decelerates until its velocity is zero; the pressure measured by a tube connected to this region is p_0. Small holes are drilled in the wall-off section, or static tube, about 10 tube diameters from the open end. The air passes externally over these holes; a tube open to the walled-off section measures p.

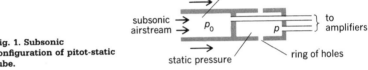

Fig. 1. Subsonic configuration of pitot-static tube.

For supersonic flow the configuration is basically the same. Provided the holes in the walled-off section are set far enough back (determined by calibration), the static pressure may be determined in the same way. The shock wave around the instrument does not permit direct measurement of p_0, however (**Fig. 2**); instead the total pressure behind the shock p_{0s} is monitored. A unique but complicated relation exists between p, p_{0s}, and p_0, thus determining the pressure terms in the above relation. SEE PITOT TUBE; SHOCK WAVE.

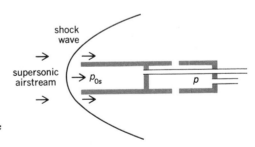

Fig. 2. Supersonic configuration of pitot-static tube.

The pressure gage on aircraft is calibrated in units of indicated airspeed and requires an additional correction for air density (or temperature).

Venturi tube. This has limited application in measuring gas velocity in industry. A reference pressure is required, either static or pitot. The tube is open at both ends and reduces in diameter sharply but smoothly from the end that heads into the airstream. The original diameter is recovered smoothly at the rear end. The rear section is three or more times longer than the front. Suction is developed at the constriction and is measured by a pressure gage connected to it by tubing. The gage is also connected to a static or pitot tube. The differential pressure measured is a function of the velocity and also of the air density. SEE VENTURI TUBE.

Bridled pressure plate. This is useful in measuring air velocities in gusts, since it has a greater speed of response than do rotating elements. The velocity pressure on the plate exposed to the wind is balanced by the force of a spring. The deflection is measured by an inductance-type transducer. The signal from the transducer is amplified and transmitted to a recorder. The indication is dependent upon the air density. Simple mechanical designs are also used to measure low air velocities.

Kata thermometer. This is used to measure low velocities in air-circulation problems. An alcohol thermometer with a large bulb is heated above 100°F (38°C), and its time to cool from 100 to 95°F (38 to 35°C), or some other interval above the ambient temperature, is measured. This time interval is a measure of the air current at the location.

Shielded thermocouple. This instrument is used for obtaining the total temperature inside a pitot tube using a thermocouple rather than a pressure-sensing device. The thermocouple must be shielded to minimize losses due to radiation and conduction. Calibration takes account of probe configurations, wall conductivity, and gas stream conditions. The same instrument may be used for subsonic and supersonic flows.

Bibliography. R. P. Benedict, *Fundamentals of Gas Dynamics*, 1983; D. Crane (ed.), *Aircraft Instrument Systems*, Aviation Maintenance Foundation, 1976; H. W. Liepmann and A. Roshko, *Elements of Gasdynamics*, 1957; L. Prandtl and O. S. Tietjens, *Applied Hydro and Aeromechanics*, 1934.

ANEMOMETER
HERBERT FOX

A device which measures the magnitude of air velocity. Anemometers are commonly known as the devices which measure wind magnitude, but they are also used to measure the rate of flow of air or of other gases in other applications, for example, in wind tunnels and on aircraft. The most common types are the cup, vane, and hot-wire anemometers. SEE AIR-VELOCITY MEASUREMENT.

Cup anemometer. This is widely used to measure horizontal wind speed, independent of direction. The use of three hemispherical cups mounted on a vertical shaft is standard in the United States. Commonly the cups rotate a worm gear which operates an electric contact every mile of wind, or fraction thereof. These contacts are recorded on a chart driven at constant speed. From the number of contacts in a selected time interval, the wind speed is deduced. Since the number of cup revolutions per mile varies with wind speed, particularly at low speeds, a correction must be made.

Wind speed practically independent of air density is directly obtained if the cups rotate an electric generator (**Fig. 1**). The generator must operate on a minimum of torque to obtain valid indication at low speeds. Its output, often amplified, drives an electric indicator calibrated in wind speed units.

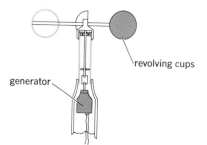

Fig. 1. Cutaway diagram of revolving-cup electric anemometer. (*After D. M. Considine, ed.,* Process Instruments and Controls Handbook, *2d ed., McGraw-Hill, 1974*)

Vane anemometer. This portable instrument is used to measure low wind speeds and airspeeds in large ducts. It consists of a number of vanes radiating from a common shaft and set to rotate when facing the wind (**Fig. 2**); a guard ring surrounds the vanes. The vanes operate a counter to indicate the number of rotations, which when timed with a stop-watch serve to determine the speed. The parts are made of lightweight materials; friction is kept low to obtain reliable measurements.

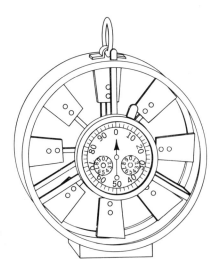

Fig. 2. Portable revolving-vane anemometer. (*After D. M. Considine, ed., Process Instruments and Controls Handbook, 2d ed., McGraw-Hill, 1974*)

Hot-wire anemometer. This is an important device used principally in research on air turbulence and boundary layers. There are two types, known as constant-voltage and constant-temperature. The cooling of an electrically heated fine wire placed in a gas stream which alters wire resistance depends on the fluid velocity. With the constant-voltage type (**Fig. 3**), as soon as the gas starts flowing, the hot wire cools off, the voltage across a Wheatstone bridge connected to the wire is kept constant, and the calibrated galvanometer shows a reading related to the

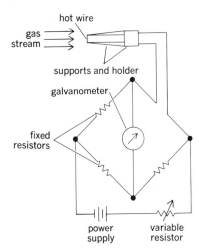

Fig. 3. Constant-voltage hot-wire anemometer.

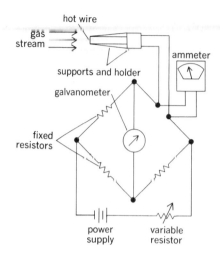

Fig. 4. Constant-resistance hot-wire anemometer.

velocity. This is useful only for very low velocities but is extremely sensitive—velocities down to 0.2 in./s (5 mm/s) are typical.

The constant-temperature, or constant-resistance, type (**Fig. 4**) is more useful. The resistance in the battery across the Wheatstone bridge is increased to such a value that the wire which was cooled by the stream is again brought to its original temperature. The current in the hot wire, read by a calibrated ammeter, gives a measure of the air velocity. Another important application is the measurement of turbulent fluctuations. These fluctuations show up as variations in wire resistance and are useful for boundary-layer studies.

Typical materials for such hot wires are platinum and tungsten with diameters of 10^{-5} to 10^{-3} in. (0.25 to 25 micrometers). S*EE* B*OUNDARY-LAYER FLOW.*

Bibliography. D. M. Considine and S. D. Ross (eds.), *Handbook of Applied Instrumentation*, 1964, reprint 1982; D. M. Considine (ed.), *Process Instruments and Controls Handbook*, 3d ed., 1985; H. W. Liepmann and A. Roshko, *Elements of Gasdynamics*, 1957; A. E. Perry, *Hot-Wire Anemometry*, 1982.

RIPPLE TANK
D*ONALD* R. F. H*ARLEMAN*

A shallow tray containing a liquid and equipped with a means for generating surface waves. Ripple tanks are used to study a number of physical phenomena which can be described in terms of wave mechanics. Water, acoustic, light, and electromagnetic waves can be investigated with equal facility.

Theory. All wave motion can be described in terms of wave period T, or frequency $f = 1/T$, and wavelength L; hence celerity C or speed of propagation of a wave is defined as $C = fL$. The speed of surface waves on the liquid in the ripple tank is dependent upon density ρ, surface tension σ, and depth of liquid h; thus Eq. (1) holds. (The equations in this article are applicable

$$C^2 = \left(\frac{gL}{2\pi} + \frac{\sigma}{\rho}\frac{2\pi}{L}\right) \tanh \frac{2\pi h}{L} \tag{1}$$

in SI units or any other coherent system of units.) In general the ripples are produced by a wave generator oscillating at a fixed frequency; hence the wavelength in Eq. (1) may be replaced by the fundamental relation $L = C/f$. Therefore Eq. (2) is valid. For a given liquid and frequency of

$$C^2 = \left[\left(\frac{g}{2\pi f}\right)C + \left(\frac{2\pi f \sigma}{\rho}\right)\frac{1}{C}\right] \tanh (2\pi f) \frac{h}{C} \tag{2}$$

ripple generation, the parenthetical terms are constants and the speed of propagation depends only upon liquid depth h. However, as the depth becomes large, the hyperbolic tangent term rapidly approaches unity and the celerity equals a constant C_0, which also depends upon the frequency and fluid properties; hence Eq. (3) is valid. A graph of ratio C_0/C versus water depth

$$C_0^2 = \left(\frac{g}{2\pi f}\right) C_0 + \left(\frac{2\pi f \sigma}{\rho}\right) \frac{1}{C_0} \qquad (3)$$

may therefore be constructed to serve as the ripple-tank calibration.

Analogy with wave phenomena. Diffractive phenomena associated with the propagation of sound, light, and electromagnetic waves can readily be studied in the ripple tank by observing that the ratio of celerities obtained above can be interpreted in the following manners: (1) For sound waves, C_0/C is the acoustic index of refraction; (2) for light waves, C_0/C is the optical index of refraction; and (3) for electromagnetic waves, C_0/C is the square root of the dielectric constant (for those materials whose magnetic permeability is near unity).

Control of water depth in the ripple tank by contouring the bottom permits the simulation of a variable or discontinuous medium for wave propagation. For example, a step change in the depth as shown in the **illustration** results in a wave diffraction in accord with Snell's law. Acoustical and optical wave diffractions have been studied in ripple tanks, and phase fronts near two-dimensional models of antenna structures and radomes have also been investigated. Ripple-tank models of large-scale water-wave motions in harbors and along seacoasts can be useful, provided the surface tension term in the celerity equations is small compared with the gravitational term. Water depths of about 1 in. (25 mm) and frequencies of 1 Hz have been used successfully in this type investigation.

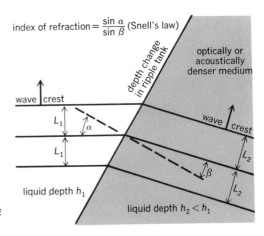

Wave refraction in accord with Snell's law at a boundary between deep and shallow sections of ripple tank.

Experimental equipment. The usual ripple tank consists of a glass or plastic plate with vertical sides to contain the liquid. Electronically driven probe vibrators are used to excite the water surface at a given frequency. Synchronously chopped light is directed through the tank to a ground-glass screen on which the phase-front shadow patterns appear stationary. Reflections from the walls of the tank are reduced to a minimum by placing folded wire screens and cloth around the boundaries. The head of a pin vibrating vertically can be made to serve as a radiating probe.

If care is taken to measure and adjust water levels accurately, depths as small as 0.004 in. (0.1 mm) can be used. At an exciting frequency of 20 Hz a range of depths of 0.2–0.004 in. (5–0.1 mm) will result in a refractive index of 1–2.5, which would also correspond to a dielectric constant of approximately 1–6. At these small depths only short paths along which the waves travel can be used because of attenuation of the wave. SEE SHADOWGRAPH OF FLUID FLOW; WAVE MOTION IN LIQUIDS.

TOWING TANK

JACQUES B. HADLER

A tank of water used to determine the hydrodynamic performance of waterborne bodies such as ships and submarines, as well as torpedoes and other underwater forms. In the narrow sense, towing tanks are considered to be experimental facilities used to measure the forces, such as drag, on ship models and in turn to predict the performance of the full-scale prototype. In general, towing tanks are rectangular in planform with a uniform cross section. Different section shapes are used, ranging from rectangular to semicircular. The cross-section dimension may vary from about 8 to 52 ft (2.5 to 16 m) in width, from about 4 to 22 ft (1.5 to 7 m) in depth, and from under 100 (30 m) to almost 3000 ft (914 m) in length; the size of the model varies in length from 4 to 30 ft (1.5 to 9 m).

Towing methods. The principal measurements made in a towing tank are force measurements, particularly drag or resistance of a towed ship model or other body. One of two principal systems for towing a model is used in most towing tanks. The simpler system consists of a gravity dynamometer and an endless cable attached to the model. A weight provides a constant towing force (**Fig. 1**). The time to traverse a fixed distance is measured when the model reaches a constant speed, thus establishing the speed-resistance relationship for the model. This dynamometer is simple and capable of high accuracy, but is limited to the measurement of the drag force of waterborne bodies. It is used in the smaller towing tanks in which the models are generally under 6 ft (2 m) in length.

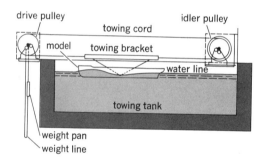

Fig. 1. Tank with model towed by falling weight.

In larger towing tanks the model is towed by a towing carriage mounted on rails at the side of the towing tank or suspended from an overhead track system (**Fig. 2**). Speed can be controlled and measured precisely on these carriages. Most carriages are equipped with a drag dynamometer as a permanent component (**Fig. 3**).

The dynamometer girder, mounted on the carriage frame, carries a long horizontal floating beam in pendulum fashion on two pairs of vertical arms terminating in flexible springs. A counterweight at the upper end of a vertical swinging arm and mounted on the girder and attached to the floating beam maintains the beam in equilibrium at any position between the limit stops. The model resistance is transmitted as a horizontal force through the upper flexible link L_1 to the lower arm of the T-shaped balance, where it is balanced by the weight W. When the model resistance is not equal exactly to a unit weight W, the remainder is taken up (or applied) by the resiliency of the whole group of flexible spring supports; the exact amount of this auxiliary load or force is recorded on the drum by the link L_2 and the recording arm shown in Fig. 3.

In modern installations, the towing force is measured at the point where the towing link is connected to the towing post in the model. The measurements are usually made by measuring the strain in a flexure with electric strain gages. This method is utilized in measuring the drag of submerged bodies.

MEASUREMENT AND DISPLAY OF PROPERTIES   **233**

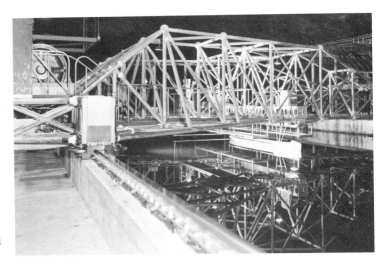

Fig. 2. Model in David Taylor Model Basin towed by carriage.

Law of similitude. Modern towing tank technology was established by William Froude in the 1870s, when he discovered the law of similitude for phenomena in which gravity is the predominating factor and established one of the essential principles of hydrodynamics for comparing model phenomena with the actual ship. There are three principal forces involved for a body moving through the water: inertia, gravity, and viscosity. The law of similitude requires that the ratio between inertia force and gravity force be the same for both the model and the prototype, and that the ratio between inertia force and viscous force be the same as well. S*EE* F*ROUDE NUMBER;* R*EYNOLDS NUMBER.*

Froude's law requires the validity of Eq. (1), where V and v represent the velocity of the

$$\frac{V}{v} = \sqrt{\frac{L}{l}} = \sqrt{\lambda} \qquad (1)$$

prototype and model, respectively; L and l represent a characteristic dimension such as length of prototype and model, respectively; and λ represents the linear ratio of prototype to model.

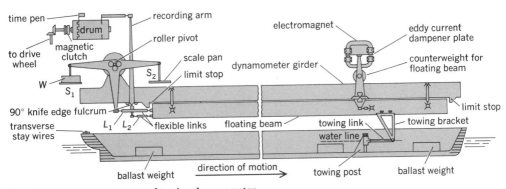

Fig. 3. Schematic arrangement of towing dynamometer.

Reynolds' law requires the validity of Eq. (2), where v_s and v_m are the kinematic viscosity

$$\frac{VL}{v_s} = \frac{vl}{v_m} \qquad (2)$$

of the fluid for prototype and model, respectively. Because model towing tanks use fresh water and ships usually operate in sea water, v_s and v_m are essentially in the same order of magnitude. Thus, Froude's law requires a model velocity which is less than that of the prototype, whereas Reynolds' law requires a model velocity which is substantially higher than that of the prototype.

Froude overcame this difficulty by dividing the ship resistance into two parts: frictional resistance, and residuary resistance consisting mainly of wave-making resistance. Frictional resistance is primarily the effect of viscosity and is thus governed by Reynolds' law. Residuary resistance is due mainly to gravitational effect and thus is governed by Froude's law. Based on this distinction, Froude developed the technique of predicting ship resistance from the resistance test of a scaled model, a technique used to this day. It can be outlined as follows. A model geometrically similar to the prototype is made and the drag or resistance measured in a towing tank at the corresponding speeds as expressed in Eq. (1). Viscous or frictional resistance r_F of the model is calculated by assuming the resistance to be the same as that of a smooth flat plate of comparable area and length. Residuary resistance r_R, which is largely a gravitational effect, is then found by subtracting the frictional resistance from the total measured resistance r_T as in Eq. (3).

$$r_R = r_T - r_F \qquad (3)$$

Residuary resistance of the ship R_R is calculated by the law of comparison at corresponding speeds as in Eq. (4). The frictional resistance of the ship R_F is calculated on the same assumptions

$$R_R = r_R \times \lambda^3 \qquad (4)$$

used in calculating the viscous resistance of the model. The total resistance of the ship is finally found by adding the residuary and frictional resistance as in Eq. (5).

$$R_T = R_F + R_R \qquad (5)$$

Besides the resistance test, there are two other important tests: the propeller open-water test and the self-propulsion test. From the measuring of the rpm, torque, and thrust of the model propeller, the shaft horsepower required to drive the prototype at designed speed can be predicted. *See* Dimensionless groups.

Force measurements. Experimentation in modern towing tanks has been extended far beyond predicting the resistance and power of a moving body in still water. Towing tanks are used to measure any combination of forces and moments upon a waterborne or submerged body under steady-state conditions. A few of the many possible types of tests involving force measurements in a towing tank are (1) lift and drag of a planing surface; (2) lift, drag, and pitch moment of a submerged hydrofoil; (3) forces and moments on a submerged body towed at an angle of attack to obtain the static coefficients of a body for equations of motion; and (4) the turning moment on a ship's rudder.

Pressure and velocity measurement. A series of experiments has evolved from the measurement of pressures and velocities. The pressure measurements may be integrated over the surface to determine the force on the surface. Typical tests are the measurement of the pressure distribution on propeller blades, the duct of a ducted propeller, and the appendages of ship models. Various velocity-measuring devices have been developed, the most common of which is the pitot tube. It is used to measure the velocity field at various locations around a model's hull, the most common of which is the plane in which the ship's propeller operates. *See* Pitot tube.

Lines of flow. A towing tank is also used to determine the lines of flow over portions of a hull; thus, appendages can be installed so that they will have minimum resistance and avoid or minimize the creation of cavitation. For surface ships, the conditions of comparable speed are maintained between model and prototype. Flow patterns on the hull are usually established by the emission of dyes upon a color-sensitive paint such as hydrogen sulfide upon a white lead-based paint. Flow patterns outside the boundary layer may be established by vanes or flags, which are free to pivot and orient themselves to the direction of flow. Typical tests are (1) establishment of the location of bilge keels; (2) orientation of shaft struts; and (3) determination of wave profile.

Wave experiments. Many towing tanks are provided with wavemakers, which extend the range of experimentation to the study of ship performance in head and following seas. At one end of the towing tank a wavemaker is installed which can generate waves of a more or less uniform profile with a predetermined height and length. At the other end of the tank a wave absorption beach is installed. Experiments are conducted on ship models at corresponding speeds in various head and following sea conditions, recognizing that the waves generated are much more regular than those encountered in the ocean. In addition to maintaining geometric similarity of the model and its prototype, the dynamic similarity of the system must also be maintained. Measurements are made primarily of the motions of the body, particularly the rotational motions, pitch and roll, and the translational motions, heave (vertical) and surge (longitudinal). Some techniques involve free-running models with sensitive accelerometers. The carriage is used to carry the recording equipment and provide power to the model through flexible cables; these flexible cables do not exert any restraining force on the model. In some instances in connnection with the measurement of motions, various components of force or moment may also be measured.

Non-steady-state experiments. With the wide availability of electronic measuring instrumentation, the study of unsteady hydrodynamic phenomena has become increasingly important in towing tanks. Such fluctuating forces or pressures are measured as (1) pressure on a model's hull from propeller blades passing in close proximity; (2) vibratory forces produced by a propeller operating in the variable velocity field behind a model; (3) route stability characteristics of a model from alternate course variations; and (4) forces and moments on a submerged body undergoing pure pitching and heaving motion. From the last measurement the coefficients for the equation of motion are obtained for a specific submarine configuration. This makes possible, through the use of the digital computer, the calculation of a submarine's motion without further experimentation and under a variety of conditions difficult to achieve in a towing tank.

Tank modifications. The towing tanks and the test equipment described have given rise to a number of special-purpose test facilities which are modifications of the tanks described above. The most important of these are the turning, rotating-arm, seakeeping, and ice tanks or basins. The turning tank is used to conduct experiments on the turning characteristics of the ship models. Again, the corresponding speed must be maintained to obtain similarity of phenomena. The rotating arm is a specialized facility used to find the so-called rotary derivatives of bodies such as submarines and torpedoes. From the measurement of the forces and moments on a body traversing a curved path, the rotary components of the equations of motion for a particular body are derived. The seakeeping basin is designed to study the motions and forces on waterborne bodies under various sea conditions. It may also be used to generate more complex seas than can be created in a towing tank with a single wavemaker. The most recent addition to the growing number of specialized towing tanks are the ice tanks. These facilities are essentially refrigerated towing tanks in which ice of any desired thickness can be frozen on the surface. They are used primarily to determine the resistance characteristics of ships which must operate in ice fields. SEE WATER TUNNEL; WAVE MOTION IN LIQUIDS.

Bibliography. R. Bhattacharya, *Dynamics of Marine Vessels*, 1978; J. P. Comstock (ed.), *Principles of Naval Architecture*, Society of Naval Architects and Marine Engineers, 1967; E. F. Gritzen, *Introduction to Naval Engineering*, 1980; K. J. Rawson and E. C. Tupper, *Basic Ship Theory*, vol. 2, 3d ed., 1983.

WATER TUNNEL
RICHARD STONE ROTHBLUM AND ROBERT J. ETTER

A hydrodynamic facility used for research, test, and evaluation, comprising a well-guided and controlled stream of water in which items for test are placed. The water tunnel is in many ways similar in appearance, arrangement, and operation to a subsonic wind tunnel. It is related to and complementary to the towing tank, in which the test item, usually a scale model of a ship or ship component, is towed through stationary water and evaluated through observation and measurement. In a water tunnel the test item is held stationary while the water is circulated. Many water tunels are capable of operation with variable internal pressure to simulate the phenomenon of cavitation. SEE CAVITATION; TOWING TANK; WIND TUNNEL.

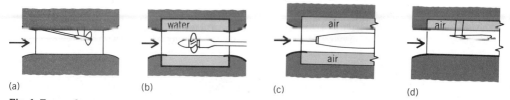

Fig. 1. Types of water-tunnel test sections with typical models. (a) Closed throat. (b) Open throat. (c) Free jet. (d) Free surface.

Classification. Water tunnels may be classified, in part, by the type of test section used. The most common section is the closed throat (**Fig. 1**a) in which the test section flow has solid boundaries. The advantage of this arrangement is its simplicity and efficiency, but the model must be small relative to the tunnel cross section to avoid large wall effects. Small wall effects are theoretically correctable. In an open-throat test section (Fig. 1b) the water jet passes through a water-filled chamber of larger diameter. This minimizes wall effects, and many tunnels dedicated to propeller testing use such an arrangement. When very low test section cavitation numbers are required, or for fully cavitating flows, a free jet (Fig. 1c) in which the water jet passes through an air-filled chamber is useful. However, capture of the free jet and removal of excess entrained air prior to recirculation is not easily achieved over a broad range of test conditions. To study cavity flows on surface-piercing components or hydrofoils which operate near a free surface, a free-surface tunnel (Fig. 1d) is required in which three sides of the water flow are bounded by solid walls and the upper surface is open to air at controlled pressure levels. This arrangement is often referred to as a water channel. SEE OPEN CHANNEL.

Another distinction among water tunnel types is whether or not they recirculate the flow. If the tunnel is nonrecirculating, water may "blow down" from a pressurized or elevated water tank or water may be diverted from a continuous source such as a waterfall or dam on a river. The blow-down tunnel has a limited test period proportional to the size of the storage tank. All water tunnels make use of transparent test-section viewing windows.

Construction. A large water tunnel planned for construction by the U.S. Navy is shown in **Fig. 2**. Typical of most designs, it is of the recirculating, closed-throat type with a carefully designed contraction ahead of the test section. An axial-flow pump located in the lower horizontal leg impels water through the circuit and test section. The strong contraction ensures good velocity uniformity in the test section, by the change of potential energy (pressure), which is uniform across the contraction entrance, to kinetic energy (velocity) of the test section. Turbulence is

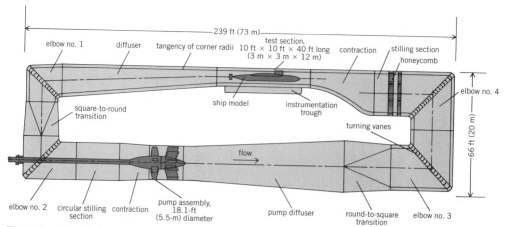

Fig. 2. Elevation of water tunnel planned by U.S. Navy at David Taylor Ship Research and Development Center, Carderock, Maryland.

reduced by passage through a stilling section containing two honeycombs. The size of a water tunnel is generally specified in terms of the dimensions of the test section where measurements are performed. The tunnel shown in Fig. 2, larger than most, has a rectangular test section of 10 ft by 10 ft by 40 ft long (3 m by 3 m by 12 m). SEE FLUID-FLOW PRINCIPLES; TURBULENT FLOW.

After passing through the text section, the velocity of the water in a recirculating tunnel is reduced to avoid cavitation and excess energy losses in other parts of the loop. Reduction of velocity takes place in the diffuser, along with an increase in pressure which helps to suppress possible cavitation of the turning vanes. Turning vanes serve to direct the flow uniformly and with minimal losses through a 90° turn at each elbow. Relatively uniform flow input to the pump is required to decrease the likelihood of undesirable vibration and pump blade cavitation.

Pressure control. In order to observe the cavitation characteristics of propellers and other test bodies, the water tunnel internal pressure must be varied independently of the velocity in the test section. Generally, for a model of a surface ship tested at speeds lower than full scale, it is necessary to decrease pressure to match full-scale cavitation conditions. If the model test speed is higher than full scale, the pressure must be increased. For submarines and similar applications which experience high full-scale operating pressures, it is also necessary to increase the tunnel pressure above the normal atmospheric level.

Closed- and open-throat tunnels. Most tunnels are of the closed- or open-throat type, operated with the entire test circuit filled with water. Consequently, they are well suited to investigations of completely submerged bodies or of conduits and turbomachinery such as pumps or turbines that are completely filled with water. Tunnels dedicated to turbomachinery are frequently referred to as test stands. Surface-ship hulls can be tested mounted flush to the top of a water tunnel test section, with the normally submerged portion of the hull exposed to the flow. The rigid wall or a specially constructed ground board takes the place of the free surface.

Free-surface tunnels. Often the behavior of the water surface around the ship has an important effect on the cavitation characteristics of the propeller or other appendages under test. For these cases, tests must be performed in the free-surface type of tunnel, or channel. The principal problem associated with free-surface tunnels is elimination of excess entrained air from the recaptured water downstream of the test section. For this reason, several free-surface tunnels have been designed to discharge water directly into a large reservoir after passage through the test section. A common practice for recirculating free-surface tunnels is to skim the top 10% of the flow from the test section and divert this to a reservoir, while recirculating the remainder after passage through a deaeration section.

Air content control. In closed or open types of tunnels, control of air content in the form of bubbles and dissolved gases is considered to be critical for the accurate replication of cavitation phenomena at model scale. Total gas content may be reduced by exposing the circulating water to reduced ambient pressure for an extended period of time. Dissolved air gradually comes out of solution to form bubbles which can be drawn off. Residual bubbles, or bubbles generated by the test body, can be removed during circulation in the tunnel by two processes: resorption and collection. Resorption takes place when part of the circuit is at a much lower height (higher static pressure) than the test section. The increased pressure forces free bubbles into solution. An alternative method is to slow the flow to a very low speed just upstream or downstream of the test section, and to collect free bubbles in trays or tubes and guide them to air evacuation points. Total removal of dissolved gas and bubbles is not desired for cavitation tests. Rather, a stable bubble population which allows good correlation of cavitation inception and cavitation patterns with realistic prototype conditions is sought. Although not a feature of most existing water tunnels, aeration devices to inject desired bubble populations are being incorporated into the design of some tunnels. Because bubbles are elastic systems, they can absorb acoustic energy. For measurements of noise generated by noncavitating bodies, it is especially important to control that part of the free bubble population which is responsible for absorption of radiated noise in the frequency range of interest.

Applications. The water tunnel is applied in much the same way as a subsonic wind tunnel, the principal difference being the possibility of vaporization of the water due to local decrease in pressure (cavitation) or the existence of a free surface and attendant interface phenomena such as waves, ventilation, and spray. Water tunnels are used to investigate the dynamics, hydrodynamics, and cavitation of submerged and semisubmerged bodies such as propellers, ships, submarines, torpedoes, and hydrofoils and of turbomachinery. They are also indispensable to research on general flow phenomena in liquids. An application of great importance has been the acoustic characterization of propellers under both cavitating and noncavitating conditions. This requires low background-noise level in the frequency range of interest. Thus, turning vanes, diffuser, and impeller must be designed to strict acoustic as well as hydrodynamic criteria. SEE HYDRODYNAMICS.

The first use of water tunnels was to identify the onset of thrust breakdown of propellers due to excessive cavitation. This is still an important application, extending to other types of marine propulsors and turbomachinery as well. In a typical test of a propeller, the thrust and torque are measured as a function of rate of rotation and speed of advance through the water, for various pressures. Cavitation inception is observed, visually using stroboscopic lighting and acoustically using hydrophones. When this information is combined with that derived from towing-tank tests of the hull alone and in combination with a propeller, the thrust, torque, and power characteristics of the full-scale ship, including the operating point of best efficiency, can then be predicted.

In sufficiently large water tunnels, such as the one in Fig. 2, the propeller can be tested with a completely appended model hull. This provides a more realistic operating environment. It also enables pressures induced on the hull by the propeller to be measured, and full-scale vibration levels to be predicted. The acoustic radiation of the propeller is also influenced by the presence of the hull. Noise and vibration characteristics of the hull-propeller combination are important for military surface ships and submarines because of the widespread use of acoustic detection techniques. For merchant ships, noise and vibration are important for habitability and avoidance of maintenance of vibration-sensitive components.

Limitations. The limitations of water tunnels in some respects are similar to those of wind tunnels. To scale dynamic effects (forces), the product of a typical dimension times the velocity must be the same in model scale as it is in full scale. Failing this, it is often sufficient that the product be above a threshold value.

This size-velocity product can be achieved by increasing model size, in which case blockage of the test section limits the fidelity of the test to full-scale conditions. Otherwise, the velocity may be increased. Besides the obvious increase in power that increased velocity requires, it becomes necessary to increase pressure to maintain cavitation similitude. Increased pressure increases the bulk of the container proportionately. This pressure constraint may be more restrictive than the power constraint, as pressure must rise with the square of the velocity. Thus water channel speeds are generally restricted to less than 50 knots (25 m/s) for relatively large tunnels. In the largest high-speed water tunnel known to be in operation, at the Swedish Maritime Center in Gothenburg, the test section is 8.5 ft (2.6 m) wide and 4.9 ft (1.5 m) deep, with a maximum velocity of 13.4 knots (6.9 m/s). SEE DIMENSIONLESS GROUPS; DYNAMIC SIMILARITY; MODEL THEORY; SIMILITUDE.

Bibliography. American Society of Mechanical Engineers, *Cavitation Research Facilities and Techniques*, 1964; W. F. Brownell, *Two New Hydromechanics Research Facilities at the David Taylor Model Basin*, DTMB Rep. 1690, December 1962; J. P. Comstock, *Principles of Naval Architecture*, 1980; H. D. Harper, A modernized control system for the DTNSRDC 36 inch variable pressure water tunnel, *Proceedings of the 20th American Towing Tank Conference*, Hoboken, 1983; R. T. Knapp, J. W. Daily, and F. G. Hammitt, *Cavitation*, 1970; H. Lindgren and E. Bjärne, *Ten Years of Research in the SSPA Large Cavitation Tunnel*, SSPA Publ. 86, 1980.

WIND TUNNEL
ROBERT G. JOPPA, RANDALL C. MAYDEW,
DONALD D. MCBRIDE, AND DANIEL J. SHRAMO

R. C. Maydew and D. D. McBride are authors of the section Hypervelocity Wind Tunnels. D. J. Shramo is author of the section Instrumentation.

A duct in which the effects of airflow past objects can be determined. The steady-state forces on a body held still in moving air are the same as those when the body moves through still air, given the same body shape, speed, and air properties. Scaling laws permit the use of models rather than full-scale aircraft. Models are less costly and may be modified more easily than aircraft, and conditions may be simulated in the wind tunnel that would be impossible or dangerous in flight.

Related aerodynamic research equipment includes ballistic guns, drop models, rockets, whirling arms, rocket-powered sleds, and flight tests. The wide range of problems studied requires a similarly wide range of specialized wind tunnels.

USES AND METHODS

Determinations commonly made with wind tunnels are as follows:

Drag of airplanes and missiles
Lifting characteristics of winged vehicles or lifting bodies
Static stability of aircraft, including missiles
Dynamic stability derivatives of aircraft
Torques required to deflect control surfaces
Pressure distributions to determine air loads
Flutter characteristics of flexible aircraft
Distribution and rate of heat transfer to aircraft parts for cooling and structural design
Performance of air-breathing engines and inlet airflow
Performance of propellers
Conditions for safe release of bombs and missiles
Effect of wind on buildings, bridges, signs, automobiles, and other nonflying structures
Nature of smoke flow from factories or ships
Nature of wind flow around geographic features

Test conditions. The tunnels in which experiments are made should closely match the conditions of full-scale flight, but it is not always possible to match all the scale parameters. An attempt is always made to match scale and the effects of compressibility of the air. SEE DIMENSIONAL ANALYSIS; DIMENSIONLESS GROUPS; DYNAMIC SIMILARITY; MODEL THEORY; SIMILTUDE.

If compressibility effects are neglected, scale is duplicated and flow patterns are similar when model tests and full-scale flight are at the same Reynolds number (Re). The Reynolds number is proportional to the product of speed and a characteristic length. SEE REYNOLDS NUMBER.

Compressibility effects are similar when the ratio of remote airspeed to the velocity of sound, called Mach number (M), is duplicated. SEE MACH NUMBER.

At flight speeds below a Mach number of approximately 0.7, one need be concerned only with the Reynolds number, since the effects of the Mach number are small and may be readily calculated. It is usually sufficient to require that the Reynolds number be greater than 1,500,000. If the flight speed is above M = 0.7, the effects of compressibility are less predictable and the Mach number must be matched. If the Reynolds number exceeds 4,000,000 the scale match is usually adequate. However, research has identified problems in the transonic speed range associated with a Reynolds number above 100,000,000. No wind tunnels are available that can match that combination of Reynolds number and M.

Special tests require the matching of other similarity parameters. Where dynamic characteristics are desired, the mass and inertia must be scaled with the forces to produce similar flight paths. If the model to be simulated is flexible, its elastic properties must be matched.

Data taken from models in wind tunnels suffer from extraneous effects due to the structure that supports the model and from the tunnel walls. The effects of support interference are minimized by design and are evaluated by careful experiments. The interference of the walls is calculated from theoretical considerations and is evaluated for each tunnel and type of model.

Measurements. Most data are required from wind tunnels through measurement of forces and moments, surface pressures, changes produced in the airstream by the model, local temperatures, and motions of dynamically scaled models, and by visual studies.

Force and moment measurements. A balance system separates and measures the six components of the total force. The three forces, taken parallel and perpendicular to the flight path, are drag lift, and side force. The three moments about these axes are moment, rolling yawing moment, and pitching moment, respectively.

Surface pressure. Surface pressures are measured by connecting orifices flush with the model surface to pressure-measuring devices. Local air load, total surface load, moment about a control surface hinge line, boundary-layer characteristics, and local Mach number may be obtained from pressure data.

Reaction of model on airstream. Measurements of stream changes produced by the model may be interpreted in terms of forces and moments on the model. In two-dimensional tunnels, where the model spans the tunnel, it is possible to determine the lift and center of pressure by measuring the pressure changes on the floor and ceiling of the tunnel. The parasite drag of a wing section may be determined by measuring the total pressure of the air which has passed over the model and calculating its loss of momentum.

Surface temperature. Measurements of surface temperatures indicate the rate of heat transfer or define the amount of cooling that may be necessary.

Dynamic measurements. In elastically and dynamically scaled models used for flutter testing, measurements of amplitude and frequency of motion are made by using accelerometers and strain gages in the structure. In free-flight models, such as bomb or missile drop tests, data are frequently obtained photographically.

Flow visualization. At low speeds, smoke and tufts are often used to show flow direction. A mixture of lampblack and kerosine painted on the model shows the surface streamlines. A suspension of talcum powder and a detergent in water does the same, is cleaner, and has a more pleasant odor.

At velocities near or above the speed of sound, some flow features may be made visible by optical devices. SEE SCHLIEREN PHOTOGRAPHY; SHADOWGRAPH OF FLUID FLOW.

TYPES

There are various types of wind tunnels, including low-speed tunnels; V/STOL tunnels; transonic, supersonic, and hypersonic tunnels; nonaeronautical tunnels; and hypervelocity tunnels.

Low-speed wind tunnel. This tunnel has a speed up to 300 mi/h (480 km/h). A low-speed tunnel has the essential features of most wind tunnels. These include a device to drive the air, a duct that provides smooth, steady, parallel flow of air in the test section, and instrumentation to measure the desired information. The tunnel may be open-circuit or return-flow (**Fig. 1**). The test section may be closed (with walls), or may be an open jet. When air-burning engines are to be tested, an open-circuit tunnel is appropriate. Most tunnels use a return duct, primarily to ensure smooth flow. The shape of the tunnel, continuously expanding from test section through the fan and around to the settling chamber, is chosen to reduce friction losses in the return circuit and to provide the powerful smoothing effect of a large contracting section just upstream of the test section. Except for special-purpose tunnels, most low-speed tunnels have a closed test section. The propulsion device (fan) causes an increase of pressure to overcome the friction losses in the tunnel circuit. Low-speed tunnels of convenient working-size test section (8 × 12 ft or 2.5 × 3.6 m) can test models at adequate Reynolds numbers with speeds up to 250 mi/h (400 km/h).

In a well-designed low-speed tunnel having a closed return, the kinetic energy of the air flowing through the test section per second may be 6 to 10 times the rate at which energy is

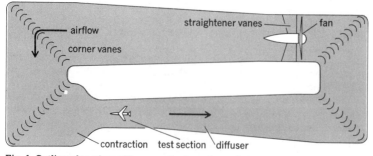

Fig. 1. Outline plan view of low-speed return-flow wind tunnel.

being supplied by the fan. This energy ratio (ER) may be used to estimate the power required to drive a tunnel by using the formula given below, where A_{ts} is test-section cross-section area in

$$\text{Kilowatts required} = \frac{13.13}{(\text{ER})} A_{ts} \left(\frac{V}{100}\right)^3 \sigma$$

square meters, V is air velocity in kilometers per hour, and σ is the ratio of the air density to the density at sea level. The energy ratio may be as low as 1 or 2 for an open-circuit tunnel and is typically about 6 to 8 for most lowspeed tunnels.

Lifting models experience interference in the tunnel because the walls restrict the downward flow behind the wing. The measured lift would only be achieved in free air at a larger angle of attack, and so a correction (typically less than 10%) is added to the angle of attack. The measured drag is reduced because of the angle of attack interference, and a correction which is a product of the lift and the angle correction is added to the measured drag. When the model has a tail, it experiences a different angle interference from that of the wing, and so a correction to the pitching moment is necessary. The drag and pitching moment corrections are of significant magnitude, but are well understood and are used confidently.

Since the model scale is usually smaller than the full-scale airplane, the Reynolds number is not matched and some scale effects appear in the data. Typically, the maximum lift of a wing is too small and the minimum drag is too large. The error can be substantial and is not subject to simple analysis. A large body of experience with these effects permits the engineer to estimate the effects of Reynolds number mismatch.

V/STOL wind tunnel. This type is a newer development of low-speed wind tunnels having a large very-low-speed section to permit testing of aircraft designed for vertical or short takeoff and landing (V/STOL) while operating in the region between vertical flight and cruising flight. In very slow flight, the flow is characterized by large downwash angles from the lifting system. The lift is usually developed by the use of power applied to a rotor, fan, jet, or jet flap. Because the rotating parts or the jet flows operate at high speeds, the flight Mach number becomes the important parameter to match. This requires that tunnel airspeeds equal flight speeds (from 20 to 100 mi/h or 32 to 160 km/h). Large test sections are required to allow the power-driven downwash to leave the model without too much distortion due to the proximity of the tunnel walls.

For economy reasons, most V/STOL tunnels are built with a standard low-speed test section (up to 300 mi/h or 480 km/h) in tandem downstream of the V/STOL test section (up to 100 mi/h or 160 km/h). This two-test-section wind tunnel provides a wide range of test conditions and has advantages in speed control at very low speeds (**Fig. 2**).

Instrumentation and test methods are similar to those of the low-speed tunnel, but models are usually more complicated because of the power systems required. Wall interference problems are yielding to research, but corrections are still difficult. Research has continued in this area.

Transonic tunnel. This type is a high-speed tunnel, capable of testing at speeds near the speed of sound, at Mach numbers from 0.7 to 1.4. In compressible fluid flow in a closed duct, sonic speed can occur only where the cross section is a minimum. Therefore, sonic speed occurs where the model is located. Additional power only causes areas of supersonic flow downstream. SEE COMPRESSIBLE FLOW.

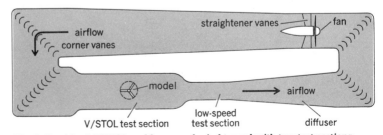

Fig. 2. Combined V/STOL and low-speed wind tunnel with two test sections.

To achieve transonic flow, the test section is surrounded by a large plenum chamber and the test-section walls are vented to this chamber by means of perforations or slots parallel to the airstream. The function of the vents is as follows. Shock waves are formed in the flow near the model when sonic speed is reached. Shock waves are reflected from a solid boundary but with a change of sign (that is, expansion waves instead of pressure waves) from a free boundary. By providing sufficient free surfaces in the form of perforations or slots, reflections which would alter the flow around the model are greatly reduced or canceled. Ventilation has proved to be useful at Mach numbers from 0.9 to 1.2, where reflections are particularly troublesome. The raw data from ventilated test sections are usually used without correction except for scale effect. SEE SHOCK WAVE.

The energy ratio for a transonic tunnel may be as large as 10 at subsonic speeds, and drop to 3 at transonic speed, leading to a power requirement of up to 5000 kW/m^2 of test section.

Supersonic wind tunnel. This tunnel is capable of test speeds corresponding to Mach numbers from 1.0 to 5. For supersonic speeds, the test section must be designed for the particular Mach number desired. When air accelerates past the speed of sound, it expands more rapidly than it accelerates, so that the tunnel must be larger downstream than it is at the minimum section, where the speed is sonic. The final Mach number is uniquely determined by the ratio of the final area to the throat area (**Fig. 3**).

Continuous-flow supersonic tunnels may require 20,000–50,000 kW/m^2 of test section. Most industrial tunnels operate intermittently from energy stored in high-pressure air tanks. The air is discharged through a fast-acting regulator valve to the tunnel and exhausted to the atmosphere (**Fig. 4**a).

The successful development of rapid recording instrumentation has made short run times possible.

The indraft, or vacuum-driven, tunnel allows atmospheric-pressure air to flow through the wind tunnel into a vacuum tank (Fig. 4b). Advantages of the indraft system include a constant (atmospheric) stagnation pressure and temperature, greater safety, and less noise and cost. Advantages of the blowdown type include higher Reynolds numbers and the ability to vary the Reynolds number over a wide range by varying the stagnation pressure.

The principal problems in the design of a supersonic tunnel are to provide sufficient pressure ratio to start and sustain the flow, and to supply adequately dry air. Drying the air to a dew

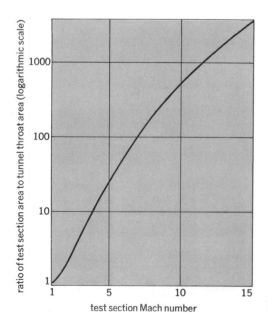

Fig. 3. Ratio of test section areas to throat area for supersonic velocity in a wind tunnel.

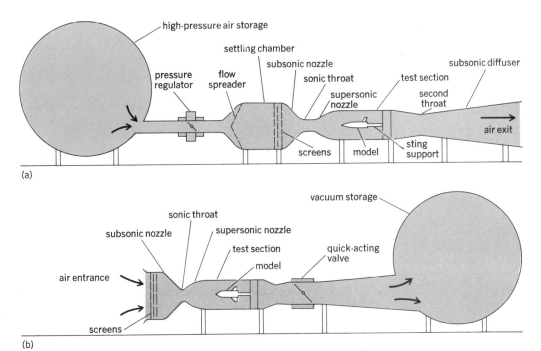

Fig. 4. Elements of (a) intermittent supersonic blowdown wind tunnel and (b) intermittent indraft supersonic wind tunnel.

point of about −40°F (−40°C) at atmospheric pressure may be required to avoid condensation shocks and uneven flow.

No wall interference effects are felt in supersonic flow, provided reflecting shock waves do not strike the model. Adjustment of results for scale effects (Reynolds number) must usually be made.

Hypersonic wind tunnel. This type is a supersonic tunnel operating in the range of Mach numbers from 5 to 15. When the Mach number is above 4 or 5, extremely high pressure ratios are required and the temperature drop in the test section is so large that liquefaction of the air would result unless the air were first heated (**Fig. 5**). Most hypersonic tunnels are intermittent types and use both high-pressure stagnation air and a vacuum-discharge tank to obtain the overall pressure ratio required (**Fig. 6**).

Heat is added by using electrical or gas-fired heaters or by passing the air through a bed of ceramic pebbles which have been previously heated. The nozzle may require cooling. The test section usually operates at a very low density, which means low Reynolds number and thick boundary layers.

Data from hypersonic wind tunnels are used directly but are subject to serious Reynolds number effects because of the low density of the flow.

Nonaeronautical wind tunnels. Testing for the effects of wind on buildings, automobiles, bridges and other nonflying objects often requires wind tunnels of special sizes and shapes.

Meteorological wind tunnels have test sections whose length is 10 to 15 times as long as they are high (**Fig. 7**). In order to develop a thick boundary layer to match the natural wind, roughness elements such as cubes or tall spirelike structures may be located on the tunnel floor upstream of the model. The roof may be adjustable to permit matching of natural zero pressure gradient. These tunnels may be used to measure the forces on buildings, and how the flow around them affects the immediate neighborhood. Dispersal of smoke and other pollutants from factories

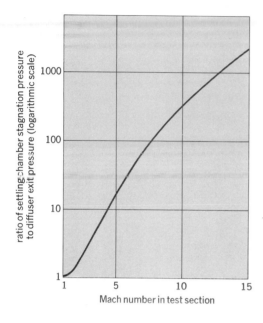

Fig. 5. Pressure ratio required to start and sustain supersonic airflow. Theoretical values assuming normal shock wave at end of test section.

may be studied. Snowdrift development, wind power generation, and dynamic testing of elastically scaled buildings or other structures are also done in meteorological wind tunnels.

Automobile wind tunnels are often very similar to aeronautical tunnels when force data are needed. For cooling tests, the tunnel may provide a jet of air not much larger than the car itself.

Special tunnels have been built for testing the dynamic character of suspension bridge models in which the test section was 50 times as wide as it was high.

Use of computers. Use of computers is rapidly changing the way in which wind tunnels are operated. The calculations involved in the conversion of raw data, in the form of forces and pressures, to final coefficient form are relatively simple but voluminous. More of this work, called data reduction, has been done by computers as their speed and memory capacity has increased. Advances in computer capability have led to changes from batch processing at the end of each day to processing at the end of each run, to on-line data reduction on a point-by-point basis. Coefficients are plotted on graphic terminals while a test is in process. The test engineer can observe the progress of the test and can recall previous tests from storage for comparison.

The computer is also used to operate the tunnel, automatically following a preprogrammed plan of test. The computer can control the tunnel speed, cause the required changes of pitch and yaw angles, and turn on the data system when specified test conditions are established.

The computer has made possible more detailed analysis of flow in wind tunnels, and has made adaptive wall tunnels possible. When the flow characteristics of a given model in free air can be calculated at a distance corresponding to the tunnel walls, the walls may be adjusted to

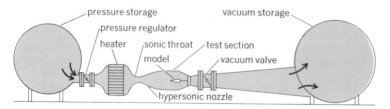

Fig. 6. Hypersonic wind tunnel with both pressure blowdown and vacuum indraft.

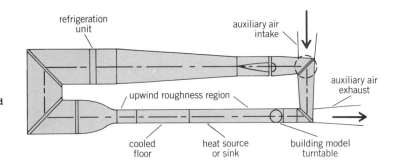

Fig. 7. Meteorological wind tunnel. (*After W. H. Rae, Jr., and A. Pope, Low-Speed Wind Tunnel Testing, 2d ed., Wiley-Interscience, 1984*)

match the free air streamlines. When this is done, the interference of the tunnel walls vanishes and no correction to the data is needed. The adjustment of the walls may be actual physical motion or virtual movement produced by injecting or extracting air through slots or holes in the walls. The techniques may be used for V/STOL or for transonic testing where corrections may be large and uncertain. Such tunnels may be called smart wind tunnels or adaptive tunnels.

Hypervelocity wind tunnels. Hypervelocity wind tunnels are tunnels possessing speed and temperature capabilities greater than those of hypersonic tunnels. They encompass the velocity range greater than 5000 ft/s (1500 m/s at Mach numbers 1 up to about 25) and are necessary for conducting aerodynamic research on vehicles such as ballistic-missile and satellite reentry bodies.

Wind-tunnel simulation of hypervelocity flight conditions requires extremely high supply (stagnation) temperatures and pressures, as may be seen in **Fig. 8**. These high supply temperatures are required for expanding the tunnel airflow through the nozzle to hypervelocities in the test section; similarly, high supply pressures are necessary to duplicate flight-altitude ambient pressure in the test section. Hypervelocity tunnels operate at considerably higher supply temperatures and pressures than do hypersonic tunnels.

The "flight corridor" indicated in Fig. 8 represents an approximate altitude-velocity band wherein flight of vehicles supported by aerodynamic lift and centrifugal force can be sustained. The upper boundary of the flight corridor is fixed by the minimum allowable atmospheric density; the lower boundary is fixed by the maximum allowable aerodynamic heating rate. This corridor

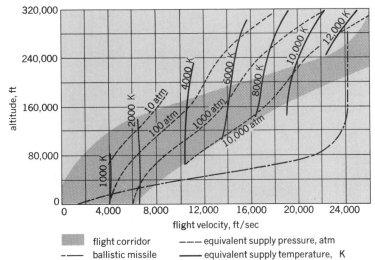

Fig. 8. Wind-tunnel simulation of hypervelocity flight requires extremely high supply (stagnation) temperatures and pressures. 1 ft = 0.3 m; 1 ft/s = 0.3 m/s; 1 atm = 10^2 kPa = 14.7 lb/in.^{2}; °F = (K × 1.8) − 460.

encompasses the flight spectrum of the proposed hypersonic transport or boost-glide vehicles. A typical trajectory of a ballistic missile from its entry into the Earth's atmosphere is shown in Fig. 8. The missile is able to tolerate the higher heating rates encountered below the corridor of continuous flight because of its short flight duration.

Complete duplication of all flight conditions in a hypervelocity tunnel is extremely difficult, if not impossible. For example, complete simulation of flight at 18,000 ft/s (5500 m/s) at 150,000-ft (45,000-m) altitude would require a tunnel supply temperature of 9000 K (16,000°F) and a supply pressure of 30,000 atm (450,000 psia or 3 gigapascals). Operation at such a high temperature is difficult, and operation at such a high pressure is not feasible at present. However, increasing the flight altitude to 300,000 ft (90,000 m) at the same velocity reduces the supply pressure required to a value that allows duplication to be achieved. Interpretation of test results is difficult, however, because the test gas is not in equilibrium owing to its rapid expansion from a highly ionized and dissociated state in the reservoir.

The major problem in designing a hypervelocity tunnel is the prevention of structural damage due to heating. Even though testing times may last only a fraction of a second, the high supply temperature (combined with the high pressure) results in a large heat flux, which reaches a maximum at the sonic throat. Damage at this point can result in large flow distortions, as may be inferred from the following example: A 0.025-in.-diameter (0.64-mm) sonic throat is required to establish a Mach 25 flow in a 115-cm-diameter (45-in.) test section. A change in throat diameter of only 0.002 in. (0.05 mm) would change the test-section Mach number by approximately 1.0.

The major types of currently operating hypervelocity tunnels are described below.

Hotshot tunnel. Mach numbers of 10–27 with testing times of 10–100 ms can be obtained with this tunnel. The principle of operation (**Fig. 9**) is to prepressurize the arc chamber to 500–10,000 psia (3.4–70 MPa) and evacuate the rest of the tunnel to a pressure of about 1 micrometer of mercury (approximately 10^{-6} atm or 0.1 pascal). The two chambers are separated by a plastic or a metal diaphragm. Electrical energy (capacitance or inductance storage of up to 100,000,000 joules) is then discharged into the arc chamber. This energy increases the air temperature and pressure, resulting in rupture of the diaphragm. The heated gas is then accelerated in the nozzle to provide hypervelocity flow. The test is usually terminated by a quick-acting dump valve which releases the stagnation chamber gas through a large port rather than requiring all the hot gases to pass through the sonic throat. This minimizes damage to both the throat and the interior of the arc chamber. The time at which this dump valve is actuated (that is, the effective run time) is influenced by a number of factors: (1) the run time necessary to obtain the desired data; (2) the time at which flow will "break down" at the model due to insufficient pressure ratio (resulting from either too large a decay in stagnation pressure or too large a buildup in vacuum tank backpressure); or (3) the maximum time the arc chamber or sonic throat can withstand the high heat-transfer rates.

Hotshot tunnels have been designed to operate at arc-chamber (supply) pressures and temperatures as high as 100,000 psia (700 MPa) and 10,000 K (18,000°F), respectively. However, melting or oxidation of the tungsten throats (diameters of 0.020–0.40 in. or 0.5–10 mm) has, in general, limited useful arc-chamber pressures and temperatures to approximately 30,000 psia (200 MPa) and 4000 K (7000°F). Nitrogen, rather than air, is used as a test medium in some tunnels to avoid

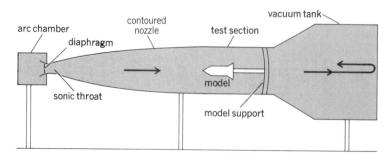

Fig. 9. Hypervelocity hotshot wind tunnel.

oxidation of the tungsten throats. Other problems are the flow contamination from solid particles (throat or electrode melting) or foreign gases (pyrolysis of plastic diaphragms and insulators) and depletion of the test gas oxygen (when air is used).

Hotshot tunnels are used for heat transfer, force, dynamic stability, and pressure model tests; data are recorded on short-response-time oscillographs or on magnetic tape. Schlieren or shadowgraph photographs and high-speed motion pictures of the luminous airflow about the model provide data on flow formation, shock waves, and flow separation.

Shock tunnel. Test Mach numbers of 6 to 25 (velocities of 4000–15,000 ft/s or 1200–4500 m/s) with testing times from 0.5 to 4 ms can be obtained in this type of facility. The major sections of a shock tunnel (**Fig. 10**) consist of the driver, a constant-area tube or driven section, and a nozzle and dump tank. A high-velocity shock wave is produced by the rupture of the diaphragm separating the high-pressure driver section, containing helium, hydrogen, or a combustible mixture, from the low-pressure driven section, containing air or nitrogen. As the shock advances, it heats and accelerates the working gas behind it in the driven tube, and a region of uniform conditions is created between the shock wave and the gas interface (the contact surface separating the driver gas and the driven gas). The shock is reflected at the nozzle throat and passes back up the driven section, and the high-pressure high-temperature driven gas ruptures the throat diaphragm and accelerates to a high Mach number in the nozzle.

In a conventional shock tunnel, the reflected shock is again reflected at the gas interface and returns downstream; the run ends when the shock arrives at the nozzle throat for the second time. Run times of 4 ms are obtained by use of the "tailored interface" technique, which consists of fixing the properties of the gas on both sides of the interface so that the primary reflected shock wave produces equal pressures and velocities on each side of the interface. When this condition is met, the shock passes through the interface without reflection.

Working-gas (supply) pressures are normally in the range of 6000–20,000 psia (40–140 MPa); while gas (supply) temperatures up to 10,000 K (17,500°F) are achievable. Initial pressure in the vacuum dump tank is about 1–5 μm of mercury (0.1–0.7 Pa).

The main advantage of the shock tunnel over the hotshot tunnel is that the test medium is less contaminated with metallic particles, although some difficulty has been experienced with driver-gas contamination due to mixing at the interface. The main disadvantage of the shock tunnel is the shorter run time.

Instrumentation with response times of less than a millisecond has been developed for measurement of heat transfer, pressure distribution, and aerodynamic forces on models. A particular problem in measuring aerodynamic forces on models arises from the necessity of making the measurements while the model is still vibrating from the impact absorbed when the starting shock passed over it. This vibration produces inertial forces which must be measured with accelerometers and accounted for in the force balance data reduction.

Graphite heater blowdown tunnel. This type of tunnel has not yet earned a nickname, so it is labeled in a very generic way. Test Mach numbers up to 14 with testing times from 0.25 to 2.5 s can be obtained with this relatively new design of hypervelocity tunnel. A Mach 18 facility has been designed but not yet built. In operation, nitrogen at room temperature is used to fill the

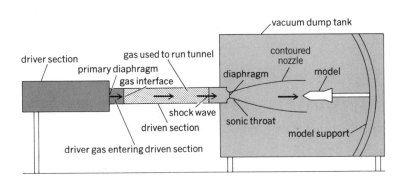

Fig. 10. Hypervelocity shock wind tunnel.

248 FLUID MECHANICS SOURCE BOOK

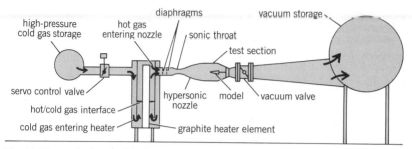

Fig. 11. Hypervelocity graphite heater blowdown tunnel.

heater (**Fig. 11**) to a pressure on the order of one-fourth of the desired test pressure. Electrical power (up to a megawatt) is then applied to a graphite heater element which heats the gas at constant volume, increasing its pressure to the desired stagnation pressure. The nitrogen is confined in the heater by a double diaphragm which is ruptured by overpressuring the interdiaphragm volume to begin the run. The decay in stagnation pressure associated with the hotshot tunnel is eliminated by introduction of controlled pressure cold gas into the bottom of the heater to push the hot test gas up, in a pistonlike fashion, out of the heater. The time over which valid data may be taken is terminated by arrival of the hot/cold gas interface at the nozzle. Control of the cold nitrogen pressure is maintained by very fast-acting electrohydraulic servo-driven control valves. Maximum supply pressures of 20,000 psia (140 MPa) and supply temperatures of 3200°R (1800 K or 2750°F) have been obtained in this type of facility.

The main advantage of this type of tunnel over the shock tunnel is the relatively long run times. During the test time available in this facility, it is possible to make a complete high speed angle-of-attack sweep from 0 to 20 or 25°, obtaining heat transfer, pressure, or aerodynamic force data over that entire angle-of-attack range. In either a shock tunnel or a hotshot, the model is preset to given angle-of-attack and data are obtained only at that condition during the run. Its main disadvantage lies in its complexity and resultant high cost.

Plasma-jet tunnel. This tunnel has the capability of developing the highest temperature (approximately 20,000 K or 35,000°F) and the longest run time (several minutes) of any hypervelocity tunnel (**Fig. 12**). It is arc-heated and utilizes either dc or ac power. The arc heater consists of a water-cooled cathode, an injection (or vortex) chamber, and a water-cooled hollow anode.

The function of the injection chamber is the generation of a strong vortex to stabilize the arc in the center of the heater. The vortex is created by the injection of high-pressure air (or other test gases) in a tangential direction at the outer edge of the vortex chamber. After injection, the test gas flows to the center of the injection chamber, through the hollow anode, and out the nozzle throat. Since the angular momentum of the test gas will be conserved as it flows to the center of the chamber, its tangential velocity will increase and its density will decrease. This effect creates

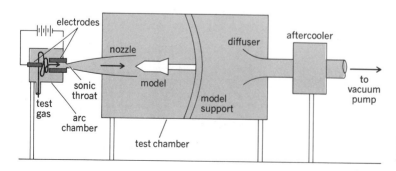

Fig. 12. Hypervelocity plasma-jet wind tunnel using dc power.

a pronounced radial density gradient in the vortex chamber and hollow anode, causing the arc to remain along the center line, where the density is lowest.

A magnetic field is used to rotate the arc so that the arc-attachment points at the cathode and anode do not become overheated. Further cooling of the anode is provided by the flow of test gas, since the swirling motion of the flow tends to keep cooler gas along the anode surface. A secondary benefit of this phenomenon is an increased heater efficiency because of reduced heat transfer to the anode and nozzle throat. Overall efficiencies as high as 60–70% can be obtained with a gas-stabilized arc heater.

The hypervelocity facility is used primarily for heat-transfer tests, reentry-shape ablation studies, and magnetoaerodynamic studies. Relatively high velocities (up to 20,000 ft/s or 6000 m/s) are achievable, but few plasma tunnels have been proposed or built for obtaining Mach numbers greater than 10. Maximum operating supply pressure is about 3500 psia (24 MPa). At high pressures the power requirements are greatly increased and arc stabilization is difficult. Plasma-jet units with power requirements of 1–60 MW are in operation or under construction; units with power requirements of several hundred megawatts have been proposed.

Disadvantages are electrode contamination and large nonuniformities in the test-gas stream, high power requirements, and operating pressure limitations. Instrumentation consists of high-speed photographic equipment, calorimeter, pitot and mass flow probes, microwave and electron beam apparatus, and a spectrograph.

Other hypervelocity wind tunnels are the gun tunnel, shock tube, expansion tube, and wave superheater tunnel.

INSTRUMENTATION

Specifically designed mechanical, electrical, and optical devices are used to measure effects of air flow across a model. Wind tunnel testing has always required instrumentation that is rugged, reliable, and accurate. Increased costs for energy and worker-power has brought demands for shorter run time and higher productivity. To fulfill these needs, modern instrumentation has been developed which can acquire accurate data quickly and provide information directly to the research engineer during the test. Measurements common to most tunnels are pressure, temperature, turbulence, and flow direction.

Pressure measurement. The earliest method used to measure a large number of pressures in tunnel tests was with multitube liquid manometer boards. A photograph of the board was taken at each special test configuration. Later, the height of each column was measured on the photo to provide the data.

Many techniques have since been developed to provide direct conversion of pressure to electrical signals. Individual strain gage, capacitance, or force balance transducers may be used, but when several hundred pressures are read this method is expensive and requires much set-up time. The mechanically scanned multiport sampling valve which switches, in sequence, 48 pressure tubes to a single transducer is often used. Since, for an accurate reading, time must be allowed for the pressure to settle after it is switched, the maximum reading rate is about 20 pressures per second.

An electronically scanned pressure (ESP) measuring system allows pressure sensors to be sampled at 10 kHz. The heart of the system is the electronically scanned pressure module (**Fig. 13**) which contains 32 silicon pressure transducers, an electronic multiplexer, and a calibration valve in a volume less than 4.2 in.^{3} (69 cm^3). The modules may be mounted within a tunnel model to avoid the slow repsonse caused by long pneumatic tube lengths. Because all transducers are periodically calibrated in place against an accurate pressure standard, the system accuracy is within ±0.15% of full scale.

Some types of testing, such as engine inlet dynamics, require measurement of pressure fluctuation up to 20 kHz. Rakes containing as many as forty 0.08-in.-diameter (2-mm) semiconductor strain gage transducers are used for this purpose. SEE PRESSURE MEASUREMENT.

Temperature measurement. Most temperature measurements in tunnel tests are made with thermocouples, thermistors, and platinum resistance temperature sensors. The airstream temperature measured is the stagnation or total temperature. The static temperature is computed from this measurement.

Heat transfer measurements may be made by constructing a thin-wall tunnel model with

Fig. 13. A 32-transducer electronically scanned pressure module. (*Pressure Systems, Inc.*)

thermocouples closely spaced on the inside surface of the wall. Individual heat transfer gages can be built by using a thin copper disk with a fine thermocouple attached, mounted in an insulated holder. The gage is installed with the sensing disk flush with the outside model surface. A heat transfer gage used in shock tunnels is constructed as a thin-film resistance thermometer. Using a 0.1-μm-thick platinum film deposited on a Pyrex glass substrate, a response time of 10^{-7} s is obtained.

Turbulence measurement. The level of velocity fluctuations is important in tunnel testing because it influences the point on a model at which the boundary layer changes from laminar to turbulent. This point of transition affects the aerodynamic drag forces on the model. Two instruments used to measure turbulence are the thermal anemometer and the laser velocimeter. SEE BOUNDARY-LAYER FLOW; TURBULENT FLOW.

The thermal anemometer is used to obtain instantaneous velocity and flow angle measurements. The sensor contains a fine wire (**Fig. 14**a) or a conducting film on a ceramic substrate (Fig. 14b) which is electrically heated to a temperature greater than the surrounding air. Since the amount of heat conducted away from the sensor depends on the local air velocity, its temperature will change accordingly. The resistance of the sensor varies with its temperature, which allows its use in a Wheatstone bridge circuit to provide a signal related to flow. SEE ANEMOMETER.

The hot wire or film by itself cannot respond to changes in fluid velocity at frequencies above 500 Hz. With electronic compensation, this response can be increased to over 1 MHz. With the constant temperature circuit, shown in **Fig. 15**, a servoamplifier is used which senses an unbalance in the bridge due to a change in sensor resistance and feeds back a current to the bridge to restore balance. In this way the wire is held at a constant instantaneous resistance, and thus a constant temperature. Therefore, no thermal lag occurs, and the feedback current is a direct measure of the turbulence.

The laser velocimeter (LV) is an optical technique that provides remote measurement of mean velocity and turbulence in the flow field of a tunnel model. Since no mechanical probe is used, there is no disturbance to the flow. As shown in **Fig. 16**, a laser beam is optically split into two equal intensity beams and focused to a point in the flow field. At this point, the light waves from each beam interfere constructively and destructively, forming a set of interference fringes. Extremely small aerosol particles (about 1 μm diameter) are injected upstream of the focal point. As each of these particles follows the airstream through the bright and dark fringes, the light that it scatters will vary in intensity. The scattered light is imaged onto a photodetector which produces an electrical signal proportional to the light intensity. The component of velocity normal to

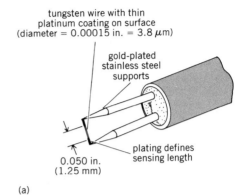

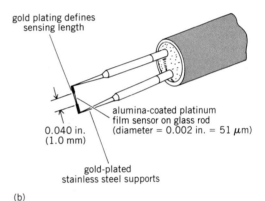

Fig. 14. Thermal anemometers with (a) hot-wire sensor and (b) cyclindrical hot-film sensor. (*TSI Incorporated*)

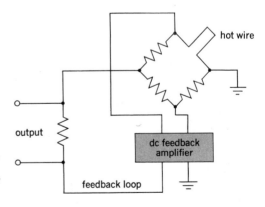

Fig. 15. Constant-temperature thermal anemometer circuit.

the optical axis in the plane of the two input laser beams can be determined from the signal frequency with the equation below, where V is the component velocity (m/s), f is the signal

$$V = \frac{\lambda f}{2 \sin\left(\dfrac{\theta}{2}\right)}$$

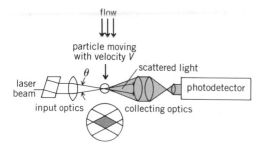

Fig. 16. Configuration of laser velocimeter.

frequency (Hz), λ is the wavelength of the laser beams (m), and θ is the angle between the two laser beams.

By using a two-color laser beam and additional optics, velocities in two directions can be measured simultaneously. A two-color laser velocimeter with blue and green beams is shown in **Fig. 17**.

A wide variety of flow visualization techniques are available for observing turbulent flow. Tufts attached to the model show flow patterns near the surface. China clay, lampblack, or fluorescent dyes supsended in viscous oil are used for studying flow in the boundary layer. Sublimation of fluorene or azobenzene is used to study extremely thin boundary layers. Schlieren photos allow the flow to be visualized in supersonic airstreams.

A method developed to visualize vortex patterns is called the laser sheet. A laser beam is fanned out into a thin vertical sheet and projected normal to the tunnel axis. A fine mist injected into the tunnel scatters the projected light and forms a visible screen. Any disturbance of the uniform flow field, such as that produced by vortices, appears as dark holes within the vapor screen.

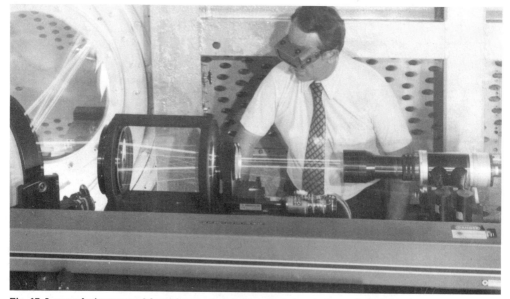

Fig. 17. Laser velocimeter used for turboprop flow-field analysis in wind tunnel at NASA Lewis Research Center.

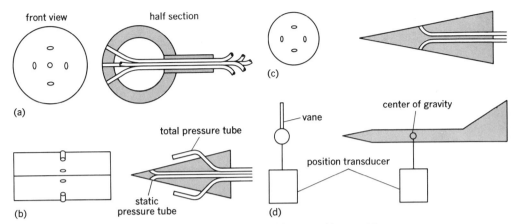

Fig. 18. Direction-measuring instruments. (a) Yaw sphere. (b) Wedge. (c) Cone. (d) Vane.

Direction probes. The flow direction and the degree to which the flow is parallel to the tunnel walls is vital in the analysis of pressure and forces on a test model in a wind tunnel. Several devices are used, depending on the flow velocity (**Fig. 18**).

Yaw sphere. In subsonic flow the pressure distribution on the surface of a sphere can be used to determine the flow direction of the air passing over the sphere. In a typical probe, yaw angle and pitch angle can be determined simultaneously. The ratio of the difference between opposite static pressures to the average of these pressures should be calibrated against the flow angles before the probe is used in a tunnel.

Wedge. In supersonic flow the sphere is not usable; instead, the wedge serves in its classical use as a means to determine the velocity of a supersonic stream and as an angle-of-flow meter. The wedge is a two-dimensional device, so that pitch and yaw direction measurements must be made separately. When the wedge is used only for directional measurement, the total-pressure tubes are not needed if the Mach number is known. The angle of flow can be obtained from tabulated data for wedges in supersonic flow. The wedge with static-pressure tubes can be used only to indicate subsonic flow angles, but first must be calibrated.

Cone. Similar to the wedge, the cone is usable in both supersonic and high subsonic flow fields to determine flow angles. The cone can be used to measure yaw and pitch angles simultaneously. However, like the sphere, it must be calibrated.

Vane. When space is not restricting, the mechanical vane can give direct-reading angles of yaw or pitch without detailed computations. The vane is mounted at its center of gravity on a calibrated position transducer and is read remotely.

Bibliography. F. Dirst, A. Melling, and J. H. Whitelaw, *Principles and Practices of Laser-Doppler Anemometry*, 1976; R. G. Joppa, Wall interference effects in wind tunnel testing of STOL aircraft, *J. Aircraft*, March–April 1969; D. B. Juanarena, *A Multiport Sensor and Measurement System for Aerospace Pressure Measurements: ISA 25th International Instrumentation Symposium*, May 1979; R. A. Kavetsky and J. A. F. Hill, *High Reynolds Number Development in the NSWC Hypervelocity Facility*, AIAA-85-0054, 1985; H. W. Liepmann and A. Roshko, *Elements of Gasdynamics*, 1957; W. C. Nelson (ed.), *The High Speed Temperature Aspects of Hypersonic Flow*, AGARDograph 68, 1964; R. C. Pankhurst and D. W. Holder, *Wind-Tunnel Technique*, 1952; P. C. Parikh, *A Numerical Study of the Controlled Flow Tunnel for a High Lift Model*, NASA Contractor Rep. 166572, 1984; C. W. Peterson, *A Survey of the Utilitarian Aspects of Advanced Flow-Field Diagnostic Techniques: AIAA 10th Aerodynamic Testing Conference* (78–796), April 1978; A. Pope and K. L. Goin, *High-Speed Wind Tunnel Testing*, 1965, reprint 1978; A. Pope and A. Pope, Jr., *Low-Speed Wind Tunnel Testing*, 2d ed., 1984.

SHOCK-WAVE DISPLAY

Eva M. Winkler

The making visible of shock waves for purposes of photography. Shock waves accompany sparks, explosions, or high-speed flows past objects in fluids. In general these waves are not directly visible, and the display of their presence requires the use of special techniques. Changes in pressure, density, and velocity of the fluid, as well as changes in its chemical composition and in the internal energy of its molecules can be used for this purpose. The method best suited to display a shock wave depends upon the strength of the shock wave to be displayed, the ambient conditions in the fluid, and the constituents of the fluid. See Shock wave.

Refraction methods. Shock-wave display techniques are used most extensively in aerodynamic test facilities. The most widely used techniques are the schlieren and shadowgraph techniques; optical interferometry is also used. All three methods are based on the variation of the index of refraction with density in the medium traversed by light. In the schlieren technique, the field displayed on the final image corresponds to the first derivative of the density normal to the schlieren knife orientation. In the shadowgraph technique, the change of illumination depends on the second derivative of the density. These two methods have the advantage of simplicity of experimental setup, and are usually preferred because they readily display even weak disturbances such as nearly sonic shocks. See Schlieren photography; Shadowgraph of fluid flow; Wind tunnel.

The interferometric technique permits a direct measurement of the density change across the shock wave. The change is presented as displacement of interference fringes (**Fig. 1**). The method is best suited for obtaining quantitative information on shock waves and flow fields. The limit of sensitivity of this technique is determined by the fringe displacement that can be measured accurately, usually considered to be 0.1 of their spacing. For air, a 4-in. (10-cm) path, and light of 500-nanometer wavelength, this limit corresponds to a density change of 1.3×10^{-6} oz/in.^{3} (2.2×10^{-6} g/cm^3). The type of instrument most commonly used is the Mach-Zehnder interferometer. Its design provides for a convenient adjustment of the virtual location of the fringes to

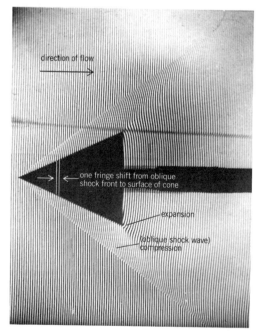

Fig. 1. Interferogram showing density change across shock wave formed around 45° cone at Mach number approximately 4. (*U.S. Naval Ordnance Laboratory*)

the object under study, and the large separation of the two interfering light beams makes it adaptable to even large test facilities.

Absorption methods. For the visualization of shock-wave phenomena in gases of low density, the above methods are too insensitive. Under such conditions, techniques utilizing the absorption of radiation or corpuscular rays provide better means for detection of shock waves. Best suited for the radiation absorption techniques are the spectral regions of strong continuous absorption, which for most gases are in the ultraviolet, or the region of soft x-rays. For investigations in air, the oxygen absorption continuum between 175 and 130 nm and soft x-rays up to 1.3 nm can be used. The x-ray technique can easily be adapted to studies in high-density fluids by utilizing x-rays of appropriately higher energy.

Corpuscular-ray absorption techniques use monoenergetic particles such as electrons, protons, or alpha particles. Their attenuation results from true absorption, scattering, and slowing down. By suitable selection of the initial energy of the particles, the technique can be adjusted to flow studies over a certain range of conditions.

Most important is the method utilizing electrons with energies of 10–30 keV. The lower limit of usefulness of the absorption techniques for shockwave display is usually correlated with a 10% change in absorption. For air and a 4-in. (10-cm) absorbing path, a shock wave of strength 6:1 can be recognized by a 10% change in absorption if the density ahead of the shock wave is about 1.4×10^{-6} oz/in.^{3} (2.5×10^{-6} g/cm^3) for 1.3-nm x-rays, 1.4×10^{-7} oz/in.^{3} (2.5×10^{-7} g/cm^3) for 10-keV electrons, and 5.8×10^{-8} oz/in.^{3} (10^{-7} g/cm^3) for the absorption at 147 nm of the oxygen molecules.

Equipment needed for absorption techniques is basically simple, consisting of a radiation or particle source with associated components, and a receiver. The range of wavelengths and energies involved requires that windows, if needed, and optical components be made of material of low absorbency, that the radiation and particle path outside the test chamber be enclosed, and that the housing be evacuated or filled with a nonabsorbing atmosphere. The receiver may be a sensitized screen, photographic film of appropriate characteristics, photoelectric receiver, ionization chamber, or Geiger counter.

Glow methods. Two other methods for visualization of especially low-density flows are the glow discharge and afterglow techniques. In these techniques the gas, prior to entering the test chamber, is subjected to a strong electrical discharge which excites it to luminescence. The light emission accompanying the discharge is used in the first method. Shock waves become visible because density changes affect the luminous intensity as well as the spectral distribution of the glow (**Fig. 2**). The useful range of the glow methods covers a density range from 1.4 ×

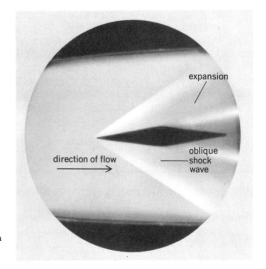

Fig. 2. Nitrogen afterglow causing shock waves to become visible around two-dimensional airfoil; Mach number approximately 2.6. (*Princeton University Press*)

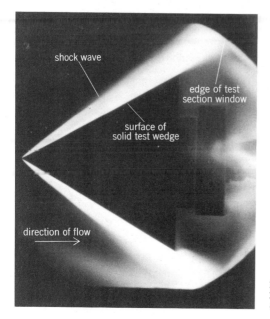

Fig. 3. Luminous flow of xenon over 60° wedge with a Mach number of approximately 12. (*U.S. Naval Ordnance Laboratory*)

10^{-4} oz/in.^{3} (2.5 × 10^{-4} g/cm^3) to 1.4 × 10^{-7} oz/in.^{3} (2.5 × 10^{-7} g/cm^3). The gas is usually excited in an insulated section which may be a separate discharge tube ahead of the wind tunnel, or the wind-tunnel nozzle proper. Depending upon the mode of excitation, which may be an electrodeless discharge, or a high-frequency condensed discharge between suitably placed electrodes, and also depending upon the operating pressure level and the type of gas to be excited, the power requirements range from a fraction of 1 kW to several kilowatts. The flow can be observed directly, or it can be photographed.

At flow conditions involving sufficiently high stagnation temperatures, the gas molecules become electronically excited and luminosity will be observed behind the shock front (**Fig. 3**). For air, this occurs at velocities approaching those of intercontinental ballistic missiles reentering the atmosphere. The luminous phenomenon is pronounced in the case of the explosion of a nuclear bomb in which a radiation front expands from the fireball. This front is characterized by a radiative process which advances faster than the pressure shock wave. This condition exists until the temperature behind the radiation front has fallen to about 540,000°F (300,000 K); thereafter, the velocity of the pressure shock wave exceeds that of the radiation front. The gas behind the pressure shock wave continues to remain luminous until, by further expansion, its temperature has decreased to only a few thousand kelvins; the luminous front then ceases to exist.

Bibliography. D. S. Dosanjh (ed.), *Modern Optical Methods in Gas Dynamic Research*, 1971; A. Pope and K. L. Goin, *High Speed Wind Tunnel Testing*, 1965, reprint 1978.

SCHLIEREN PHOTOGRAPHY
James D. Trolinger

Any technique for the photographic recording of schlieren, which are regions or stria in a medium that is surrounded by a medium of different refractive index. Refractive index gradients in transparent media cause light rays to bend (refract) in the direction of increasing refractive index. This is a result of the reduced light velocity in a higher-refractive-index material. This phenomenon is exploited in viewing the schlieren, with schlieren photographs as the result. Modern schlieren systems are rarely limited to photography; electronic video recorders, scanning diode array cameras, and holography are widely used as supplements.

Knife-edge method. There are many techniques for optically enhancing the appearance of the schlieren in an image of the field of interest. In the oldest of these, called the knife-edge method (see **illus.**), a point or slit source of light is collimated by a mirror and passed through a field of interest, after which a second mirror focuses the light, reimaging the point or slit where it is intercepted by an adjustable knife edge (commonly a razor blade is used). The illustration shows the "z" configuration which minimizes the coma aberration in the focus. Mirrors are most often used because of the absence of chromatic aberration.

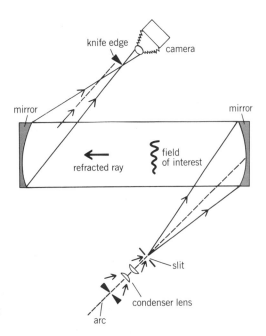

Knife-edge method of viewing schlieren, employing the "z" configuration.

Rays of light that are bent by the schlieren in the direction of the knife edge are intercepted and removed from the final image of the region of interest, causing those regions to appear dark. Consequently, the system is most sensitive to the density gradients that are perpendicular to the knife edge. The knife edge is commonly mounted on a rotatable mount so that it can be adjusted during a measurement to optimally observe different gradients in the same field of interest. The intensity in the processed image is proportional to the refractive index gradient. A gradient in the same direction as the knife edge appears dark. Gradients in the opposite direction appear bright. This method, employed with arc light sources, is still one of the simplest ways to view refractive index changes in transparent solids, liquids, and gases. A well-designed schlieren system can easily detect the presence of a refractive index gradient that causes 1 arc-second deviation of a light ray.

Color schlieren. Except for locating and identifying schlieren-causing events such as turbulent eddies, shock waves, and density gradients, schlieren systems are usually considered to be qualitative instruments. Quantitative techniques for determining density are possible but are much more difficult to employ. The most common of these is color schlieren. The knife edge is replaced with a multicolored filter. Rays of light refracted through different angles appear in different colors in the final image.

Use of coherent light. The availability of lasers and new optical components has expanded the method considerably. When a coherent light source such as a laser is used, the knife edge can be replaced by a variety of phase-, amplitude-, or polarization-modulating filters to produce useful transformations in the image intensity. For example, replacing the knife edge with a 180° phase step produces the Hilbert transform. Adding a graded-amplitude wedge to this performs differentiation of the image. Phase-shifting only the central focused spot produces phase-contrast interferometry. A variety of filters produce images referred to as schlieren interferometry. These include the insertion of stops, gratings, and Wollaston prisms in place of the knife edge.

Applications. Schlieren systems are common in aerodynamic test facilities and optical test laboratories. A schlieren system was used in the space shuttle *Space Lab 3* flights to observe fluid densities around crystals growing in zero gravity. SEE SHOCK-WAVE DISPLAY; WIND TUNNEL.

SHADOWGRAPH OF FLUID FLOW
RICHARD F. CHANDLER

A simple method of making visible the disturbances that occur in fluid flow at high velocity. The three principal methods of optical fluid flow measurements, schlieren, interferometer, and shadowgraph, depend on the fact that light passing through a flow field of varying density is retarded differently through the field, resulting in a turning of the wavefronts, that is, a refraction of the rays, and in a relative phase shift along different rays. The first of these, the refraction of the rays, is the basis for shadowgraph flow visualization.

Figure 1 shows light crossing a flow field of varying density in the y direction, as represented by the lines of constant density. The light can be considered to be acting as a wave entering at the left. The lines marked w are the wavefronts at successive times, and the lines orthogonal to them and marked r are the rays of light. Because large density gradients have a greater effect on the velocity of the light, the wavefront turns as shown. The ray of light is turned through the same angle. It can be shown that for plane flow, where conditions are the same along any x direction, the deflection of the ray is proportional to the density gradient.

For three-dimensional flow, as in a wind tunnel, the final deflection depends upon all of the density gradients encountered.

The application of the shadowgraph to a wind tunnel is shown in **Fig. 2**. The rays enter normal to the sidewall through the window at the left. (The window is optical glass of high quality to reduce refraction and to assure that observed effects are caused by the flow in the test section.)

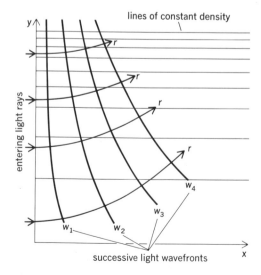

Fig. 1. Density gradient in fluid flow results in a turning of the wavefronts, that is, a refraction of light rays.

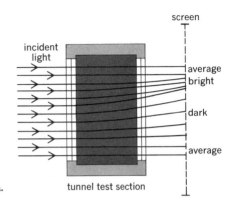

Fig. 2. Shadowgraph produced by light passing through fluid in test section indicates flow pattern.

As they pass through the test section, they encounter a change in density of the fluid in the tunnel and are deflected. The light rays then fall on a screen, where they are observed or photographed. Where the rays have crowded together, the screen is brightly illuminated, and where the rays diverge, the screen is dark. Where the spacing is unchanged, the illumination is normal, even though there has been a change of density along the path of the ray.

The light need not be parallel when it enters the test section, so the slit source and lens systems of the other methods are not required. This simplicity makes the shadowgraph system considerably less expensive than other methods; it is often used where the finer resolution of the other systems is not required or desirable. SEE SCHLIEREN PHOTOGRAPHY; SHOCK-WAVE DISPLAY.

CONTRIBUTORS

CONTRIBUTORS

Allan, William. Retired; formerly, Dean, School of Engineering, City College of City University of New York.

Arnstein, Dr. Karl. Retired; formerly, Vice President in Charge of Engineering, Goodyear Aircraft Corporation, Akron, Ohio.

Bloomfield, Prof. Philip E. Department of Physics, University of Pennsylvania; City College of City University of New York.

Bradner, Mead Technical Director, Foxboro Company, Foxboro, Massachusetts.

Brower, Prof. William B., Jr. Division of Fluids, Rensselaer Polytechnic Institute.

Bryson, Prof. Arthur E., Jr. Chairperson, Department of Aeronautics and Astronautics, Stanford University.

Cambel, Dr. Ali B. Executive Vice President for Academic Affairs, Wayne State University.

Catchpole, John P. Admiralty Materials Laboratory, Dorset, England.

Chandler, Richard F. Mechanical Engineer, Air Force Missile Development Center, Holloman Air Force Base, New Mexico.

Chen, Dr. Shih-Yuan. Department of Environmental Sciences, Rand Corp., Santa Monica, California.

Cornish, Dr. Joseph J., III. Chief Engineer, Technology, Lockheed-Georgia Company, Marietta, Georgia.

Cox, Dr. Charles S. Department of Oceanography, University of California, San Diego.

Curtis, Prof. Charles F. Department of Chemistry, University of Wisconsin.

Daugherty, Prof. Robert L. Division of Engineering, California Institute of Technology.

Dobrowolski, Dr. Zbigniew C. Chief Development Engineer, Kinney Vacuum Company, Canton, Massachusetts.

Emerson, Lewis P. Engineer in Charge of Slow Measurement, Foxboro Company, Foxboro, Massachusetts.

Etter, Robert J. David W. Taylor Naval Ship Research and Development Center, Bethesda, Maryland.

Fox, Dr. Herbert. Department of Aerospace Technology, New York Institute of Technology.

Fulford, Prof. George D. Department of Chemical Engineering, University of Waterloo, Ontario, Canada.

Hadler, Jacques B. Webb Institute of Naval Architecture, Glen Cove, New York.

Hansen, Arthur G. President, Georgia Institute of Technology.

Harleman, Dr. Donald R. F. Department of Civil Engineering, Massachusetts Institute of Technology.

Harris, Dr. J. Postgraduate School of Studies in Chemical Engineering, University of Bradford, England.

Hirschfelder, Dr. J. O. Department of Chemistry, University of Wisconsin.

Joppa, Dr. Robert G. Department of Aeronautics, University of Washington, Seattle.

Kochendorfer, Fred D. Manager, Communications Satellite Program, Philco WDL.

Krause, Lloyd N. Consultant, Westlake, Ohio.

Landweber, Dr. Louis. Iowa Institute of Hydraulic Research, University of Iowa.

Lapple, Charles E. Consultant, Fluid and Particle Technology, Air Pollution and Chemical Engineering, Los Altos, California.

Leadon, Dr. Bernard M. Research Professor, Department of Aerospace Engineering, University of Florida.

Lin, Dr. Shao-Chi. Department of Applied Mechanical and Engineering Sciences, University of California, San Diego.

McBride, Dr. Donald D. Experimental Aerodynamics, Sandia National Laboratories, Albuquerque, New Mexico.

Markovitz, Dr. Hershel. Department of Chemistry, Carnegie-Mellon University.

Maydew, Dr. Randall C. Aerodynamics Department, Sandia National Laboratories, Albuquerque, New Mexico.

Menkes, Joshua. Science and Technology Divisions, Institute for Defense Analysis, Arlington, Virginia.

Murphy, Dr. Glen. Department of Nuclear Engineering, Iowa State University.

Nachtrieb, Prof. Norman H. Chairperson, Department of Chemistry, University of Chicago.

O'Connell, Prof. John P. Department of Chemical Engineering, University of Florida.

Pai, Dr. Shih I. Institute of Physical Science and Technology, College of Engineering, University of Maryland.

Rockett, Frank H. Engineering Consultant, Charlottesville, Virginia.

Ross, Dr. Robert S. Goodyear Aerospace Corporation, Akron, Ohio.

Rothblum, Richard Stone. David W. Taylor Naval Ship Research and Development Center, Bethesda, Maryland.

Schoenherr, Dr. Karl E. Retired; formerly, Technical Director, Hydromechanics Laboratory, Naval Ship Research and Development Center, Washington, D.C.

Scott, Dr. John E., Jr. Department of Aerospace Engineering and Engineering Physics, University of Virginia.

Sell, Dr. Heinz G. Metals Development Section, Westinghouse Lamp Divisions, Bloomfield, New Jersey.

Sellars, Dr. John R. Manager, Engineering Mechanics Operations, TRW, Inc., Redondo Beach, California.

Shramo, Daniel J. Lewis Research Center, National Aeronautics and Space Administration, Cleveland, Ohio.

Sprenkle, Raymond E. Retired; formerly, Director of Education, Bailey Meter Company, Cleveland, Ohio.

Steward, Dr. John W. Department of Physics, University of Virginia.

Streeter, Prof. Victor L. Department of Civil Engineering, University of Michigan.

Talbot, Prof. Lawrence. Department of Mechanical Engineering, University of California, Berkeley.

Trolinger, Dr. James D. Vice President, Spectron Development Laboratories, Costa Mesa, California.

Walsh, Dr. Peter J. Department of Physics, Fairleigh Dickinson University.

Walter, William C. President, Frontier Systems, Rolling Hills, California.

Whitaker, Prof. Stephen. Department of Chemical Engineering, University of Houston.

White, Prof. Frank M. Department of Mechanical Engineering, University of Rhode Island.

Wilkinson, Dr. W. L. Postgraduate School of Studies in Chemical Engineering, University of Bradford, England.

Winkler, Eva M. U.S. Naval Ordnance Laboratory, Silver Spring, Maryland.

Zifcak, John H. Foxboro Company, Foxboro, Massachusetts.

INDEX

INDEX

Asterisks indicate page references to article titles.

Absolute viscosity 15
Acoustic waves 134-136
Aerodynamic heating 103-104
Aerodynamics 99-101*
Aeromechanics 11*
  see also Aerodynamics; Aerostatics
Aerostatics 11*
  Archimedes' principle 8*
  buoyancy 7*
Aerothermodynamics 101-104*
  aerodynamic heating 103-104
  external flow 102-103
  internal flow 101-102
Air-velocity measurement 226-228*
  anemometer see Anemometer
  bridled pressure plate 228
  kata thermometer 228
  pitot tube 225-226*, 227
  shielded thermocouple 228
  venturi tube 223-224*, 227-228
Alfvén wave 82
Anemometer 228-230*
  cup anemometer 228-229
  hot-wire anemometer 229-230

Anemometer (*cont.*):
  vane anemometer 229
Archimedes' principle 8*, 61
Axial meter 218

Barometer 194, 198*
Bernoulli's theorem 65-66, 106-107*, 184-185
Berthelot, P.E.M. 189
Bingham plastics 119
Borda mouthpiece 224-225*
Boundary-layer flow 20, 26-32*, 62-64, 86-87
  boundary-layer control 31-32
  boundary-layer separation 27-28
  critical Reynolds number 29
  laminar and turbulent layer 28
  laminar layer stability 29-30
  layers in compressible flow 31
  skin friction 30-31
Bourdon-spring pressure gage 195, 198-200*
Bridgman, P.W. 159
Bridled pressure plate 228
Buckingham. E. 159

Buoyancy 7*, 9
  horizontal 7
  stability 7

Capacitive pressure transducer 203
Capillary wave 140-141*
Catchpole, J.P. 158
Cavitation 184-188*
  Bernoulli's principle 184-185
  cavitation number 185
  effects 187-188
  physical causes 186-187
  types 185-186
Collisionless flows 88
Compressible flow 43-44*
Conservation of energy: flowing gas 78-79
Continuity equation: gas dynamics 76-77
Contrapropagating meter 218
Correlation meter 218
Creeping flow 126*
Critical Reynolds number 29
Crocco's equation 112*
Cup anemometer 228-229

D'Alembert's paradox 107*
Damköhler's ratio 79

Darcy-Fanning equation 92
Deflagration wave 81-82
Deflection meter 218
Detonation wave 81-82
Diabatic internal gas flow 84-85
Differential-pressure measuring devices 209-211
Dilatant fluids 119
Dimensional analysis 152-157*
 applications 154-157
 examples of dimensional formulas 153-154
 theory 152-153
Dimensionless groups 158-172*
 classification 159-160
 general nomenclature 171-172
 named groups 160-171
 named groups, history of 158
 sources and uses 158-159
Doppler meter 218
Doublet flow 50-51*
Ducted flow 39-40*
Dynamic similarity 157-158*

Elasticity, nonlinear 129-130
Electrolytic tank 110*
 apparatus 110
 applications 110
Electromagnetic flowmeter 212
Energy equation: gas dynamics 78-79
Engel, F.V.A. 159
Euler, L. 54
Euler's momentum theorem 106*
External flow *see* Boundary-layer flow; Wake flow

Fanno flow 85-86
Flow measurement 203-221*
 accuracy 220-221
 air velocity *see* Air-velocity measurement

Flow measurement (*cont.*):
 anemometer *see* Anemometer
 axial meter 218
 contrapropagating meter 218
 correlation meter 218
 deflection meter 218
 differential-pressure secondary devices 209-211
 differential-producing primary devices 205-209
 Doppler meter 218
 electromagnetic flowmeter 212
 flow correction factors 214
 fluidic-flow measurement 217-218
 fluidic-flow sensor 218
 fluidic-oscillator meter 217
 heat-loss flowmeter 216
 hot-wire anemometer 216
 laser Doppler velocimeter 220
 mass flow rate meters 215-216
 metering orifice 221-223*
 noise-type meter 218
 nuclear magnetic resonance flowmeter 220
 open-channel meter, ultrasonic 218
 open-channel meters 211
 pitot tube 225-226*, 227
 positive-displacement quantity meters 212-214
 quantity meters 212-214
 rotating impeller meters 214
 target meter 211
 thermal devices 216
 Torricelli's theorem 221*
 tracer method 212
 ultrasonic 218
 variable-area meters 211-212
 venturi tube 223-224*, 227-228
 vortex flowmeters 216-217
Flow nozzle 207
Flow tube 207

Fluid dynamics 54*
 see also Aerodynamics; Gas dynamics; Hydrodynamics
Fluid flow 14-20*
 compressibility of fluids 16
 compressible flow 19-20
 density and specific weight 15-16
 dynamic similarity 157-158*
 measurement *see* Flow measurement
 non-newtonian *see* Non-newtonian fluid flow
 principles *see* Fluid-flow principles
 shadowgraph of 258-259
 steady flow 18-19
 uniform flow 18
 unsteady flow 19
 viscosity 14-15, 58
Fluid-flow principles 54-66*
 angular momentum principle 57-58
 Archimedes' principle 61
 Bernoulli's equation 65-66
 boundary-layer flow 62-64
 conservation of mass 55
 density 55-56
 Euler's laws of mechanics 54-55
 fluid characteristics 58
 incompressible flow 59-60
 inviscid flow 64-65
 irrotational or potential flow 65
 laminar flow 60-61
 linear momentum principle 56-57
 Navier-Stokes equations 59
 Newton's law of viscosity 58-59
 static fluids 61
 turbulent flow 60-61
 uniform, laminar flow 61-62
 viscosity 58
Fluid statics 6-7*
 see also Aerostatics; Hydrostatics

Fluidic-flow measurement 217-218
Fluidic-flow sensor 218
Fluidic-oscillator meter 217
Fluids 14, 178-182*
 compressibility 16
 interfaces 180-181
 mixtures 180
 newtonian fluid 118*
 non-newtonian fluid 118-123*
 transport properties 181-182
 vapors and liquids 178-180
Free-turbulence problem 24
Free vortex 46
Friction factor 35*
Friction flow: internal one-dimensional gas flow 85-86
Froude, W. 233
Froude number 174*, 233-234
Fulford, G. 158

Gas 188-192*
 empirical equation of state 189-190
 principle of corresponding states 190-191
 theoretical considerations 191-192
 see also Fluid entries
Gas dynamics 76-89*
 continuity equation 76-77
 detonation and deflagration waves 81-82
 dimensionless parameters 79
 energy equation 78-79
 external flow 86-87
 hydromagnetic (Alfvén) waves 82
 internal one-dimensional flow 83-86
 low-pressure see Low-pressure gas flow
 Mach number functions 82-83
 momentum equation 77
 rarefied gas 87-89

Gas dynamics (cont.):
 scope of subject 76
 shock wave 80-81
 speed of sound 79-80
Gas kinematics 75*
Gas kinetics 99*
Gas mechanics 10*
Gukhman, A.A. 159

Head meters 205-209
Heat-loss flowmeter 216
Helical impeller 214
Herschel, C. 224
Horizontal buoyancy 7
Hot-wire anemometer 216, 229-230
Hydraulic analog table 110-112*
 applications 111-112
 theoretical basis 111
Hydraulic gradient 38*
Hydraulic jump 143-144*
Hydraulics 74-75*
 applications 75
 fluid properties 75
Hydrodynamics 67-74*
 flow field 67
 hydrokinematics 67*, 68-69
 hydrokinetics 70-72
 irrotational flow 72-73
 towing tank 232-235*
 viscous flow 73-74
 water tunnel 235-238*
Hydrokinematics 67*, 68-69
Hydrokinetics 70-72, 74*
Hydromagnetic wave 82
Hydromechanics 8*
Hydrostatics 9-10*
 applications 9
 Archimedes' principle 8*
 buoyancy 7*, 9
 pressure-measuring instruments 10
 pressure transmission 10
 stability of body 9-10
Hypersonic wakes 87
Hypersonics 101, 102

Ideal fluid flow 14, 16-17
Ideal gas 16
Inclined-tube manometer 198
Incompressible flow 42-43*, 59-60
 mass, momentum, and energy relations 42-43
 speed ranges in gas dynamics 42
Internal one-dimensional gas flow: diabatic flow 84-85
 flow with friction 85-86
 variable area flow 83-84
Internal wave 141-143*
Inviscid flow 64-65
Inviscid fluid: Bernoulli's theorem 106-107*
 Crocco's equation 112*
 D'Alembert's paradox 107*
 Euler's momentum theorem 106*
 Kelvin's circulation theorem 108*
 Kelvin's minimum-energy theorem 109*
 Laplace's irrotational motion 108-109*
Irrotational flow 72-73
 Kelvin's minimum-energy theorem 109*
 Laplace's irrotational motion 108-109*
Isentropic flow 44-45*

Jet flow 38-39*
Kármán vortex street 47-48*
Kata thermometer 228
Kelvin's circulation theorem 108*
Kelvin's minimum-energy theorem 109*
Kinematic viscosity 15
Klinkenberg, A. 159
Knudsen, V.O. 95
Knudsen number 79, 87, 98-99*
Kolmogoroff, A.N. 24

Laminar flow 17-18, 21*, 60-61
  aerothermodynamic considerations 101 102
  in boundary layer 28
  low-pressure gas flow 93-94
  in pipes 33
  Reynolds number 173
  time-independent 123-125
  uniform flow 61-62
Langhaar, H.L. 159
Laplace's irrotational motion 108-109*
Laser Doppler velocimeter 220
Leibnitz, G.W. 55
Linear viscoelasticity 127-129
Liquid 178-180, 182-184*
  cavitation 184-188*
  theoretical explanations of liquid state 183-184
  thermodynamic relations 182
  transport properties 183
  wave motion see Wave motion in liquids
  see also Fluid entries
Liquid-sealed drum-type gas meter 214
Lobed impeller 214
Low-pressure gas flow 89-98*
  conductance 91
  design criteria for operating systems 91
  flow regimes 89
  laminar flow 93-94
  molecular flow range 95-98
  nomenclature and units 91-92
  through nozzles and orifices 94-95
  pressure measurement 89-90
  pumping speeds 90
  throughput 90-91
  transition range between laminar and molecular flow 95
  turbulent flow in circular conduits 92-93

Mach number 174*
  branches of aerodynamics 101
  functions for perfect gas 82-83
  gas dynamics 79
  shock wave 145-150*
Magnetic pressure transducer 201-202
Magnetohydrodynamics 25-26
Manometer 10, 194, 196-198*
  inclined-tube 198
  micromanometer 198
  U-tube 194, 197
  well-type 197
Mass flow rate meters 215-216
Metering orifice 221-223*
Micromanometer 198
Model theory 174-176*
  gas dynamics 77
Mooy, H.H. 159

Navier-Stokes equations 59, 66-67*, 77
Newtonian fluid 118*
Newton's law of viscosity 58-59
Nonlinear elasticity 129-130
Nonlinear viscoelasticity 121-122, 130-131
Non-newtonian fluid 118-123*
  nonlinear viscoelasticity 121-122
  time-dependent fluids 119-121
  time-independent fluids 119
  viscoelastic fluids 121
Non-newtonian fluid flow 123-126*
  laminar flow, time-independent 123-125
  thixotropic fluids in pipes 125-126
  turbulent flow, time-dependent 125
  turbulent suppression 125
Normal shock wave 146-148

Nozzle 40-42*
  aerothermodynamic effects 102
  design considerations 41
  energy exchanges 40
  isentropic flow 44-45*
  low-pressure gas flow 94-95
  variable area internal gas flow 83-84
  in wind tunnel 40-41
Nuclear magnetic resonance flowmeter 220
Nutating-disk meter 212-214

Oblique shock wave 148-150
Open channel 36-38*
  hydraulic jump 37
  stream gaging 37-38
Open-channel meter 211
  ultrasonic 218
Orifice plate 206
Oscillatory waves 138-139
Oseen's flow 126
Oval-shaped gear meter 214

Pankhurst, R.C. 159
Pascal's law 7*
Piezoelectric pressure transducer 202-203
Pipe elbow 208-209
Pipe flow 33-35*
  laminar flow 33
  pipe elbow 208-209
  thixotropic fluids in pipes 125-126
  turbulent flow 33-35
  turbulent low-pressure gas flow in circular conduit 92-93
  water hammer 144-145*
Pitot tube 207-208, 225-226*, 227
Poiseuille's law of flow 93
Positive-displacement quantity meters 212-214
Potential flow 109*
Prandtl, L. 26

INDEX

Prandtl-Meyer expansion fan 45-46*
Pressure measurement 194-196*
  barometer *see* Barometer
  Bourdon pressure gage *see* Bourdon-spring pressure gage
  expansible metallic-element gages 194-196
  liquid-column gage 194
  manometer *see* Manometer
  measurement standards 196
  pitot tube *see* Pitot tube
  pressure transducer *see* Pressure transducer
Pressure transducer 200-203*
  capacitive 203
  electrical 196
  magnetic 201-202
  piezoelectric 202-203
  resistive 201
  resonant 203
  strain-gage 201
Pseudoplastic fluids 119, 121

Quantity meters 212-214
Rarefied gas dynamics 87-89
  collisionless flow 88
  flow regimes 87-88
  slip flow 88-89
  transition flow 89
Rayleigh flow 84-85
Real fluid flow 14, 17
Real gas 16
Resistive pressure transducers 201
Resonant pressure transducer 203
Reynolds, O. 29, 126, 173
Reynolds number 172-173*
  boundary-layer flows 29
  laminar and turbulent flow 173
  model testing 173
Rheology 126-131*
  linear viscoelasticity 127-129

Rheology (*cont.*):
  models and properties 127
  nonlinear elasticity 129-130
  nonlinear viscoelasticity 121-122, 130-131
Rheopectic fluids 120
Ripple *see* Capillary wave
Ripple tank 230-231*
Rotary-abutment meter 214
Rotary-vane meter 214
Rotating impeller meters 214

Schlieren photography 256-258*
Shadowgraph of fluid flow 258-259*
Shock wave 80-81, 145-150*
  aerothermodynamics 102
  bomb blast 150
  normal 146-148
  oblique 148-150
Shock-wave display 254-256*
  absorption methods 255
  glow methods 255-256
  refraction methods 254-255
Similitude 152*
Sink flow 50*
  doublet flow 50-51*
Skin friction 32-33*
  boundary layers 30-31
Slip flow 88-89
Solitary waves 139
Source flow 48-50*
  characteristics of a source 48-49
  doublet flow 50-51*
  flow about a body 49-50
  sink flow 50*
Standing waves 139
Stokes, G.G. 158
Stokes stream function 51-52*
Strain-gage pressure transducer 201
Streamline flow 21-22*
  laminar *see* Laminar flow
Subsonics 101, 102
Supersonics 101, 102

Thermal flow measurement device 216
Thixotropic materials 119-120
  fluid flow in pipes 125-126
Throttled flow 42*
Time-dependent fluids 119-121
  turbulent flow 125
Time-independent fluids 119
  laminar flow 123-125
Torricelli's theorem 221*
Towing tank 232-235*
  force measurements 234
  law of similitude 233-234
  lines of flow 234
  non-steady-state experiments 235
  pressure measurement 234
  tank modifications 235
  towing methods 232
  velocity measurement 234
  wave experiments 235
Transition flow (rarefaction) 89
Turbine: aerothermodynamic effects 102
Turbine meter 214
Turbulent flow 18, 22-26*, 60-61
  aerothermodynamic considerations 101-102
  in boundary layer 28
  diffusion 25
  jet mixing and wakes 24
  local isotropy 24-25
  logarithmic velocity profile 23
  low-pressure gas flow in circular conduit 92-93
  magnetohydrodynamics 25-26
  in pipes 33-35
  random nature 22
  Reynolds number 173
  semiempirical theories 22-23
  statistical theory 24
  time-independent 125
  turbulent stresses 22

U-tube manometer 194, 197
Ultrasonic flow measurement 218

van der Waals, J. 189
Vane anemometer 229
Variable-area internal gas flow 83-84
Variable-area flowmeter 211-212
Venturi tube 206, 223-224*, 227-228
Viscoelastic fluids 121
Viscoelasticity: linear 127-129
  nonlinear 121-122, 130-131
Viscosity 14-15, 58, 114-117*
  D'Alembert's paradox 107*
  flow behavior of complex fluids 117
  liquids 115-116
  measurement 116-117
  molecular basis in gases 115
  Newton's law of 58-59
Viscous fluid *see* Newtonian fluid
von Helmholtz, H. 158
von Kármán, T. 47, 80
Vortex 46-47*
  arrays 47
  free 46
  Kármán vortex street 47-48*
  vortex tube 46-47

Vortex cage meter 214
Vortex flowmeter 216-217
Vortex tube 46-47

Wake flow 24, 26*, 87
Water hammer 144-145*
  boundary conditions 145
  magnitude of disturbance 145
  propagation 144
Water tunnel 235-238*
  applications 237-238
  classification 236
  construction 236-237
Wave motion in fluids 134-137*
  acoustic waves 134-136
  applications 134
  internal wave 141-143*
  liquids *see* Wave motion in liquids
  seismic waves 137
  shock wave 145-150*
  wave classification 134
  waves of larger amplitude 137
  zone of action and zone of silence 136
Wave motion in liquids 137-140*
  oscillatory waves 138-139
  ripple tank 230-231*

Wave motion in liquids (*cont.*):
  solitary waves 139
  standing waves 139
  surges 140
Weber, M.G. 158
Well-type manometer 197
Wind tunnel 238-253*
  computer use 244-245
  direction probes 253
  graphite heater blowdown tunnel 247-248
  hotshot tunnel 246-247
  hypersonic 243
  hypervelocity 245-249
  instrumentation 249-253
  low-speed 240-241
  measurements 239-240
  nonaeronautical 243-244
  nozzle use 40-41
  plasma-jet tunnel 248-249
  pressure measurement 249
  shock tunnel 247
  supersonic 242-243
  temperature measurement 249-250
  test conditions 239
  transonic 241-242
  turbulence measurement 250-252
  types 240-249
  uses and methods 239-240
  V/STOL type 241

3 0020 00142 7840